# Breeding Plants for Disease Resistance
Concepts and Applications

# Breeding Plants
# For Disease Resistance
## Concepts and Applications

## Edited by R. R. Nelson

The Pennsylvania State University Press
University Park and London

Second printing, 1977


Library of Congress Cataloging in Publication Data


Nelson, Richard Robert, 1926-
  Breeding plants for disease resistance.

  1. Plants—Disease and pest resistance.     2. Plant-
breeding.     I. Title.
SB750.N44            631.5'3             71-128175
ISBN 0-271-01141-6


Library of Congress Catalog Card Number: 71-128175
International Standard Book Number: 0-271-01141-6
©1973  The Pennsylvania State University
Printed in the United States of America
Designed by Maria DeFebo

*To Dr. C. C. Wernham*
*whose concern for and dedication to the*
*improvement of crop plants through disease resistance*
*inspired his students and his colleagues*

# Contents

# Preface

The old adage that "nations swarm for food" bodes ill to civilizations unable to sustain themselves. Food-deficient nations house a majority of today's people, and their starving millions attest to the urgent need for greater quantities of quality food throughout much of the world.

A number of proposals have been suggested as means to alleviate the world's food problem. Some, such as population controls, would necessitate what many consider to be impossible changes in moral, religious, and philosophical ethics. Others, such as "a meal in a pill" and "down to the sea for algae," undoubtedly would test the psychological fiber of many of us to the breaking point.

There are some viable alternatives. The reclamation of nonarable land currently is converting deserts into farmlands and abandoned mining tracts into orchards. The continued development and increased use of synthetic materials will lessen the need to devote arable land to fiber crops.

Productivity may be the prime scientific key to the world's food needs. A greater yield of better quality food per acre of arable land is the avowed mission of agricultural sciences. That mission remains undiminished despite the tendencies of urban societies to provide fewer funds and less encouragement to the "wetbacks" of science. And yet, if we don't swim, they won't eat. Agricultural research on food productivity is a composite of science, sense, and society. It is time for society to acquire more sense about science.

This book focuses attention on and extols the virtues of what many believe is the key of all keys to greater food production. Disease resistance in plants is well documented as a spectacular means of increasing productivity, as well as an able, if not consistent, defender against epidemic losses. Many of its prime credentials are assembled herein.

Part I summarizes my general considerations of the concepts, principles, and terminology that are germane to controlling plant diseases by resistance. Some of the concepts and several of the terms are controversial in their interpretation, and my treatment of them may well be deemed the same. I firmly believe that the principles and concepts pertinent to disease resistance are applicable to the control of most diseases of most crops. That the amount of scientific facts doubles every seven years while the number of concepts and principles remains fairly constant lends credibility to that belief. Part II treats many of the world's important crop species and ably demonstrates the role that resistance has played in their continued prominence. Each of the crop chapters has been written by one or more of the world's recognized leaders in the field. Their names and contributions to science are well known to many. My thanks go to each of them for their contributions and their patience with me during the editing phase. The authors were asked to restrict the number of their citations. Any omissions by them of pertinent references should not be construed as a minimization of their importance, or as an oversight on the part of the authors. If this book is a worthy contribution, it is because of them. In fact, some of my colleagues who followed the book from its inception refer to it as one "where the fiction comes first and the facts last."

Certain important crop species are not treated here; the reasons for their absence are varied.

This book is dedicated to Dr. C. C. Wernham. In a very real sense, it is his book. His entire professional career was dedicated to the proposition that disease resistance in plants was essential to the well-being of man. With that dedication to guide him, he set out to compile this book upon his retirement. His untimely death left it in its incipient stages. It was finished out of respect for him.

R. R. Nelson

*University Park, 1973*

# Part I
## General Considerations, Concepts, and Terminology

# 1 Introduction

## R. R. Nelson[1]

## The Past

*If men could learn from history, what lessons it might teach us. But passion and party blind our eyes and the light which experience gives is a lantern on the stern, which shines only on the waves behind us. . . .*

*Coleridge*

Plant diseases have been one of the greatest hazards to crop production almost since man began to domesticate plants. Diseases became epidemic and inflicted serious losses in ancient times, as evidenced by Biblical references to severe blasting, blighting, and rusting of plants.

Man has attempted to explain the occurrence of plant diseases for thousands of years. The accepted thinking and philosophies of the times tempered such explanations along certain lines. Prior to 1800, for example, the theory of spontaneous generation was accepted generally to explain the origin and occurrence of most living things, including plant diseases. This theory was advanced by Hebrew writers before 600

1. Professor of Plant Pathology, The Pennsylvania State University.

B.C. The idea of Aristotle (384-322 B.C.) that living organisms were formed by the union of a passive principle "matter" with an active principle "form" influenced the minds of men and scholars for more than 2000 years. During the Greek and Roman empires, the occurrence of plant diseases was thought to be punishment for displeasing the gods, and elaborate ceremonies were held to regain their favor. From then until early in the 19th century, beliefs concerning the causes of plant diseases were not appreciably different from those of the Greeks and Romans. The 50 years preceding the acceptance of the germ theory were dominated by the autogeneticists, who believed that fungi were the products of morbid sap produced by the living cells of the host.

Numerous achievements prior to 1850 appreciably aided the transition from the general ignorance of the Dark Ages to an awareness of the causes of plant diseases. The invention and later improvement of the microscope, the systematic treatment of the fungi by Fries and Persoon, the early work on the infectious nature of diseases of wheat by Tillet and Fontana, and the brilliant research of Prevost in 1807 on the cause of bunt of wheat are some of the milestones in the development of the germ theory. Despite these scientific facts and tools, general acceptance of the germ theory still was lacking by 1850.

The Irish famine, which resulted from 5 epidemics of potato late blight, may well have been the final contributor to a general acceptance of the germ theory of plant disease. Rapid increases in population, dependence on one crop for food, and widespread starvation resulting from plant diseases dictated that man must acknowledge the germ theory or die. The classic work of de Bary proved the validity of the germ theory once and for all.

Progress on the cause and control of plant diseases during the last 100 years has been spectacular. Yet the heavy toll that diseases still reap in all areas of the world and the approaching problem of feeding the world's expanding population clearly illustrate the immediate need for additional knowledge of the nature and control of plant diseases.

The impact of plant diseases on crops is still grossly underestimated in most agricultural areas. This is particularly true of those crops that are rarely, if ever, subjected to severe epidemics of a plant disease. The dramatic losses from stem rust of wheat, coffee rust, late blight of potatoes, and black shank of tobacco; the near elimination of the American chestnut; the ever-increasing losses from oak wilt; and the 1970 epidemic of southern corn leaf blight in the United States are graphic examples of the effect of plant disease epidemics. Spectacular losses create spectacular concern about diseases, although intense concern may be short-lived. In most crops, however, reductions in yield and quality are brought about in less spectacular fashion by diseases that occur annually or nearly so in less than epidemic proportions. Total losses from these diseases over a period of years can more than equal the losses incurred during a single epidemic year.

There are many methods of effective disease control. Such measures as chemotherapy, cross-protection, and systemic fungicides are of recent origin. Others came into play only after the acceptance of the germ theory, which postulated that organisms can cause disease. Seed treatment of cereal grains preceded this era, but it was used empirically. Disease control by resistance probably was the first method used to combat plant pathogens. The selection and use of disease resistant plants may have been a conscious or unconscious part of man's philosophy from the time he began to cultivate crop plants: he certainly must have been concerned with plant improvement at an early date. There is no reason to assume that the tendency to select the most productive types of plants is the singular prerogative of modern man. Indeed, the ancient civilizations of Greece and Rome recognized differences among plants in their susceptibility to diseases. Natural selection and survival of the fittest probably provided early man with some base of resistance in the wild plants that served as the progenitors of cultivated crops.

Recent history and our understanding of basic differences between wild and cultivated species provide excellent insight into many of the reasons for man's early confrontation with diseases of crop plants as he began to domesticate plants for convenience and for greater quantities of food and fiber. Wild species typically are comprised of populations of plants exhibiting considerable genetic diversity for many traits; man tends to emphasize homogeneous and uniform populations of cultivated or domesticated plant species to facilitate cropping and harvesting practices. Wild species typically exhibit a variety of traits or characteristics which may be beneficial or necessary to survival, but which definitely are not desirable to man in his domesticated crops. Wild species typically grow at random in nature as single plants or groups of few plants. For convenience and to preserve land, man crops his domesticated species as large populations of plants in relatively small areas.

Recent attempts commercially to domesticate wild rice in northern Minnesota illustrate the significance of these basic differences between wild and cultivated species. Wild rice plants grow randomly in scattered groups along banks of streams or in wet, boggy areas. Most of the plants shed their kernels at maturity as a natural response to wind or rain, a trait that man refers to as "shattering." The obvious benefit to the plant of quick release of seed is that the species sows its seed for reproduction. The obvious disadvantage to man of shattering plants of domesticated species is loss of yield. One of the first steps in domesticating wild rice in Minnesota was the selection of nonshattering plants that occurred in low frequencies among natural populations. Necessary as this selection was, it served to reduce the genetic diversity of the populations. The domestication of wild rice proceeded with the intensive cropping of the species in prepared rice paddies. Almost immediately the cultivated wild rice was subjected to a leaf spot disease which rapidly reached epidemic proportions and now threatens the domestication of wild rice until resistance or some other appropriate control measure is obtained. The domestication of coffee and bananas in large plantations created a situation of intense cropping which no doubt contributed significantly to epidemics of coffee rust and banana wilt. Early man probably faced similar instances of disease problems as he domesticated wild species. To what extent early man selected for disease resistance and how efficient he was in doing so are not known.

Precisely when the use of disease resistance as a means of controlling plant diseases had its modern beginning depends largely on how one defines the term "modern." For example, Foex is credited with saving the grape industry of France a century ago by importing American varieties resistant to downy mildew. Considering that significant progress in selecting for disease resistance occurred near the beginning of this century, it may be appropriate to designate the era around 1900 as the modern beginning of disease control by hereditary methods. Certainly the belated appreciation about 1900 of Mendel's contributions provided a firm scientific basis for exploiting the inherent resistance present in most crop species to a multitude of causal agents.

Initial efforts to identify and obtain disease resistant plants utilized the process of selection from already available lines or varieties. In the United States Orton initiated a program of selecting cotton for resistance to fusarium wilt. Plants were tested in heavily infested soil; from those that appeared healthy, seed was saved for testing in the same soil the following season. Persistence in this method of selection and progeny testing led to the production of several varieties of Sea Island and Upland cotton. Success with the cotton wilt program was paralleled by progress with cowpeas resistant to wilt and to the root knot nematode. In each of these efforts Orton selected resistant cultivars from varieties already in cultivation.

The technique of within-variety selection was used by Bolley in North Dakota to seek resistance against fusarium wilt of flax at about the same time Orton was working with cowpeas and cotton. Flax grown in successive years in the same soil showed such a high incidence of *Fusarium lini* that only a few plants of a tested variety managed to survive. Progeny testing of these surviving individuals on "flax-sick" soil led to the production of several resistant varieties. But the resistance of the flax varieties proved to be unstable, and many investigations were initiated to pinpoint the underlying causes. Out of these studies came two well-established concepts: (1) the almost limitless abilities within the flax wilt organism to generate new races capable of attacking presumably resistant flax varieties; and (2) the dependence of the host on specific homozygous genotypes to combat specific strains of the wilt pathogen.

In order to obtain wilt resistant watermelons Orton had to resort to hybridization. The cultivated watermelon did not yield resistant individuals for progeny testing, but the citron melon, grown only for livestock, was remarkably resistant. The $F_2$ generation segregates produced resistant watermelon types, and subsequent selection and crossing to watermelon led to the introduction of the wilt resistant watermelon Conqueror. This is probably the first use of the back-cross method in breeding for disease resistance.

As early as 1901 Farrer in Australia reported on the breeding of wheat varieties resistant to bunt or stinking smut. Biffen, in England, crossed two wheat varieties, one resistant and one susceptible to stripe rust. Cognizant of Mendel's work, Biffen established that resistance was conditioned by a simple recessive factor and that resistance was inherited independently of other parental traits. His findings contributed to the reasoning that resistance could be combined with other characteristics desired in commercial varieties. Thus, a new dimension was added to the value of hybridization between different genotypes; a dimension that was to be used extensively in subsequent years.

## The Present

*If I have seen a little farther than others it is because I have stood on the shoulders of giants. . . .*

*Newton*

It is likely that more than 75 per cent of the current agricultural acreage in the United States is planted with varieties resistant to one or more plant diseases. When Coons (1953) estimated a one-half to three-quarters of a billion dollars financial benefit to growers in the United States from using resistant varieties in 1953, he estimated that 50 per cent of all crop acreage utilized resistant varieties. Compounded the world over today, benefits accrued from the use of resistant varieties certainly would amount to many billions of dollars annually.

Another means of viewing the economic value of disease resistance was illustrated succinctly by Stakman and Harrar (1957) when they stated, "were potatoes resistant to the principal pathogens that attack them in certain areas of the northern United States, the cost of production could be reduced at least $50 an acre." Similarly, Reitz (1954) discussed the value of breeding for disease resistance in the United States and estimated that "the increased income from wheat alone at $2 per bushel would return four-fold the annual cost of all tax-supported agricultural research." Although the development of resistant varieties is not inexpensive, the benefits more than justify the costs. Furthermore, resistance is the only means of disease control that does not add directly to cost of production, although seed of newly released resistant varieties may command a premium price until ample stocks are available.

Resistance enables crop plants to defend *themselves* against their pathogens or against a level of disease deleterious to the crop. All other control measures dictate man's continued involvement in performing one or more of a variety of operations before, during, and/or after crop production. Frequently, the relative success of other control measures is determined by the preciseness of their use. The success of resistance is not so subject to such external variation.

The genetic control of plant diseases is isolated in this book in order to focus attention upon the impact of accomplishments that have occurred and upon the methods that have made this approach a viable and dynamic part of modern agriculture. And yet the threads of scientific pursuit are so interwoven that one principle of disease control is seldom employed exclusively. Combinations of principles creep unobtrusively into almost every facet of plant disease control. The integrated use of disease resistance with principles of exclusion, eradication, physical or chemical protection, avoidance, therapy, cross-protection, and biological control is commonplace in modern agriculture, and it further illustrates the importance of genetic control of diseases. Combinations of principles are used to combat the same or different diseases. The value and/or stability of resistance to a particular disease is enhanced by eliminating weed or alternate hosts from cultivated areas, for example virus infected *Sorghum halepense* from corn fields, and barberries and buckthorn from areas grown intensively to wheat and oats. Turf grasses with partial resistance to *Helminthosporium* leaf spot diseases commonly are sprayed with fungicides to control the same diseases.

Similarly, potato varieties exhibiting some degree of resistance to the late blight pathogen are protected further by fungicides. The net result in both instances is either more effective control or adequate control at a reduced cost. Disease resistance often is more effective when crop rotation or sanitation is utilized, since both of these cultural practices frequently tend to reduce the amount of initial inoculum available for disease onset. The stability of disease resistance is enhanced by quarantine restrictions regulating the introduction of germ plasm of hosts and pathogens from other geographic areas. The inadvertent introduction of a new race of a pathogen can be disastrous when current resistant varieties have not been evaluated for their response to the race. Plowing-under of green cover crops reduces the incidence of scab of potatoes, presumably by creating a soil environment more conducive to certain microorganisms that are antagonistic to *Streptomyces scabies.*

Combinations of control measures to combat different diseases are virtually standard practices. It is unlikely that a single variety of any crop species will be developed with resistance to all races of all its pathogens. Resistance to one disease would be meaningless should the crop succumb to another disease. Most crops are subject to attack by one or more pathogens from planting until harvest. Thus it is commonplace to treat seeds of blight resistant maize hybrids with fungicides to combat damping-off and seedling blights. Wilt resistant vegetables are often sprayed for foliar diseases. When several diseases are potentially common threats to a crop, resistance to some diseases may make fungicidal control of the remaining ones an easier and economically feasible task.

The crop often determines the method of disease control to be employed. Crops of high unit value, such as orchard fruits, vegetables, glasshouse crops, and ornamentals, are commonly protected from disease by fungicides. Their value justifies the cost. Crops of low unit value have been protected largely by hereditary control. Wheat, oats, barley, corn, flax, forages, and sugarcane are not normally protected from disease principally by fungicides. Their value does not justify the cost. However, application of fungicides on wheat at critical periods in the development of the crop has shown considerable value in protecting the crop against a significant increase in disease. The principle of using fungicides to reduce the rate at which disease develops, rather than as a protectant in the classical sense, is virtually unexplored and may provide a new dimension to disease control. This would certainly be the case should breeding for disease resistance shift primary goals toward accenting field resistance rather than race-specific resistance.

# The Future

*Man has become the potential master of his own fate in his struggle for existence. Thanks to science and technology, he has the means to restrict his numbers simply and humanely and to increase his food supplies quickly and substantially. But to become the actual master of his fate he must have the will and wisdom to utilize fully his present means and the wit to devise better ones for the future. . . .*

*Stakman*

Our forefathers in the American colonies, as well as our ancestors in England, traditionally built villages and towns about a commons. Implicit in this arrangement was the understanding that all the families living in the village or town had the right to graze their cattle on the grass in the commons. This situation progressed beautifully until with time, as the number of families increased, there were too many cattle for all to graze on the commons. This parable has come to be known as the "Tragedy of the Commons." Today, with our advanced technology and a 20-20 vision gained from hindsight, we are prone to view our forefathers with a jaundiced eye for having planned so poorly for the inevitable problems of the future.

Any time a major agricultural crop of any nation is inundated by a widespread epidemic of a plant disease, that situation qualifies imminently as a modern-day tragedy of the commons. The 1970 epidemic of southern corn leaf blight in the United States is the most recent example of such a tragedy. The extreme susceptibility of corn hybrids produced in male-sterile cytoplasm and which occupied 90 per cent of the nation's acreage, and a virulent race of the fungus with an extraordinary appetite for them, formed the nucleus for the epidemic and for a retrospective look at science and crop production. It is clearly evident that the world's agriculture must return to basic principles. Male-sterile hybrids were indeed a triumph of scientific engineering, yet their very success has led to their demise. The knowledge that male-sterility could be incorporated into hybrids literally paved the way for a monoculture of our nation's most important agricultural crop. An epidemic was predictable years ago.

Scientists know that wild plant species have survived their evolutionary "struggle for existence" because of their diversity. Populations of plants find safety from widespread disease when all plants are not alike. Finding resistance genes in plants is one thing; using them wisely is another. Placing the same kind of resistance into all hybrids or varieties of a crop creates an instant monoculture. Placing different genes for resistance into different varieties creates diversity and safety from a widespread epidemic.

Man has the wherewithal to avoid future epidemics of plant diseases. The idea that epidemics of plant diseases are inevitable can no longer be entertained, and scientists have an obligation to use their knowledge to make certain that epidemics are avoided. Making that knowledge work for the world's benefit requires support in the form of money for research, and agencies able to provide that kind of support must review their role as an obligation just as intense as the obligation of scientists. If the recent epidemic of corn blight has taught us anything, it is that we can ill afford another similar "tragedy of the commons."

The control of plant diseases by genetic means has undergone extensive modification during the last 70 years. Techniques, goals, and philosophies are ever changing. This book highlights past accomplishments and points to the success and limitations of disease resistance. The past serves both as a source of encouragement and as a lesson in futility. The future is one of promise and challenge. The next 70 years should find disease resistance clearly established as the most effective and ideal means of controlling plant diseases.

Fortunately for man, most species of cultivated plants possess adequate resistance to most of their parasites. If they do not, their wild relatives do, and man can and does take advantage of the situation. The question we are faced with, and the one to which much of this book addresses itself, is concerned with how best to use what resistance we can find. It is inevitable that new races of plant pathogens will continue to appear from time to time, and they will do so irrespective of the relative resistance or susceptibility of currently grown cultivars. The best use of resistance genes would negate the consequences of newly appearing races by curbing their increased frequency among populations of pathogens.

The development of agronomically adaptive crop varieties with sufficient resistance to major diseases has enabled many developing countries to begin to meet their ever-growing demands for food. The continued success of the "green revolution" will hinge largely on our ability to maintain adequate resistance to an array of plant pathogens which actually hold the key to continued crop production. The current national concern about the use of pesticides places a great emphasis on disease control by genetic means. An understanding of the biochemical bases of resistance and the nature of gene action should add a further perspective to the artful manipulation of hereditary material. The extent to which plants can "live with" a certain amount of disease without deleterious effects must be investigated, for such knowledge could shift the emphasis from the often unstable specific resistance against specific races to a more generalized resistance conferring an acceptable level of protection against a broad spectrum of a pathogenic species. The stability of disease resistance will be enhanced by an increasing awareness of pathogen variation and by the means to combat or negate the consequences of such enormous plasticity.

An understanding of the genetic basis for the interaction of host and parasite is imperative for the most effective use of genetic material.

As is the case with much scientific endeavor, past approaches to the control of disease by resistant varieties often have been empirical. We know that certain end results will be obtained if certain procedures are followed. A better understanding about *why* certain varieties are resistant and others are susceptible is vital to the future success of resistance as an effective and lasting means of disease control.

Authors who have contributed chapters on specific crops have identified future needs and challenges for their crops, while the chapters in Part I are concerned with future needs in general. It is readily apparent that much remains to be done. Hopefully, it will be equally apparent that these goals can be accomplished.

## References

Coons, George H. 1953. Breeding for resistance to disease, p. 174-192. *In* Plant disease, the yearbook of agriculture. U.S. Government Printing Office, Washington, D.C.

Reitz, L. P. 1954. Wheat breeding and our food supply. Econ. Bot. 8:251-268.

Stakman, E. C., and J. George Harrar. 1957. Principles of plant pathology. The Ronald Press Company, New York. 581 p.

# 2 The Meaning of Disease Resistance in Plants
## R. R. Nelson

There is a tremendous amount of misunderstanding about the term "disease resistance," although misunderstandings are not so readily apparent when some general parameters of disease resistance are considered. For example, few would disagree that resistance is a means of disease control by natural, rather than by physical or chemical means, or that resistance connotes a hereditary struggle against some specific causal agent. An appreciation of the true nature of disease resistance results from a general understanding that resistance implies an active, dynamic response of the host to a parasite. The means by which plants *avoid* disease by avoiding infection should not be viewed as mechanisms of resistance. A host is actively involved when it resists the attempts of a parasite to derive all or part of its food supply from the host. Host resistance comes into play after infection is initiated. The normal, passive means by which plants avoid disease by avoiding infection exist before infection and are independent of any parasite.

Even if universal accord could be reached on the general parameters of disease resistance, for example that resistance is an active, dynamic response of the host to infection, the meaning of disease resistance would still be hazy. The principal problem hinges on the use of certain scientific terms and the meaning each conveys or is meant to convey. Many terms are used in speaking and writing of disease resistance. Several of these terms are borrowed from medical science and are used in their original sense or modified slightly so they apply more directly to disease in plants. Some terms apply to the interaction between host and causal agent, some apply specifically to the host, and others designate attributes of the causal agents. Since terms are symbols or abbreviations intended to convey specific thoughts, the need to equate the appropriate symbol with the appropriate thought is imperative, but this need is frequently not met. This chapter discusses the meanings of several important terms associated with disease resistance. No single definition or meaning of most of the terms treated herein would be universally accepted, and what follows reflects my concept of these terms. Other contributing authors to this book or workers in the field may apply different meanings to these terms, either by specific definition or by thought. Such differences seem inevitable.

## Resistance and Susceptibility

The terms "resistance" and "susceptability" are used to characterize those situations in which some degree of host-parasite interaction is evoked. Each term is relative, and the two terms are relative to one another. They represent a continuum of interactions on a single scale or spectrum. Because they are relative terms they are often used ambiguously and arbitrarily. When, for example, is a plant resistant and when is it susceptible? Plants are resistant or susceptible as different investigators deem them so.

The degree of resistance of a plant frequently is related to the relative incidence of infection and the relative extent of pathogenesis, and in this sense the terms are used erroneously. The resistance of a host is characterized by the amount of disease that it sustains or by the damage that it incurs. Disease, however, is not a character of the host, but rather is a product of the interrelationship of host and parasite under a given environment. A relative level of incompatibility between host and parasite usually results in a low level of disease, and from this comes the conclusion that the host is resistant. In fact, a low level of disease can be accounted for by two different phenomena: either the host has a sufficient capacity to defend, or the pathogen has an insufficient capacity to attack. Nonetheless, we do speak of genes for resistance in the host, and we do "look" for host genes that condition disease responses. It is in this context that disease resistance is discussed herein.

## Two Kinds of Disease Resistance

Disease resistance to plant pathogens is evidenced by one of two major kinds of host responses. The host either (1) resists the establishment of a successful parasitic relationship by restricting the infection site and the infection process, or (2) it resists the colonization and growth of the parasite subsequent to a successful infection, even though the infection process, culminated by reproduction of the parasite, is completed. Resistance is considered in this context as an active, dynamic response of the host to a parasite and it excludes such passive phenomena as immunity, klenducity, or disease escape, which are discussed later in this chapter. Resistance to the successful establishment of a viable parasitic relationship classically is referred to by such terms as hypersensitivity, specific resistance, nonuniform resistance, vertical resistance, or major gene resistance. Resistance to colonization and growth subsequent to infection is a host response characterized erroneously by the term tolerance and variously by the terms field resistance, generalized resistance, nonspecific resistance, partial resistance, uniform resistance, horizontal resistance, multigenic or polygenic resistance, and minor gene resistance. The fact that resistance is relative places the two kinds of resistance on a continuous scale. The difference between a pinpoint fleck reaction and a small necrotic lesion is a relative indication of host response to a parasite. The difference between a small necrotic lesion and a larger one is just as relative. Hypersensitivity at the cellular level, for example, technically can be viewed as resistance to growth. The main point is that two major kinds of resistance phenomena are evidenced in host reactions.

The terms vertical resistance and horizontal resistance have been popularized by Van der Plank (1968). They are probably not the best terms to use, because neither succinctly describes the kind of host response incited by a parasite or the relative effectiveness of the resistance to different races of the parasite. Although the terms race-specific resistance and nonspecific resistance may not be totally acceptable, they do delineate major host responses. These terms will be used to characterize the two main kinds of host reactions.

Race-specific resistance has been a major tool in efforts to control disease by genetic means. The merits of this kind of resistance have long been extolled; its shortcomings are becoming increasingly more clear. As the term implies, resistance of this type is dramatically effective against one or more races of a pathogen and dramatically ineffective against other races. It is an all-or-nothing resistance. Classically, race-specific resistance is expressed in the form of a hypersensitive response to the race or races against which the resistance is effective. It is a resistance against the establishment of a successful infection site. The resistance prevents the continued colonization of the parasite and consequently curbs the subsequent production of inoculum.

Genetically, race-specific resistance usually behaves as a single-gene trait. More often than not race-specific resistance is dominant over susceptibility. A single gene may confer resistance to one or many races of a pathogen. It is by far the most dramatic resistance reaction elicited by plants against their pathogens. The acceptance of race-specific resistance probably results from the fact that man is philosophically and scientifically schooled to react most favorably to the most dramatic host response. His search for the dramatic host response usually begins when current resistance has been matched by a new race of the pathogen. The search for a source of resistance to match the new race is accomplished by matching the race against many potential bases of resistance in field or greenhouse programs. This approach normally can yield only race-specific resistance. Race-specific resistance is easily recognized and readily attained because of its relatively simple inheritance.

The inability of races of a pathogen to overcome race-specific resistance usually is a dominant trait, while the ability to overcome the resistance is recessive. Because race-specific resistance functions only when a "resistant" host interacts with an "inable" pathogen, Flor (1955), working with flax rust, suggested a gene-for-gene relationship. Briefly, the theory states that for each gene conditioning race-specific resistance in the host there is a specific and related gene in the pathogen that conditions the ability to overcome that resistance.

While race-specific resistance reacts specifically against specific races, it is incorrect to assume that a gene-for-gene system must apply in all cases of this kind of resistance. Race-specific resistance may operate on a one-for-oneness and indeed does in some well-documented instances. However, the test for genes for race-specific resistance is not a one-for-oneness, but a differential reaction to different races. A differential reaction to different races would be the sole criterion for race-specific resistance even if it were shown that *all* examples of this kind of resistance functioned on a gene-for-gene basis.

It is generally agreed that resistance of any kind is subject to break-down when new races with novel pathogenic capacities arise by mutation, heterocaryosis, or genetic recombination. The pathogen nor-mally requires but a single genetic change to overcome a single gene for resistance. The instability of race-specific resistance in cereals to stem rust and in potato to late blight are ample testimony to the relative ease by which effective genetic changes occur among populations of plant pathogens. The relative instability of race-specific resistance can be ex-pected if for no reason other than the sheer probabilities that single-gene changes occur more often than multiple-gene changes.

From the standpoint of the onset and subsequent increase of disease among plant populations, race-specific resistance functions by reducing the initial amount of inoculum available for disease onset. For example, if races 1 and 2 comprised equal amounts of the initial available inoculum, and the host possessed a specific gene for resistance against race 1, the amount of functional initial inoculum would be reduced by one-half. The effective reduction of functional initial inoculum is one of the significant features of multiline varieties. The increased probability that portions of the available inoculum will come to rest on components of the multiline with race-specific resistance to the race(s) is directly proportional to the number of component lines comprising the multiline. It is probable that a synthetic, a variety of a cross-pollinated species produced by combining selected lines of plants with subsequent normal pollination, functions in a manner similar to that of a multiline variety by selectively restricting the impact of initial inoculum.

Following the initial establishment of successful infection sites, pathogens proceed further to colonize their hosts. A period of colonization is followed in most instances by a period of reproduction, which from a disease standpoint represents the production of inoculum for subsequent infection cycles and generations of disease. The ex-ceptions to this total pattern are those pathogens whose primary in-fection and subsequent colonization are restricted to one generation of disease. Smut species colonizing kernels or seeds of cereal crops are examples of pathogens that accomplish the disease cycle in a single in-fection process.

Plant species may possess several types of resistance mechanisms that come into play after infection sites are established. These mechanisms tend, in general, to restrict the extent to which pathogens can colonize host tissue and the relative degree to which pathogens are able to produce inoculum for subsequent infection or disease cycles within a single growing season. Resistance of this nature tends to reduce the amount of disease that develops, usually by reducing the amount of disease that occurs within a single infection cycle and by reducing the rate at which disease develops from one infection cycle to another.

Plant resistance to colonization and reproduction of pathogens essentially interferes with several different phases of pathogenesis. The incubation period of a pathogen may be defined as the time required from the initial stages of infection to the production of inoculum for a subsequent infection cycle. Plants may express resistance to pathogens by increasing the time required to complete the incubation period. Resistance also may be expressed by a restriction of the amount of tissue that is colonized from a single infection site. The end results may be smaller pustules or lesions, for example. A reduction in the amount of inoculum produced in colonized tissue frequently is a characteristic of plant species possessing post-infectional resistance. The time required for sporulation to occur may be extended in colonized tissue that has reached a stage where sporulation is possible and normally expected. These major effects by plants restricting the development of their pathogens all function to decrease the rate of infection within and among populations of plants.

Certain general characteristics of resistance mechanisms affect the rate of infection of pathogens. To the casual observer, it is not a spectacular form of resistance. Disease is present, and the impact of the resistance on the progression of disease is not evident when viewed at a single point in time. There are virtues to resistance to disease development that are becoming patently evident. Resistance to spread is not race-specific, at least in the sense of all-or-nothing effects on different races, as we have noted for race-specific resistance. It can be termed appropriately as nonspecific resistance if certain general parameters are understood. There is a growing misconception that nonspecific resistance reacts uniformly against all races of a pathogen. While the gross effects are the same, the degree to which nonspecific resistance functions against different races can vary to a considerable extent. A brief discussion of a hypothetical model will illustrate the point. Assume that there are only three races of a pathogenic species. These races are capable of inciting disease on a variety with no genes for race-specific resistance and no genes for nonspecific resistance. The variety is "susceptible" to the three races but not equally so, because of inherent differences in virulence among the races. The most virulent race is able to incite more disease and does so in a shorter period of time. A second variety has no genes for resistance that are race-specific to the three races, but it does possess resistance genes that reduce the rate of disease development by reducing infection rates within and among different plants. The incubation period of two of the three races is lengthened by three days on the second variety as compared to their capabilities on the first variety. The incubation period of the third and most virulent race is lengthened by only one day on the second variety as compared to the first variety. The variety with genes for resistance reacts uniformly against the three races in that the incubation period of the three races is lengthened, and yet it reacts differentially to the three races with respect to its relative ability to lengthen the incubation period of the three races. In the broad sense, these resistance genes confer nonspecific resistance to the variety because disease increase is reduced against all races of the pathogen. That the nonspecific resistance is not uniformly effective against all races should be expected in light of the known abilities of races or different isolates of races to differ in virulence.

Van der Plank (1968) portrays horizontal resistance as reacting uniformly against all races of a pathogen. In that sense, there appears to be no evidence that horizontal resistance to any plant pathogen exists in any plant species. There are, however, many examples in which plant selections or cultivars express a level of resistance to colonization or spread of disease. Whether this kind of resistance is effective against all races and uniformly so against all races are the issues at stake when horizontal resistance is discussed. Since there is no evidence that these criteria can be met in any host-pathogen situation, it seems advisable either to discard the term horizontal resistance or to modify its meaning.

Nonspecific resistance usually is conditioned by the combined action of several genes. Its polygenic nature probably accounts for the relative stability of the resistance over long periods of time. The resistance is stable because races with the necessary genes to overcome it are less likely to arise. Whereas single genetic changes in the pathogen often are sufficient to overcome race-specific resistance conditioned by a single resistance gene, several different genetic changes probably are needed by the pathogen to overcome resistance that is polygenic in nature. The probability that any given race can acquire, accumulate, and maintain all the necessary genetic changes is lessened. Therefore, stability of nonspecific, polygenic resistance seems to be based on probabilities of sequential events occurring in the pathogen.

The "loss" of polygenic, nonspecific resistance usually is gradual and is seldom complete. The stepwise loss of resistance can be subtle and may be evident only to the keen observer. Whereas "complete" nonspecific resistance may function to increase the incubation period of a pathogen by three days, for example, "partial" nonspecific resistance, resulting from the loss of one or more of the components of the polygenic system, may be effective in increasing the incubation period by two days. Similar relative efficiencies of complete and partial polygenic resistance are evidenced in the effects of resistance on lesion size and number and on the various aspects of the sporulation process.

## Hypersensitivity

"Hypersensitivity" is a ubiquitous term in the literature on parasitism and disease resistance. Since Ward (1902) first focused attention on the hypersensitive reaction (HR) as a potentially significant defense mechanism in brome to brown rust, the phenomenon has been the subject of extensive investigation and interpretation. Muller (1959) described HR as "all morphological and histological changes that, when produced by an infectious agent, elicit the premature dying off (necrosis) of the infected tissue as well as inactivation and localization of the infectious agent." Fungi, bacteria, viruses, and nematodes have been reported to elicit the HR. The sequence of events which take place, the time required for the initiation and culmination of the phenomenon, and the extent to which various pathogens are inactivated vary for different host-pathogen interactions.

Hypersensitivity has come to imply death of the host cells and the pathogen in a majority of host-parasite interactions. A notable exception seems to be the viruses, which can be recovered from the hypersensitive reactions. Apparently the assumption that fungal pathogens are permanently inactivated is no universally valid. It is now known that certain of the cereal rust pathogens that invoke the hypersensitive response are not dead but merely quiescent, and they can resume colonization under proper environmental conditions after long periods of time. The author has recovered *Helminthosporium carbonum* and *Helminthosporium turcicum* from hypersensitive reactions several weeks after the response was elicited on certain corn inbreds. In fact, the hypersensitive reaction to *H. carbonum* induced in young corn plants with certain genotypes often remained as such until the plants approached maturity, at which time the necrotic flecks began to enlarge to form small lesions. The physiological phenomena associated with senescence apparently lessened the resistance of the host. A long latent period of infection is associated with physiological activity of the host. The anthracnose pathogens of banana and tomato infect green, immature fruits, but symptoms of the disease are not evident until the fruits are

ripe. Potato clones carrying *R* genes from *Solanum demissum* that confer a hypersensitive response to certain races of the late blight pathogen
often develop lesions on the older, lower leaves while the young leaves
still react by hypersensitivity. Host defenses involved in an active,
dynamic resistance probably are related to the presence and availability
of specific metabolites. Plants approaching maturity or senescence may
fail to produce the metabolites at all, or they may produce them in insufficient amounts. Sporulation by isolates of the southern corn leaf
blight pathogens is more abundant in lesions on older leaves than in
those on younger ones. Defense, once again, seems to be chemically
founded.

It is standard procedure to evaluate the hypersensitivity of plants
within a few days after inoculation. The occurrence of the HR is often
considered to be the ultimate, irrevocable response of the host to the
pathogen. However, in those instances in which the reaction actually
results in a temporary inactivation of the pathogen, the HR is the
response of the host to the initial stages of the infection process. The subsequent response of the host would never be recognized by some of the
current methods of evaluating the HR. The initial evaluation in those
instances is in a sense an artifact. Hypersensitivity should not be viewed
as the permanent inactivation of the pathogen until appropriate studies
have shown this to be the case in each specific host-parasite interaction.

## Tolerance

Tolerance may be defined as the inherent or acquired capacity to endure disease. Little is known practically or theoretically about tolerance,
the mechanisms underlying the phenomenon, or its inheritance. It is difficult to classify precisely the position that tolerance occupies among the
many kinds of host response to parasites. From the standpoint of
relative sensitivity to disease-inciting agents, tolerance and hypersensitivity appear to represent the two extremes in host responses to infection. However, to equate tolerance and hypersensitivity to opposite
ends of a single, continuous spectrum of host reactions dictates that
tolerance be characterized as a resistance phenomenon. Theoretically, at
least, tolerant plants are not resistant to a parasite. Tolerant plants are
susceptible to infection and susceptible to spread and colonization of the
parasite, but they exhibit tolerance of the parasite by resisting the impact
of disease. Thus, tolerance accomplishes the same net result as active
resistance mechanisms by obviating the consequences of infection.

Tolerance negates infection by a desensitization of the plant, whereas hypersensitivity, for example, presumably negates infection by localizing infection sites.

Theoretically, tolerance may be operative when losses from disease are disproportionately less than what the occurrence of disease would suggest. Disease losses are measured most effectively by assessing the yield of the product.

Certain plants do seem capable of enduring more disease than other plants. Certain maize inbreds and hybrids exhibit a high degree of susceptibility to maize dwarf mosaic virus, as judged by symptom expression, but suffer remarkably little loss of yield in field plot yield trial studies. Tuthill and Decker (1941) studied the effect of leaf roll virus on the yield of Cobbler and Chippewa potato varieties. Regardless of the percentage of leaf roll, the yield of Cobbler was reduced more than that of Chippewa. With 100% leaf roll infection, Cobbler exhibited a 53.0% reduction in yield, whereas the yield of Chippewa was reduced by 40.7%. Chippewa appeared to be more tolerant of the virus and the effects of the disease.

Simons (1969) has studied tolerance in oats to crown rust and has concluded that the trait is quantitatively inherited. Several selections of different species of *Avena* have been reported to exhibit tolerance to the barley yellow dwarf virus. Schafer (1971) recently has prepared an excellent factual and conceptual review of tolerance to plant disease.

The concept of tolerance cannot be applied to diseases that affect the end-product of the host. There can be no tolerance in the cereals to smut pathogens that attack the kernels. Cultivars of alfalfa can exhibit no tolerance to foliage-attacking pathogens if the crop is grown and harvested for hay. The same rationale applies to the various diseases of fruits, vegetables, and so forth.

The tendency to equate tolerance to a form of nonspecific resistance seems erroneous. Although tolerance and nonspecific resistance may function similarly to minimize losses incurred from disease, they do so by entirely different means. Nonspecific resistance reduces losses by reducing the amount of disease and the rate of disease increase, features that are not associated with tolerance. To equate the two would be analogous to equating immunity to disease escape on the basis that disease is absent in each case.

One underlying philosophy supporting the value of nonspecific resistance is that plants can "live with" a certain amount of disease without incurring appreciable ill-effects. If a similar philosophy can be extended to tolerance, the need to explore the traits conferring the capacity to endure disease and the hereditability of those traits seems imperative.

# Disease Escape

The term "disease escape" is used to characterize those situations in which inherently susceptible plants do not become diseased because they do not become infected. Escape from disease is a hazard of all programs concerned with development of resistant varieties. Unless all factors favoring disease development are operating to give a finite and discrete evaluation to genetic differences among plants, true plant reactions are obscured. Disease ratings often vary from year to year, from plot to plot, from region to region, and from researcher to researcher. Variation may be caused by differences in the selection pressure placed upon plant populations. The hazards of disease escape in breeding programs are discussed in greater detail in Chapter 3.

Disease escape also is a useful and practical means of avoiding disease under natural conditions. Disease escape can be related to factors associated with the pathogen, its host, or the environment. Susceptible plants frequently escape disease or escape damaging levels of disease when the inoculum of a pathogen is absent or below a critical level, or when the pathogen is in an inactive stage. Nonexposure of susceptible plant populations to their pathogens is fairly common. Wheat planted early in certain areas of Washington tends to escape infection by the bunt pathogen because soil moisture and soil temperature are unfavorable for the germination of chlamydospores of the smut fungus. Potato varieties susceptible to powdery scab can be grown in warm climates. Sweet corn varieties susceptible to bacterial wilt escape the disease in Canada and New England, where low winter temperatures are not favorable for the survival of the flea beetle vector. Sugar beet varieties susceptible to the curly top virus can be grown in areas where the leaf hopper vector does not occur. Early planted cabbage often escapes the full brunt of fusarium wilt in northern areas of the United States. In this case, early planted varieties develop a larger root system and a more advanced kind of vascular root system by the time climatic conditions are most favorable for disease onset and progression. Early planted cabbages do not escape infection, but they do escape the full brunt of the disease. Similarly, early maturing potatoes susceptible to late blight escape the full impact of the disease because temperature-moisture relationships do not favor rapid blight development early in the season.

Examples of disease escape can be attributed variously to morphological, physiological, or functional properties of the host. Certain corn hybrids produce ears that turn outward and hang down and thus prevent the occurrence of favorable micro-environmental conditions necessary for certain ear-rotting organisms. It is possible to select lines of barley that retain the panicle in the boot until self-pollination is completed, thus preventing exposure of the pistil to inoculum of the loose smut organism. Although van der Plank (1968) characterizes such barley varieties as having "horizontal resistance amounting to practically absolute immunity," the closed inflorescences enable those susceptible varieties to escape disease rather than to resist the parasite in an active, dynamic manner so typical of reactions attributed to horizontal resistance.

Varieties of raspberries susceptible to the virus causing raspberry mosaic frequently escape the disease because the plant is a nonpreferred host of the insect vector. Where a choice exists, the vector prefers to feed on a more delectable host. The term "klenducity" is used commonly to characterize the type of disease escape accomplished by the avoidance of the insect vector.

## Immunity

Immunity in plants is absolute (100%) freedom from disease. Plants are immune to disease because they are immune to infection. Immune plants may or may not be immune to the active penetration or the passive ingress of an organism, both of which are preinfectional processes.

It is widely accepted that immunity is the usual condition in plants, whereas some measure of host response to parasites is a rarity. Such a conclusion is developed from the reasoning that apples, for example, are immune to virtually all plant pathogens while interacting actively with varying degrees of resistance or susceptibility to relatively few pathogens. While it is true that plant species are parasitized by relatively few causal organisms, the belief that they are immune to the remaining parasites places the concept of immunity into the realm of academic interest rather than of practical importance. To be a meaningful concept of significance to disease control by hereditary means, immunity should be viewed as follows: immunity to members of a specific pathogenic species should not be ascribed to a member or members of a plant species until it has been demonstrated that other members of the same plant species are not immune to members of the same pathogenic species. Such a concept of immunity has meaning in our consideration of disease control. It would be of extreme practical significance if it were determined that a clone of *Solanum tuberosum* was immune to all isolates of *Phytophthora infestans*, the incitant of late blight of potatoes. It would be of little significance if it were determined also that the same clone or all clones of *S. tuberosum* were immune to all isolates of *Puccinia graminis tritici*, the causal agent of stem rust of wheat, since the rust fungus is not a parasite of potatoes. To put it another way, the knowledge that *S. tuberosum* is immune to *P. graminis tritici* is of no practical assistance in developing immunity or resistance in potato to any of its known pathogens.

Immunity as characterized herein probably is a rarity in plants. To find a tomato line immune to all members of even one of the tomato pathogens is an unlikely quest. Because immunity is absolute freedom from disease, the use of the terms "virtually immune," "almost immune," and "near immunity" should be avoided. The term "immune response" occurs frequently in the literature. Its use is improper because immunity is a passive, static characteristic of a plant species. A pathogenic entity can evoke no response from a passive agent. Immunity to disease is a highly desirable trait, but it probably plays a minor role in the heritable control of plant diseases.

# References

Flor, H. H. 1955. Host-parasite interaction in flax rust — its genetics and other implications. Phytopathology 45:680-685.

Muller, K. O. 1959. Hypersensitivity, p. 469-519. *In* J. G. Horsfall and A. E. Dimond, [ ed.] Plant pathology — an advanced treatise, Vol. 1. Academic Press, New York.

Schafer, John F. 1971. Tolerance to plant disease. Annu. Rev. Phytopathol. 9:235-252.

Simons, M. D. 1969. Heritability of crown rust tolerance in oats. Phytopathology 59:1329-1333.

Tuthill, C. S., and P. Decker. 1941. Losses in yield caused by leaf roll of potatoes. Amer. Potato J. 18:136-139.

Van der Plank, J. E. 1968. Disease resistance in plants. Academic Press, N.Y. 206 p.

Ward, H. M. 1902. On the relations between host and parasite in the Bromes and their brown rust, *Puccinia dispersa* (Erikss.). Ann. Bot. (London) 16:233-315.

# 3 The Detection and Stability of Disease Resistance

R. R. Nelson
D. R. MacKenzie[1]

Assuming that a plant species possesses genes for resistance, or that resistance factors are present in a related, cross-compatible species, the selection and development of disease resistant plant populations usually is a relatively straightforward process when disease is present and constantly exerting an adequate selection pressure on host material heterogeneous for disease reaction. A notable exception to this generality often is apparent in cases where nonspecific or generalized resistance is sought in order to control disease by curbing disease increase. Detecting and incorporating nonspecific resistance theoretically is a straightforward program, but in practice it is difficult to accomplish for reasons discussed in Chapter 2.

Genetic units conditioning plant responses to pathogens can be identified and manipulated only when disease symptoms point to their presence. The relative resistance or susceptibility of a plant is its maximum response to representative members of a pathogen evaluated under epiphytotic conditions over a sufficient period of time to reveal the true limits of its reaction. The maximum pathic response is evaluated

1. Rockefeller Foundation Post-Doctoral Fellow in Plant Pathology.

comparatively with a known susceptible genotype whose prime function is to point to the presence or absence of conditions suitable for the occurrence of maximum disease.

The true limits of disease reaction frequently are not attained when one or more factors operate suboptimally to create an inferior selection pressure. The failure to detect the maximum pathic response inevitably promotes serious consequences because, in such cases, the plant's true maximum response to disease is poorer than presumed. The most obvious consequence of an inferior selection pressure is the instability of the presumed resistance.

Disease escape can be disastrous in programs designed to detect and comparatively evaluate levels of resistance among populations of plants. Selecting for resistance in the presence of an inadequate amount of disease creates an instant artifact. The time required to incorporate promising genetic material into commercially or agronomically desirable cultivars often spans many years, and to begin the process with less resistance than is either presumed or necessary is a misuse of time and funds.

Few programs of selection for resistance among variable plant populations can rely solely on natural infection. The benefactors of the exceptions to this general rule are most fortunate. Artificial inoculations are the backbone of many selection programs. The methods of inoculation, the timing of inoculation, and the number of times that inoculum should be applied vary for each disease problem. The desired end result, however, is the same in all cases, that is, adequate disease for optimum selection.

It is possible to have adequate disease and still lack an adequate selection pressure. The contributing factor in such a case is the pathogen. Populations of a plant pathogen are as variable as or more variable than populations of their host. Different isolates or populations of isolates can and usually do exhibit marked differences in pathogenicity, virulence, aggressiveness, or fitness. The selection of the most resistant plants from a disease nursery must always be tempered by the knowledge that the selected plants exhibit known resistance only to what they were evaluated against. To what extent the isolate or isolates used as inoculum are representative of the range of variation within the species is a key and critical factor. It may not be possible to formulate a general rule regarding numbers and sources of isolates needed to constitute an adequate and representative inoculum, but one can speak to the question in terms of probabilities. Inoculum comprised of a single isolate is less likely to be representative of the range of variation that exists within a species than inoculum comprised of 50 isolates; 50

isolates collected from one location will be less representative than 50 isolates from 50 locations; and so on. The crucial principle is, the more representative the inoculum, the greater the chance for stability of the selected resistance.

The maintenance of inoculum for subsequent use is an important aspect of screening for disease resistance. Fungal and bacterial pathogens are notorious for their tendencies to lose pathogenicity or virulence when maintained in vitro by certain means. Prolonged culturing on agar substrates is risky, particularly when sporulation capacities are considered. There are countless examples in which the relative ability to incite a given amount of disease is correlated directly with sporulation. Spores usually are more infective agents than mycelia. Mutations in vitro affecting sporulation inevitably result in a reversion to mycelial types with markedly reduced sporulation capacities. Mycelial colonies exhibit faster growth rates than sporulating colonies. Consequently, mutations to mycelial types within an in vitro culture quickly result in a largely mycelial culture. A less effective inoculum is the end result.

Inoculum should be maintained in a manner that tends to keep the pathogen in an inactive or dormant state. Lyophilization and storage under liquid nitrogen are two effective means of storing most bacteria, sporulating fungi, and certain viruses in a dormant state for prolonged periods of time. Cultures can be recovered in their original state, usually with a high degree of viability. Cultures of soil pathogens are maintained effectively in vials of soil. Many fungi and bacteria are preserved readily in water blanks stored at temperatures slightly above freezing. Fungal pathogens inciting foliar diseases are maintained in diseased tissue stored under dry conditions to prevent saprophytic colonization.

Susceptible check lines in a breeding nursery can point to an insufficient level of disease, but they cannot identify the reason(s). With the variety of effective storage techniques available, the reason for insufficient disease need not be an ineffective inoculum.

The kind of resistance sought dictates the methods used for its detection. Plant selections invoking a hypersensitive response to a race of a pathogen can be evaluated readily from the results of primary inoculations, usually in a short period of time. Seedling tests often are reliable in these cases and are amenable to greenhouse facilities. Reaction type usually is a valid criterion of host response.

Resistance factors that serve to reduce the amount and rate of disease development cannot be detected by methods of screening that use host response to primary inoculation as a criterion for resistance or susceptibility. Resistance of this type must be evaluated over time. Seedling tests are of no value, and greenhouse screening is not possible unless facilities are available to reproduce climatic regimes necessary for generations of disease over time.

Suitable resistance in plants may not be detected when disease evaluations are made within certain parameters. Differences in the nature or magnitude of host reactions at different physiological ages of plants often complicate the detection of resistance or restrict the periods at which disease evaluations can be made. Selection for disease resistance in the seedling stage has a number of obvious advantages. Large numbers of seedling plants can be screened in a relatively small area in a greenhouse or growth chamber. Critical control of environmental conditions often is possible in enclosed facilities. Differences in the relative virulence among pathogenic isolates are detectable more readily under controlled conditions. However, seedling tests are of value only when seedling reaction is essentially similar to the response of the plant under field conditions at the physiological stage of maturity of the plant when it is subject to attack by its pathogen(s). Plants frequently are very susceptible to attack in the young seedling stage but demonstrate adequate resistance as they reach certain stages of maturity. Many workers have noted a decrease in the amount of rice blast as the leaf stage increases. For example, there is more disease at the 5-leaf stage than at the 10-leaf stage. It is unusual to find resistance in corn seedlings to *Helminthosporium turcicum,* the incitant of northern leaf blight. Inherently resistant lines of corn begin to exhibit their true resistance to the parasite after the 7-9-leaf stage. Detectable resistance in corn seedlings possessing the dominant gene *Ht* for resistance to northern leaf blight often is displayed when the gene is in an appropriate genetic background.

Altering the nutrition of a plant can alter its disease expression. In general, this effect is evident only on plants with intermediate levels of resistance or susceptibility. Highly susceptible plants usually exhibit a similar degree of susceptibility under varying nutritional conditions. A plant responding to a parasite by a hypersensitive reaction probably would do so under vastly different nutritional regimes.

Disease expression can be altered by excesses or deficiencies of major or minor elements, or by an imbalance of nutrients. Apparently, no general rule is applicable to a given element and its effect on disease. High levels of nitrogen, potassium, or phosphorus result in increased disease for some host-parasite systems and decreased disease in others.

Temperature can alter disease expression. Some resistance genes are temperature-sensitive. In wheat and oats, certain genes associated with resistance to stem rust or leaf rust are functional under given temperature regimes and are less effective against certain races under other regimes. There is a general tendency of temperature-sensitive genes to be less effective at higher temperatures.

Light intensity or day length can alter disease resistance. Soil moisture content can affect host response. Frequently, increasing the water content of soil results in increased disease, a phenomenon which may be related to increased succulence of host tissue. It is patently clear that external factors can influence host response to infection and colonization and thereby affect the true measure of resistance or susceptibility that a given plant inherently may possess. These factors can complicate the selection and development of resistant cultivars. Regional testing for disease resistance frequently results in dramatically different results among locations. Needless to say, promising selections should be evaluated under the different environmental conditions that prevail in the areas of potential use of the cultivar.

## Epidemiology

Plant pathology was largely a descriptive science in its tender years. The separation of plant pathology from mycology was academic. As information accumulated and as new concepts were introduced it was inevitable that the study of plant disease should assume new postures. Today there are many branches of plant pathology, one of which is concerned with the increase of disease. This discipline is called *epidemiology*. Epidemiology may be defined as a study of the factors that affect the rate of increase of disease among populations of plants over time.

The essence of epidemiology is the disease cycle. The disease cycle has no beginning and no end: one segment leads to the next. The particulars differ among pathogens and their hosts, but the overall cycle remains the same. Propagules depart from infected tissue, travel to susceptible tissue, enter, colonize, and then produce more propagules, continuing the cycle. The faster the cycle moves, the greater the amount of disease.

The rate at which the disease cycle occurs is of concern to epidemiologists and is becoming of increased concern to breeding for disease resistance. The application of epidemiology to breeding for disease resistance has been evident in two major areas. The now classical failure of hypersensitive, race-specific resistance to withstand continual race changes of the pathogen has caused an increased attention to the less ephemeral forms of resistance. From an understanding of the fundamental types of resistance that function primarily or exclusively to reduce the rate of disease increase and the different techniques needed to identify them, new hope has been added to overcoming many disease problems that heretofore seemed hopeless.

The second contribution of epidemiology to breeding for disease resistance has been an increased knowledge of how to create an epidemic at will. The ability of any breeding program successfully to identify resistant progeny depends upon the development of substantial levels of disease. An awareness of the factors contributing to rapid increase of disease allows a choice of locations, environmental factors, dates of planting, and so forth which will contribute to intense levels of disease.

The growing awareness of resistance mechanisms that reduce the amount of disease by curbing the rate of disease increase is accompanied by a similar awareness of the factors that influence increase of disease. It is on the latter point that research on disease resistance has made a major contribution to epidemiology. Historically, epidemiological research to monitor increases of disease has evaluated the impact of time and climatic parameters on various phases of the infection cycle. Primary attention has been given to the influence of temperature, relative humidity, light, and dew periods on sporulation, spore dispersal, infection, and colonization of tissue. Little attention has been given to the remaining members of the disease triad, the host and the pathogen. Interpretative and predictive epidemiology can be misleading when the assumption is made that different populations of a pathogen race and different varieties of a susceptible complex act as constants and are prone to respond similarly to various climatic regimes. Studies with different isolates of race T of *Helminthosporium maydis,* incitant of southern corn leaf blight, and different susceptible varieties in male-sterile cytoplasm are illustrative. Just as a little increase in dew can account for a great increase in disease, so can a little decrease in virulence or epidemiological fitness of race T account for a great decrease in disease. The decrease in disease is essentially similar to the decrease associated with decreasing dew periods or decreasing temperatures for colonization. Similarly, a slight decrease in susceptibility of a susceptible corn hybrid can account for a substantial decrease in disease. The main effect is on the increased time required for colonization and the increased time for lesion development to reach a stage conducive for maximum sporulation. The decrease in disease is essentially similar to the decrease incurred by decreasing colonization temperatures 5 C below the optimum range.

Nonspecific resistance or field resistance functions to reduce disease by reducing the rate of increase. The potential value of this kind of resistance is of sufficient significance to warrant a brief treatment of the underlying principles of epidemiology.

The components of an epidemic are susceptible tissue, pathogenic organism, environmental influences, and time. None of these components is independent. To return to the general disease cycle, it must be stressed that each segment is critical in the development of the epidemic. Each must be reviewed in relation to the necessary components. For example, the landing of viable spores on resistant material slows the increase in disease. Propagules traveling through dry air may not be viable upon landing. Colonization and sporulation are known to be affected by several environmental factors including soil fertility, air temperature, and moisture. Time is a factor common to all phases of the development of an epidemic.

# The Mathematical Expression of an Epidemic

## Model I: Disease as a Function of Time

Weekly observations in a wheat field might give the following observations on stem rust during the course of an epidemic. A trace (0.05%) of pustules was observed in a large planting of wheat known to be susceptible to the prevalent races of stem rust. At the end of the first week the frequency of pustules approached 2%. Curiosity turns to apprehension as the percentage of infected pustules progresses from 2 to 5, 10, 20, 50, 95 and 100% infection after 2 months observation (Figure 1). Although the data have been fabricated to illustrate several points, the curve does show what might happen during an epidemic. With an understanding of the disease cycle, we can consider the sequential steps as one and go directly to an inquiry of what has happened.

Let us first consider the nature of the epidemic curve. When the plotted values are connected with a smooth line, a sigmoid or S-shaped curve results. From 0 to 2 weeks the curve is slow to develop. Next the epidemic increases rapidly, and the curve flattens out as it approaches 100% after 8 weeks of observation. There are reasons for this type of curve. If 1 lesion gives rise to 10 lesions, then 10 will give rise to 100. These 100 lesions will give rise to 1,000, and so on. Progress is slow in the beginning but, as in the case of interest being added to an investment, the amount grows in size and a fixed rate of interest brings increasingly greater returns on the investment. Consider one dollar invested at 10% per annum versus one million dollars invested at the same rate. The principle is the same for disease as for money: the larger investment earns more return. This is reflected in the first half of the sigmoid curve. The flattening out of the curve does not represent a decrease in the rate of interest paid, but a limit on the ability to pay. As susceptible tissue runs out, the epidemic comes to a slow but inevitable halt.

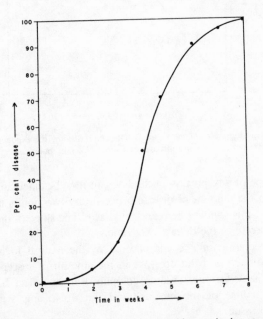

Figure 1.   The increase of disease in time. The sigmoid curve is characteristic of plant disease development. Data have been fabricated to demonstrate the general disease curve.

Sigmoid curves are, by their nature, difficult to characterize. However, they can be straightened by expressing the vertical axis in a logarithmic scale (Figure 2). Portions of the previously sigmoid curve can now be expressed quite readily in mathematical terms. The poor fit of the line, particularly at the higher values, will be explained later.

Let the proportion of disease be represented by $X$. Subscripts can then denote at which point in time we are referring to a given quantity of disease, so that $X_2$ is the amount of disease present after 2 weeks of observation, and $X_5$ is the amount present after 5 weeks.

Let a point in time (in weeks) be represented by $t$. Subscripts can then denote the point in time to which we are referring, so that $t_2$ represents 2 weeks and $t_5$ represents 5 weeks.

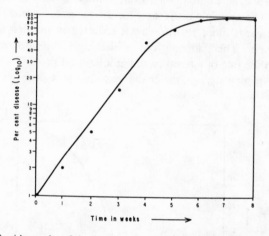

Figure 2. Semi-$\log_{10}$ plot of the Figure 1 data demonstrating the exponential nature of the relationship between disease increase and time. Note the poor correction, especially near the end of the epidemic.

The slope of any portion of the line can now be calculated from the increase in the amount of disease between any two points in time. Let the slope of this segment be represented as $r$. The slope of the line is the rate of increase of the disease over time.

The calculation of $r$ is complicated by the natural logarithmic increase of disease and the decrease in the amount of susceptible tissue. Therefore, any formula used to compute $r$ must reflect both the exponential nature of the increase and the decreasing availability of susceptible tissue.

The exponential nature of the increase of disease can be accounted for by using the natural log of the proportion of disease. Expressed mathematically this is $\log_e X$. However, the expression does not allow for the decrease in the amount of susceptible tissue as the pathogen progresses through the epidemic. The direct relationship between diseased tissue and healthy susceptible tissue is as $X$ (proportion of disease) is to $1 - X$ (proportion of healthy but susceptible tissue). It is the ratio between these two that is of importance. Therefore, $X/1 - X$ is the ratio between what is now diseased and what is still available to the pathogen. The natural logarithm of this relationship, $\log_e X/1 - X$, corrects the faults encountered in the plot shown in Figure 2. The corrected version is shown in Figure 3.

Now that biases have been corrected, we can proceed to calculate the slope ($r$) of the line. The desired value is the relationship between the amount of disease at two points in time. Let one point in time be $t_2$ and a second $t_5$. This is the portion of the epidemic of most confidence, the exponential phase of the epidemic. The amount of disease at each point in time is, by our notation, $\log_e (X_2/1 - X_2)$ and $\log_e(X_5/1 - X_5)$. The difference between these two values is the increase in disease. The difference between $t_2$ amd $t_5$ is the amount of time elapsed during this increase. The ratio between the increase in disease and the amount of time is $r$, that is, the increase in disease per unit time. The general formula for the calculation of $r$ is expressed as:

$$r = \frac{1}{t_2 - t_1} \left[ \log_e \left( \frac{X_2}{1 - X_2} - \frac{X_1}{1 - X_1} \right) \right]$$

where appropriate subscript numerals can be substituted for the time to be studied. For our data, $r = 0.182$ per unit per day.

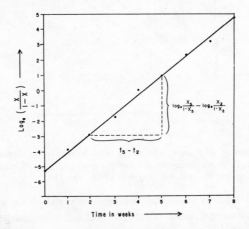

Figure 3.   Data of Figure 1 corrected for exponential nature of disease increase and the diminishing availability of susceptible tissue. See text for explanation of the values $t_5-t_2$ and $\log_e \frac{X_5}{1-X_5} - \log_e \frac{X_2}{1-X_2}$ and their application to computing epidemic rates.

Further complication of this relationship is not within the scope of this chapter. It must be stressed, however, that this mathematical treatment of an epidemic is done with full knowledge of the effects of the environment on the rate of increase of disease. As knowledge of these effects is gathered, epidemiologists will become more able to predict changes in $r$.

Breeders must be concerned with the effects of resistance on disease development ($r$). These effects can be of two types, and both affect the curve presented in Figure 3. The first effect is exemplified by hypersensitive resistance. Races of many pathogen species have been reported. When a new race appears that can overcome hypersensitive resistance, the new race functions as if the hypersensitive gene were not present. The slope of the epidemic curve ($r$) is unchanged (Figure 4). Mixtures of races are most commonly found in nature. Hypersensitive hosts selectively screen out those races able to produce disease. It is these races that are able to increase and their rate of increase ($r$) is not affected by the hypersensitive genes.

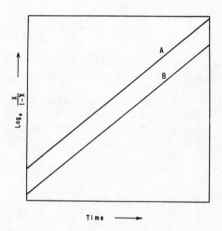

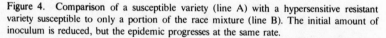

Figure 4.   Comparison of a susceptible variety (line A) with a hypersensitive resistant variety susceptible to only a portion of the race mixture (line B). The initial amount of inoculum is reduced, but the epidemic progresses at the same rate.

The second type of resistance has been referred to as nonspecific, generalized, horizontal, field, polygenic, multigenic, minor gene and so forth. The variety of names reflects the present state of confusion concerning this form of resistance. The single identifying characteristic has been, until recently, that this form of resistance lasted for years with "no deterioration" of resistance resulting from new races. Van der Plank (1963; 1968) has proposed that another identifying characteristic can be used to demonstrate what he calls horizontal resistance. This trait is the rate ($r$) of increase of disease in time. Graphic comparison of the two types of reactions is given in Figure 5, where a susceptible variety offering no resistance to disease increase is compared to a variety with nonspecific resistance. Races of a pathogen are said to react similarly to this form of resistance. No new races will arise to overcome this form of resistance. Resistance, as the argument continues, should be everlasting.

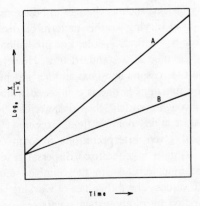

Figure 5. Comparison of a susceptible variety (line A) and a variety with general resistance (line B). The rate of epidemic increase is less for the general resistant variety regardless of the race composition of the pathogen population.

## Model II: Disease as a Function of Distance

The mathematical description of disease is an attempt to express a multidimensional system in a two-dimensional scheme. The faults of expressing the increase of disease as a function of time lead to the in-

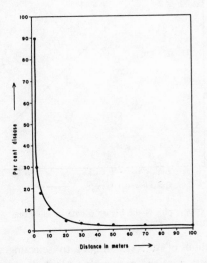

Figure 6. The dispersal of disease over distance. Data have been fabricated to demonstrate the relationship of the dispersal of plant pathogens and, therefore, of plant disease.

tegration of other factors, such as changes in host resistance as the plant matures, the effects of varying weather patterns on disease increases, and so forth, which are not entirely visible. For prediction into the future it is reasonable to use time as a standard unit. However, for other purposes it can be just as reasonable to use another independent variable of which disease is a function. If one is concerned with the rate at which disease spreads through a variety, then distance can be used as the unit of increase of disease at any point in time. Gregory (1968) has reviewed the literature and given interpretations of plant disease dispersal gradients. Figure 6 gives a generalized dispersal gradient from a local source of disease inoculum. Again the data are fabricated, but they represent realistic values. The difficulties of working with disease curves can again be avoided by plotting data points on log-log paper, which straightens the relationship between disease and its dispersal from a point source (Figure 7). This is justifiable from a wealth of knowledge accumulated in the study of particle dispersal patterns.

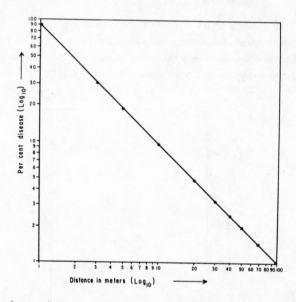

Figure 7.   $Log_{10}$ x $log_{10}$ plot of the Figure 6 data to show one method of straightening disease dispersal curves.

The application of such techniques to selecting resistant varieties awaits the rejection of the notion that plant disease is a stagnant affliction of a singular plant. Little progress can be expected with such antiquated views of plant disease, regardless of disease severity, that is, endemic or epidemic.

# References

Gregory, P. H. 1968. Interpreting plant dispersal gradients. Annu. Rev. Phytopathol. 6:189-212.

Van der Plank, J. E. 1963. Plant diseases: epidemics and control. Academic Press, New York. 349 p.

————. 1968. Disease resistance in plants. Academic Press, New York. 206 p.

# 4 Pathogen Variation and Host Resistance
## R. R. Nelson

Plant pathogenic microorganisms are notorious for their ability to vary. Their capacities to generate increased or novel parasitic abilities are the principal reasons for man's continuing efforts to control plant diseases by genetic means. Stable resistance would be commonplace in an era of stable pathogens.

Plant pathogens effectively exercise the conventional mechanisms of genetic variation and employ other means which appear to be unique to certain of their members. Most plant parasites change rapidly and dramatically by mutation and by genetic recombination through "conventional" sexual processes. Although mutations constitute the ultimate source of genetic diversity, the reassortment of genetic material during the sexual process is a significant mechanism contributing to pathogen variation within many fungal pathogens. Variation in certain parasitic attributes may be dependent upon the combined or additive effects of two or more genes. In such cases, single mutations within populations would not improve or change a particular attribute. The recombination of different mutant genes into a single genotype during the sexual process would provide the appropriate gene combination to create a new pathogenic variant.

Genetic reassortment via parasexuality or mitotic recombination is an acknowledged means of variation within some fungal pathogens. Heterocaryosis, the coordinated capacities of genetically dissimilar nuclei in vegetative or asexual systems, appears to be an important mechanism for variation within certain fungal parasites. Heterocaryosis may, in fact, constitute the most important means of variability in fungal species that lack a functional sexual process. Although the mechanisms controlling the parasexual cycle, mitotic recombination, and hetero-caryosis are poorly understood, the consequences of the variation derived from these phenomena place additional stresses on those con-cerned with disease control by hereditary means. It is not the purpose of this chapter to discuss how these and the conventional mechanisms func-tion, nor to document them as functional phenomena for given pathogens. Instead, we will focus on the kinds of variation that occur and discuss their consequences relative to disease control.

# *Terminology*

Although relatively few terms are used to designate specific attributes of plant pathogens, the fact that each of these specific terms is used by dif-ferent researchers to connote different attributes has created confusion among plant pathologists regarding their meaning and the intent that in-dividual authors mean to convey. Terminology relating to disease-inciting agents is important to our consideration of disease resistance, and consequently those terms will be treated briefly.

## *Pathogenicity*

"Pathogenicity," simply defined, is the ability of an organism to incite disease on given members of a host species. Conversely, an organism is nonpathogenic when it is unable to incite disease on given members of a host species. The appropriate antonym of pathogenic is nonpathogenic.

A species is not an organism or an entity of any kind, but is rather an abstraction created artificially by man to facilitate classification and nomenclature. Isolates of a species are living entities, and pathogenicity should rightly be considered as their individual ability to incite disease on given members (genotypes) of a host species. Pathogenicity is not an attribute of a species, even though it is often stated as such. Pathogenicity is a qualitative attribute. An organism (isolate) is either pathogenic, or it is nonpathogenic.

## Virulence

The amount of disease that an isolate incites on a given member of a host species is a measure of its virulence. Virulence is relative. The virulence of one isolate pathogenic to one or more specific host genotypes cannot be measured. Virulence can be evaluated only when two or more isolates pathogenic to the same specific host genotype are compared for the relative amount of disease that each incites. Virulence of an isolate is not determined by the relative resistance or susceptibility of its host, because virulence is not an attribute of the host. For example, an isolate pathogenic to cultivars A and B of the same host species incites more disease on A than on B. The virulence of the isolate cannot be measured, because virulence is a relative attribute of the pathogen. In this example, all that can be said is that cultivar A is more susceptible than B to the isolate or, conversely, that B is more resistant than A.

The term "avirulence" appears commonly in the literature. Its use should be avoided. Because virulence is relative, there can be no zero virulence or avirulence. An isolate incapable of inciting disease on a given genotype is nonpathogenic rather than avirulent.

Virulence can be measured in a number of ways, although all methods use the amount of disease incited as the prime index. In the case of pathogens inciting leaf spots of foliage, for example, virulence commonly is evaluated by the relative number or size of leaf spots incited on a specific host genotype by two or more isolates.

## Aggressiveness

There are instances in which two pathogenic isolates exhibit the same virulence on a host by virtue of inciting the same amount of disease. Similarly, there are instances when two such isolates require different amounts of time to incite the same amount of disease. The isolate causing the same amount of disease in less time often is considered to be more aggressive. Time aspects of parasitism probably should not be considered as a component of virulence.

Disease is a product of the interaction of the host, its parasite, and the environment. Changes in any of the three factors can result in less disease or in no disease. An organism pathogenic to a given plant under certain climatic conditions may be nonpathogenic to the same plant under other conditions. The same applies to the relative virulence or aggressiveness of a pathogenic isolate. Thus, the characterization of the pathogenicity of an isolate or an assessment of its virulence and aggressiveness must be made in context with a given host and given environmental parameters.

# Kinds of Variation

Variation within populations of plant pathogens essentially encompasses changes in five principal attributes. These attributes are: (1) the ability to incite disease; (2) the ability to incite a greater amount of disease; (3) the ability to incite disease more effectively or more efficiently; (4) the ability to incite disease under a wider range of environmental conditions; and, (5) the ability better to persist. The first two attributes are concerned with the relative ability of a species to create new and significant populations. The latter three attributes are concerned primarily with the relative ability of new populations to become more dominant members of the species. Populations acquiring new pathogenic capacities or increased virulence would pose relatively few problems to man if the populations were unable to increase in frequency among other populations; they certainly would not have a profound effect on the stability of disease resistance.

An epidemic of a plant disease is incited by a virulent race that is epidemiologically able to cause widespread disease among populations of plants. Fitness attributes are the keys to epidemic populations. The improved ability to persist in the nonparasitic stage enables populations of races of nonobligate parasites to generate greater amounts of initial inoculum for onset of disease. The ability to incite initial and subsequent infections under a wide range of environmental conditions often results in earlier disease onset, which in turn lengthens the time available for disease increase. An improved infection efficiency (the number of infections resulting from a given inoculum) and a superior disease efficiency (the number of lesions or pustules resulting from a given number of infection sites) contribute to increased disease. A shortened generation cycle (the time required from infection to production of subsequent inoculum) permits greater numbers of generations of disease over a fixed time. Several ramifications of inoculum production directly affect parasitic fitness. The capacity to generate inoculum under a broad range of climatic regimes is a significant component of fitness. Greater inoculum production per given area of diseased tissue promotes greater disease increase. An increase in the number of cycles of inoculum production from the same diseased tissue is an obvious attribute. Populations of a virulent race may possess all or none of these fitness capacities, if we assume that fitness properties are conditioned by genetic factors different from those controlling pathogenicity or virulence. Existing evidence strongly supports that assumption.

The most conspicuous consequence of pathogen variation is the loss of resistance of a currently grown cultivar that had been presumed to possess resistance to prevailing populations of the pathogen. Loss of resistance in such instances occurs most often in cultivars utilizing race-specific genes for resistance. Presumably, isolates of a pathogen usually attain their increased parasitic capacities for race-specific varieties by acquiring one or a very few "new" genes. The relationship between the breakdown of race-specific resistance and pathogen variation is discussed in greater detail in Chapter 2. It is sufficient to say here that efforts to utilize race-specific, hypersensitive resistance have made variation worthwhile to plant pathogens.

Spectacular increases in disease can occur for different reasons or combinations of reasons. The prime requisite is the presence of a population of the pathogen with the necessary genetic capacity to incite a spectacular level of disease. Such a population must possess the necessary fitness attributes to incite disease under climatic regimes wherever the cultivar is grown. Finally, the susceptible cultivar must occupy vast acreages of the crop. Monoculture is a significant key to epidemic diseases.

Variation resulting in an improved capacity to survive is a significant phenomenon influencing the relative dominance among populations of plant pathogens. The ability of a population to survive better than other populations is a prerequisite to its increased frequency. Superior survival qualities of nonobligate parasites are concerned with the nonparasitic and parasitic stages of their life cycles. Nonparasitic survival can be equated to all situations in which the pathogen is not functioning as an active parasite of its living host. No special reservations are necessary to discuss the different conditions under which nonparasites survive or their different survival structures. The same principles apply, for example, to survival in soil and survival in plant debris, or to survival as sclerotia or chlamydospores and as mycelial strands. The relative extent of survival under different conditions or by different structures may vary for a given pathogen; for example, foliar pathogens may survive better on debris above ground than below ground, or a given pathogen may persist better by sclerotia than by mycelial fragments. Nonetheless, the principles of survival and their importance to population dynamics are the same.

The ability of a biotic plant pathogen to survive saprophytically in a given area may be a significant epidemiological factor conditioning the frequency, prevalence, and severity of the disease that it incites. The dependence on introduced inoculum from other areas for primary infection and disease onset influences the regularity with which the disease occurs from year to year as well as the severity, at least in instances where disease severity is correlated with time of disease onset. Yield losses often are conditioned by the stage of maturity of the plant at the time disease becomes well established. Losses from most foliar diseases of maize are greater when infections are heavy before tasseling, while grain yields are seldom affected when disease does not become severe until the plants are reaching maturity. As a general rule, disease losses should be greater when primary inoculum is local rather then introduced.

The nature of the host may affect the relative ability of plant pathogens to survive saprophytically. Studies by Nelson and Scheifele (1970) on the overwintering of *Trichometasphaeria turcica (Helminthosporium turcicum)*, incitant of northern leaf blight of maize, suggest that more susceptible lines of maize may serve as poorer hosts for the overwintering phase of the pathogen. Similar results were reported by Nelson (1971) for the overwintering of *Helminthosporium maydis,* incitant of southern leaf blight of maize. The explanation in both cases may be nutritional. In the presence of a pathogen, more susceptible cultivars should sustain more disease. When there is more diseased tissue, there is less disease-free tissue. From the onset of the overwintering phase to the time when climatic regimes suitable for disease onset are present in the subsequent season, several climatic periods probably occur that are sufficient to activate the dormant pathogen to grow and sporulate. Presumably the pathogen draws nutrients during these periods of activity from disease-free tissue surrounding the diseased tissue. Available nutrients in the tissue would be depleted as the pathogen initiates additional cycles of growth and sporulation. The pathogen presumably cannot be activated in the absence of a nutritional substrate.

Overwintering tissues are colonized abundantly by saprophytic species apparently because of the ability of these species to utilize the available degradation metabolites produced by pathogen colonization and saprophytic growth. Their external presence on tissues materially reduces the ability of pathogens to sporulate and, in that sense, restricts their relative survival.

Different isolates and populations of isolates of a pathogen exhibit differential survival capacities in their nonparasitic phase. The studies by Nelson and Scheifele (1970) and Nelson (1971) demonstrated that certain isolates of *H. turcicum* and *H. maydis* dominated the overwintering populations on different maize lines. Nonparasitic survival was not correlated with abilities to cause disease in the parasitic phase. Fitness for one asset is independent of fitness for another. Differential survival among populations of plant pathogens should be expected in light of their known capacities to vary for almost any identifiable trait.

Populations of nonobligate parasites that exhibit poor survival qualities in nonparasitic situations would have to possess superior abilities to increase during parasitic phases to remain dominant members of a species.

Pathogen variation frequently results in an extension of the host range of members of a pathogenic species. At least it is unlikely that a pathogen with a wide host range possessed all of the necessary genes for parasitism at its point of origin. When the inability to cause disease is a dominant trait, as so often is the case, mutations from dominance to recessiveness theoretically can contribute to new parasitic abilities. It is known that different genes condition the pathogenicity of isolates to different host species.

Although pathogens may not always incite serious levels of disease on all of their hosts, extended host ranges provide additional sources of overwintering for nonobligate parasites. Secondary hosts may also serve as contributors of primary or secondary inoculum. In such instances, some of the usual benefits of crop rotation could be minimized.

The differential suitability of host cultivars as overwintering substrates and the differential ability of pathogen populations to survive saprophytically may have important implications to field programs concerned with breeding for disease resistance. It is not an uncommon practice to plant border rows in disease nurseries with highly susceptible lines and collect the diseased material for use as inoculum for the subsequent year's nursery. Should the susceptible line(s) prove to be a poor host for the fungus to survive saprophytically, or should the host serve as a selective medium for strains with better survival abilities, the amount or nature of the available inoculum may not be appropriate for selecting maximum resistance. There is no evidence available from which to presume categorically that better survival abilities are associated with increased virulence.

Survival qualities are key epidemiological traits of plant pathogens. The identification of these traits among populations of pathogens is an important tool for early prediction of disease occurrence and severity. Too little research has been devoted to this area.

It is not uncommon for new, virulent races to remain in low frequencies for many years. Their belated rise to dominance among races probably can be attributed to the acquiring of traits which enabled them to compete favorably with other races. Their rise to dominance usually threatens currently grown cultivars. Extensive efforts are devoted to the monitoring of new races, and similar efforts should be directed constantly towards their relative prevalence and distribution. Monitoring for improved fitness may avert epidemic occurrence of some diseases.

The ability of a pathogen to incite disease under a wide range of climatic conditions promotes its increased frequency. All pathogens operate under definable climatic controls. Each has an optimal, minimal, and maximal climatic regime in which it can infect, colonize, and sporulate. Expanding the optimum, lowering the minimum, or raising the maximum results in more disease or a more rapid disease increase.

Some recent unpublished research by the author and coworkers on the epidemiology of southern corn leaf blight illustrates the above points. The 1970 epidemic of the disease in the United States was caused by a race, designated race T, of *Helminthosporium maydis* with an extraordinary capacity to parasitize corn hybrids produced with male-sterile cytoplasm; hybrids which occupied approximately 90% of the 1970 acreage. It is now known (Nelson et al., 1970) that isolates of race T have existed throughout many areas of the world for many years, the earliest collection having been obtained in North Carolina in 1955.

The pathogen requires dew for infection and sporulation. Higher temperatures favor more rapid colonization and increased numbers of infections under shorter dew periods. Sporulation is more abundant in darkness and at higher temperatures, to a point.

Comparative studies with the 1955 isolate of race T and a representative isolate of the 1970 population revealed some substantial differences in certain fitness attributes. The relative ability to cause infection was studied at dew temperatures of 12-28 C, at 2-degree intervals, with 2-12-hour dew periods, at 2-hour intervals. Pertinent to our discussion is that the 1970 isolate could consistently cause infection with 4 hours of dew at 12 and 14 C, while the 1955 isolate failed to do so in most of the cases. Furthermore, colonization was more rapid by the 1970 isolate at temperatures of 20 and 25 C. Finally, the 1970 isolate required shorter dew periods for sporulation and produced up to 15 times more spores per given area under climatic regimes favorable for both isolates. By these criteria, the 1970 isolate was more fit to cause an epidemic. Our studies prompt us to speculate that the 1955 isolate could not have caused an epidemic, even in the presence of vast acreages of susceptible hybrids. Its inability to operate efficiently under climatic regimes that occur during the initial portions of the growing season in many areas would tend to delay disease onset and restrict disease increase. And, an epidemic is time.

# References

Nelson, R. R. 1971. Studies and observations on the overwintering and survival of isolates of *Helminthosporium maydis* on corn. Plant Dis. Reporter 55:99-103.

————, and G. L. Scheifele. 1970. Factors affecting the overwintering of *Trichometasphaeria turcica* on maize. Phytopathology 60:369-370.

————, J. E. Ayers, H. Cole, and D. H. Petersen. 1970. Studies and observations on the past occurrence and geographical distribution of isolates of race T of *Helminthosporium maydis*. Plant Dis. Reporter 54:1123-1126.

# 5 The Use of Resistance Genes to Curb Population Shifts in Plant Pathogens

## R. R. Nelson

Attempts to control plant diseases by the use of resistant cultivars have resulted in frequent and sometimes spectacular shifts in the racial composition of populations of plant pathogens. Such shifts have led to the increasing prevalence of races capable of attacking widely used and presumably resistant germ plasm. The cyclic rise and fall of resistant cultivars is well known and needs no specific documentation; the economic consequences of their demise are equally well known.

Fortunately for man, most species of cultivated plants or their wild relatives possess adequate resistance to most of their parasites. The issue at hand is how best to use what resistance we can find. It is inevitable that new races of plant pathogens will continue to appear from time to time. The best use of resistance genes would encompass gene usage that negates the consequences of new races by curbing their increased frequency among existing populations. This chapter evaluates the prospects for stabilizing racial populations of pathogens by use of resistance genes. Several potential approaches have been suggested. Some of them are controversial and, therefore, my examination of their credentials may be deemed controversial.

# Two Kinds of Disease Resistance

There are two major kinds of disease resistance in plants. The host either resists the establishment of a successful parasitic relationship by restricting the infection site and the infection process, or it resists the colonization and growth of the parasite in the host subsequent to infection. Resistance to the successful establishment of a viable parasitic relationship has been referred to by such terms as hypersensitivity, specific resistance, nonuniform resistance, vertical resistance, or major gene resistance. We will use the term vertical resistance (VR), because one of the concepts under scrutiny in this chapter uses that term. Resistance to colonization and growth subsequent to infection is a host response characterized variably by the terms field resistance, generalized resistance, nonspecific resistance, uniform resistance, horizontal resistance, and minor gene resistance. We will use the term horizontal resistance (HR), because it is the suggested alternate of VR.

Cultivars are considered to possess genes for VR if they react differentially to races of a pathogen. While VR genes confer an effective degree of resistance against a certain race or races, they are ineffective against other races; plants lacking VR to a given race usually are highly susceptible to that race. Genes for VR function against epidemic development of plant diseases by reducing the amount of effective inoculum available for disease onset. Races lacking virulence genes to match VR genes are essentially disqualified from epidemic involvement. From a genetic standpoint, VR is usually, but not necessarily, conditioned by a single gene. More often than not VR is dominant over susceptibility.

In contrast to VR, HR functions by reducing the amount of disease that develops and the rate of disease increase. Reduction in the amount and rate of disease increase can be accomplished by host mechanisms that retard penetration, increase the incubation period, restrict the amount of tissue that is colonized from a single infection site, and reduce the amount and duration of inoculum production or sporulation.

Some or all of the effects of HR are imposed upon all races of a pathogen. Resistance to disease increase is not race-specific, although the degree to which HR functions against different races varies considerably. From a genetic standpoint, HR usually is polygenic in inheritance.

# Origin and Nature of Pathogenic Races

Analysis of population shifts of plant pathogens and the role that resistance genes may play in curbing such shifts have become subjects of increased interest to those concerned with disease resistance. Several points are germane to the issue. The increased frequency of virulent races and the concurrent susceptibility of a host cultivar usually occur when cultivars have been developed with VR to a specific race. In all probability, single-gene resistance can be overcome by a single gene change in the pathogen. Because the ability of a race to match a VR gene usually is recessive, a mutation from dominance to recessiveness renders a race virulent against a cultivar with a particular VR gene. Races of plant pathogens arise by mutation, genetic recombination, genetic reassortment via mitotic recombination, and heterocaryosis. Races arise totally independently of the relative resistance or susceptibility of their hosts. To assume any degree of dependence upon their hosts would necessitate accepting the idea of "directed" origin, for which there is no evidence. Host genotypes may influence the ultimate frequency or the sustained presence of new races, but not their origin.

A race that increases in distribution and frequency among populations of a plant pathogen by virtue of its ability to match a particular gene for VR must exhibit two fundamental assets. The race must, of course, possess the necessary gene(s) for virulence that negates the gene for VR. It also must possess the necessary abilities that enable the race to tend to become a more dominant member of the species. These abilities can be characterized in a general way as "fitness attributes" and include: (1) the ability to attack a suitable host under different environmental regimes in which the host is grown; (2) the ability to produce greater amounts of inoculum over a longer period of time; (3) the ability to become disseminated over wide geographic areas; and (4) the ability to persist or survive in the absence of the primary host, either as a parasite on alternate or secondary hosts, as a saprophyte, or in a resting or dormant stage.

Other pertinent attributes may not directly affect the ultimate frequency of a race, but do directly influence the rate at which that frequency is attained. Races with the ability to cause a greater amount of disease in a shorter period of time with the resulting production of greater amounts of inoculum for subsequent generations of disease would assume a more dominant position among populations of races in a shorter period, provided they possessed the other necessary fitness attributes.

# Curbing Population Shifts by Horizontal Resistance Genes

More often than not, polygenic HR has remained stable and effective for long periods of time. The stability of HR appears to rest on genetic probabilities. A strain of a pathogen probably would have to acquire several new genetic abilities to overcome resistance conditioned by several genes. Whether the strain would be required to match the host gene-for-gene is not important: each genetic "improvement" in the pathogen probably is attained independently of other improvements involved in overcoming polygenic resistance. Thus, the probability that any given strain can acquire and then maintain all the necessary genetic changes is lessened. The stability of HR, then, seems to be based on the probability of occurrence of necessary sequential events in the pathogen.

Certain potato (*Solanum* spp.) lines have exhibited stable HR in Mexico to a variety of races of *Phytophthora infestans* (Mont.) de Bary, the late blight organism. However, there is some recent indication of a gradual erosion of the HR in some potato lines. The fact that the erosion is gradual supports the idea that the pathogen needs to acquire several new genes to completely overcome HR. The future use of HR to curb population shifts seems very promising. The problems inherent in detecting genes for HR and incorporating them into acceptable cultivars are another story.

# Curbing Population Shifts by Vertical Resistance Genes

Four primary uses of genes for VR have been suggested as potential means of curbing population shifts. Three of these will be treated briefly, while the fourth, the concept of stabilizing selection, will be probed in depth because of its controversial nature.

*Multilines.*   Multilines are belends of different genotypes, each of which, in the simplest case, contains a different gene for VR. The excellent and comprehensive review by Browning and Frey (1969) on multilines as a means of disease control is recommended highly. Multilines exhibit VR by reducing the amount of initial inoculum available for successful infection. Inoculum of races lacking virulence genes to match particular VR genes is ineffective when it lands on components of the multiline containing those VR genes. Multilines also exhibit HR by curbing subsequent increase of disease for the same reason: they reduce the amount of effective inoculum of the right races at the right places.

Can multilines curb racial shifts? Theoretically, the answer probably is a qualified yes. The qualifications are concerned with the number of VR genes available in different host species, the number of VR genes necessary in a given multiline to confer adequate protection to new races, and with what constitutes "adequate" protection. Genetic probabilities again are an issue. For example, it is unlikely that a race would acquire all of the necessary virulence genes to match all VR genes in a multiline with seven components possessing seven VR genes. It is probable that a race could acquire a few of the seven necessary virulence genes and, as a consequence, some disease would occur. Whether that level of disease is acceptable or whether new VR genes could be substituted for the ineffective ones are important questions. Some have considered multilines to be a conservative approach to disease control. In a sense that is true, particularly if one is a devotee of the race-specific resistance syndrome where disease is virtually absent during the "good years."

*Gene deployment.*    Historically, a new VR gene which confers protection against the prevalent pathogen race is incorporated rapidly into cultivars wherever the crop is grown. The "instant" monoculture invites new races to increase in frequency, and therein lies the frequent demise of the VR gene. As an alternative, it has been suggested that different, effective VR genes be deployed in and restricted to different geographical regions of a country in which the crop spans a sizable area. What we are considering, essentially, is a geographical multiline, and the usefulness of gene deployment to curb population shifts can be considered similar to that of a multiline. The geographical deployment of three VR genes, for example, risks severe disease losses in one area with the occurrence of a race that has acquired the matching virulence gene, while protecting the two remaining areas from the wrath of the new race. I would consider the merits of gene deployment to be less than those of a multiline in curbing population shifts, although the mechanics of gene deployment are easier.

*Pyramiding vertical resistance genes.*    It has been suggested that several or many VR genes be incorporated into a single genetic background to curb population shifts. Thus, as opposed to our seven-component multiline, for example, all seven VR genes would be used in a single genotype. Admitting bias, the author thinks pyramiding VR genes to be a most promising approach. The author believes, with some evidence (Nelson, MacKenzie, and Scheifele, 1970), that VR genes functioning collectively in a single genetic background can contribute to or confer horizontal resistance. There are those who express great concern over "spending" all of our VR genes at one time. Nonetheless, this author casts his vote for the pyramid and considers that approach superior to gene deployment or multilines.

*The concept of stabilizing selection.*    Can genes for race-specific or VR contribute in another way to stabilizing population shifts in pathogens? Van der Plank (1968) thinks so. His concept of stabilizing selection explains how VR genes influence the selection and survival of races within variable populations. The concept revolves about genetically simple and complex members of a host and its parasite. A simple host cultivar possesses few, if any, genes for VR, whereas a complex cultivar possesses many VR genes. A simple race of the pathogen is characterized by its ability to attack only simple host cultivars and, as such, it would possess few genes for virulence. A complex race is able to attack complex cultivars with many VR genes by virtue of possessing many virulence genes. Parasite complexity is relative to host simplicity. Complex races, therefore, have unnecessary genes for virulence as they attack simple cultivars. Simple races attacking simple cultivars have few, if any, unnecessary genes for virulence.

Stabilizing selection means that simple races are more fit to survive than complex races. The rationale for the improved survival ability of simple races is different for obligate parasites and nonobligate parasites. Simple races of obligate parasites are considered to be more fit to parasitize simple cultivars than are complex races, whose collection of virulence genes for complex cultivars is unnecessary against a simple host. The simple host is the stabilizing factor for obligate parasites. Data from research and observations with stem rust (*Puccinia graminis* Pers. f. sp. *tritici* Eriks. & E. Henn.) of wheat (*Triticum aestivum*) in Australia in 1958 (Watson, 1958) and late blight of potato (Graham, 1955) have been cited (van der Plank, 1968) in support of the concept and are summarized briefly herein. Many races of stem rust occur throughout the eastern Wheat Belt of Australia. The relative simplicity or complexity of the races, with respect to their genes for virulence, can be determined readily by the reaction of standard and supplemental differential wheat cultivars, several of which possess known specific genes for VR. Stem rust in the southern zone of the Wheat Belt has not posed a sufficiently serious threat to warrant breeding cultivars with specific resistance genes to counteract specific races. These common cultivars are assumed to lack VR genes and are further assumed to qualify as simple cultivars according to van der Plank's definition. Stem rust in the northern zone periodically reached serious proportions, necessitating the breeding of cultivars with VR to a particular race. During the 1957-58 survey of the eastern Wheat Belt, the most commonly grown wheat cultivars in the northern zone contained a single gene for VR.

Simple races dominated the parasitic population in both zones. In the South, where cultivars contained no VR, the race with no unnecessary genes for virulence comprised 62% of the population. Races with one, two, three, and four unnecessary virulence genes comprised 22, 13, 3, and 0.2% of the population, respectively. In the North, where the popular cultivars contained a single VR gene, the race with only the necessary gene to match the VR gene comprised 68% of the racial population. Other races declined in frequency in proportion to the number of unnecessary virulence genes; the most complex race formed only 1% of the population.

A similar story exists for late blight of potatoes. It is appropriate because of man's efforts to combat races of the pathogen with VR genes transferred from *Solanum demissum* Lindl. to the cultivated potato, *Solanum tuberosum* L. As VR genes were incorporated into *tuberosum*, new races of *Phytophthora infestans* were detected with virulence genes to match the resistance genes. Single VR genes and combinations of VR genes were effective against some races and ineffective against others, establishing the vertical nature of their effects.

The simple races of *P. infestans* appeared most fit to survive on simple cultivars under natural conditions and in greenhouse studies. The race with no virulence gene to match any of the VR genes from *S. demissum* was identified most frequently on older cultivars of *S. tuberosum* without VR genes. Races with just enough virulence genes to match the VR genes of different cultivars were more common on those cultivars than races with both necessary and unnecessary virulence genes. In greenhouse studies, Thurston (1961) passed a simple race without virulence genes to match any VR genes in various combinations with races havinh 1-3 virulence genes specific for different VR genes for 1-8 generations on a cultivar having no VR genes. The simple race dominated each mixture at the conclusion of the varying numbers of generations on the simple cultivar.

Other data support the idea that simple races of obligate parasites are more fit than complex races to survive on simple cultivars. Some of these are discussed by van der Plank (1968), and others are reviewed or cited by Browning and Frey (1969) and Watson (1970). They will not be reviewed here because of space limitations. Their omission should not be construed as a minimization of their significance nor of their support for the concept.

The stabilizing of races of obligate parasites in favor of races with fewer virulence genes has focused on the impact of simple and complex cultivars. Simple cultivars stabilize racial populations at reduced levels of virulence, whereas complex cultivars exert a selection pressure in favor of complex races. Stabilizing selection for obligate parasites can operate only in the presence of two different host genotypes. The assumption is correctly made that all plants of a genetically homogeneous cultivar exhibiting VR to a pathogen will be resistant. Some plants of a cultivar cannot serve as a source of inoculum to other plants of the same cultivar. For stabilizing selection to function, the inoculum must come from a second cultivar; a simple cultivar without the VR exhibited by the first. Races able to match virulence genes against the VR of the first cultivar will be less fit to survive on the simple cultivar, thus effectively reducing their potential inoculum. The bulk of the inoculum produced on simple cultivars will be that of simple races unable to attack the cultivar with VR.

Van der Plank (1968) has given recent attention to stabilizing selection as it applies to nonobligate parasites. Nonobligate parasites exist or persist in a nonparasitic phase as active saprophytes or in a resting stage, and they provide primary inoculum to reinitiate the parasitic phase.

Van der Plank (1968) suggests that the prevalence in nature of simple races over complex races of nonobligate parasites is a result of their superior capacities to exist in the nonparasitic phase. That is, it is the selection for fitness to compete on the medium on which the pathogen exists during its nonparasitic phase that stabilizes the races.

A single valid example is cited by van der Plank (1968) in support of his contention that the saprophytic medium is the stabilizing force for nonobligate parasites. It has to do with fusarium wilt of tomatoes (*Lycopersicon esculentum*) incited by *Fusarium oxysporum* f. sp. *lycopersici*. Cultivars carrying the single vertical gene *I* are resistant to race 1 but susceptible to race 2 of the pathogen. Although race 2 was first reported in 1945, it did not increase substantially in nature and had not posed a threat to widely grown tomato cultivars susceptible to race 2. Van der Plank attributes the apparent curbing of race 2 to effective stabilizing selection caused by the low ability of race 2 to exist or persist saprophytically in the soil. The current status of race 2 will be discussed later. There are instances in which VR genes have not stabilized racial populations. Races of the wheat stem rust fungus ably exist in Australia with virulence genes effective against VR genes *Sr* 8 and *Sr* 15, even though gene *Sr* 8 has never been used in commercial wheat cultivars in that country, and gene *Sr* 15 has been used only sparsely. The race of *P. infestans* with the virulence gene necessary to overcome the potato resistance gene *R* 4 is as fit to survive on cultivars lacking VR genes as is the simple race without virulence genes. Complex races of the flax (*Liinum usitatissimum*) wilt pathogen (*Fusarium oxysporum* f. sp. *lini*) have proven equally fit to survive saprophytically in the soil. In fact, races with virulence genes effective against VR genes were present even before the VR genes were utilized.

Van der Plank (1968) accounts for the aforementioned exceptions to stabilizing selection by distinguishing VR genes as either "strong" or "weak" genes. Vertical resistance genes are strong if races able to match them are stabilized on simple cultivars (for obligate parasites) or in their nonparasitic phase (for nonobligate parasites). Vertical resistance genes are weak if races able to match them are equally fit to survive on simple hosts, in the case of obligate parasites, or in their nonparasitic phase(s), in the case of nonobligate parasites. Thus, Van der Plank (1968) says that genes *Sr* 8 and *Sr* 15 in wheat, gene *R* 4 in potato, and the VR genes in flax are weak genes; races with virulence genes to match these VR genes are equally fit to survive. Other VR genes in wheat including gene *Sr* 6, certain *R* genes in potato, and *I* gene in tomato are said to be strong genes: races with virulence genes to match these VR genes are less fit to survive.

Strength is relative. Some strong VR genes are stronger than others. The relative strength of VR genes can be estimated by determining the "half-life" of a complex race on a simple host. The half-life of a complex race of an obligate parasite is the number of generations (time) required for the complex race to decrease to 50% of its original frequency when it is cultured in combination with a simple race on a simple cultivar. The fewer generations required for each half-life, the stronger the VR gene.

How do strong VR genes stabilize races? Van der Plank (1968) offers a speculative rationale. He suggests that, as simple races become more complex by acquiring virulence genes to match VR, their normal parasitic processes are upset. Virulence against VR genes requires alternative processes or pathways to accomplish effective parasitism. The adoption of these substitute parasitic pathways renders the races less fit as parasites on simple hosts or as survivors in a nonparasitic phase.

Are there practical implications to stabilizing selection? Van der Plank (1968) is confident that there are when certain prerequisites are satisfied. The prerequisites are concerned with strong VR genes, their deployment and nondeployment, and with epidemic diseases that follow regular patterns. Stem rust of wheat in the United States and late blight of potatoes in Europe are appropriate illustrative examples. Stem rust of wheat in the United States follows an epidemic pattern according to van der Plank (1968). For all practical purposes, the rust overwinters on wheat in Mexico and Texas. It blows northward in the spring and finally reaches spring wheat in the northern states and Canada. In the fall, northern winds carry the rust back to Texas and Mexico. Such an epidemic pattern sets the stage for strong VR genes. The suggestion is to use the strong VR genes in spring wheat cultivars in the North but not in winter wheat cultivars in the South. Inoculum produced on wheat in the South, then, will be stabilized in favor of simple races without virulence genes to match the VR genes in the North.

Early- and late-maturing potato cultivars in northwestern Europe present a comparable epidemic system. Much of the inoculum reaching late-maturing cultivars originates on early-maturing cultivars. Incorporating strong VR genes in the late cultivars and keeping them out of early cultivars should find racial populations stabilized by the early, simple cultivars.

## An Evaluation of the Stabilizing Selection Concept

The concept of stabilizing selection has met with mixed reactions from plant pathologists. Those who express some concern about the universality of the concept have done so primarily out of a reluctance to accept the irrevocable, universal axiom that fitness equals simplicity when strong VR genes are matched by complex races. Van der Plank's strongest argument in support of stabilizing selection in obligate parasites was derived from existing data on stem rust of wheat in Australia where, in certain sections of the country, simple races predominated over complex races on simple cultivars. Recent evidence suggests that fitness per se is not necessarily associated with simplicity. Certain races with unnecessary genes have not dropped out of the current Australian populations. Watson (1970) reports that one of the most prevalent strains of *Puccinia recondita* in Australia is a race with virulence genes not needed to attack the commonly grown wheats. Work by Brown and Sharp (1970) with *Puccinia striiformis* demonstrated that a particular complex race had greater survival abilities than a simple race when compared in mixtures on a simple cultivar. Watson and Luig (1968) likewise demonstrated the predominance of a complex race of wheat stem rust after four uredial generations. In other words, an induced mutant strain of wheat stem rust with three additional genes for virulence was competitive with the simpler strain from which it was obtained after eleven uredial transfer generations on seedlings. One of the additional virulence genes matched VR gene *Sr* 6, which van der Plank considers to be a strong VR gene.

Ogle and Brown (1970) compared the relative survival ability of wheat stem rust races 21-2,7 and 21-2,3,7 for several serial transfer generations on eight different wheat cultivars in the greenhouse. Both races possessed virulence genes to match VR genes *Sr* 11 and *Sr* 15, while race 21-2,3,7 carried an additional virulence gene to match VR gene *Sr* 9. Four of the wheat cultivars had no VR genes, two cultivars carried the VR gene *Sr* 11, and two the VR gene *Sr* 15. Race 21-2,3,7, the race with the greater number of unnecessary virulence genes, predominated in mixtures after one generation on all cultivars and comprised more than 90% of the population on all cultivars after three generations.

The effect of generation time (the period from inoculation to production of inoculum for the next generation) was evaluated by Ogle and Brown (1970). Generation times of 10, 14, and 18 days were studied for three of the eight wheat cultivars. Fewer generations were required to reach the maximum frequency of the more complex race 21-2,3,7, when the generation time was 18 days. These results suggest that the superior survival ability of race 21-2,3,7 can be accounted for, in part, by its ability to produce greater amounts of inoculum over a longer period of time. The results also point out that different methods used in serial transfer studies, including sampling techniques, generation times, and so forth, could result in different patterns of survival. Certain survival trends obtained by manual transfer from host to host could indeed be artifacts.

While the simple race (0) of *P. infestans* predominated over different complex races on a simple potato cultivar in greenhouse mixtures by Thurston (1961), some of his field studies yielded different results. Different adjacent plots of a simple potato cultivar were inoculated separately with one of five races: simple race (0); races (1) and (4) with one unnecessary virulence gene each; and races (2,4) and (1,2,4) with 2 and 3 unnecessary genes, respectively. The relative fitness of the races was determined by comparing disease spread within the plots. Race (2,4) spread most rapidly, followed by race (1). Thurston (1961) commented on these results by suggesting that different isolates of a race may vary in aggressiveness. His results indicate not only that simple races are not always the most fit, but they also raise a question about the use of greenhouse studies to evaluate fitness. Concentrations of inoculum used in the greenhouse often may be considerably higher than those under natural conditions. One or more facets of a given sampling technique may preferentially bias the study in favor of one component of the mixture under artificial parameters.

Martens et al. (1970) evaluated the relative prevalence of races of *Puccinia graminis* Pers. f. sp. *avenae* Ericks. & E. Henn., incitant of stem rust of oats (*Avena sativa* L.), during a recent ten-year period in Canada. A high prevalence of virulent races with unnecessary genes for virulence prompted them to conclude that relative competitive ability was not reduced by unnecessary virulence genes.

It is possible that certain of these recent findings could be disqualified as negating the concept of stabilizing selection by concluding or assuming that weak VR genes were involved. The same conclusion certainly would not be appropriate to many of the recent studies.

Some recent information from nonobligate parasites raises a question about the validity of stabilizing selection as a universal phenomenon. Van der Plank (1968) cites race 1 vs. race 2 of the tomato wilt pathogen in support of the saprophytic medium as the stabilizing factor for nonobligate parasites. Briefly repeated, the VR gene *I* in tomato confers resistance to race 1 but not to race 2 of the wilt pathogen. Although race 2 had been detected at various times in various areas, it had never posed a threat to cultivars with the *I* gene. Van der Plank attributed the curbing of race 2 to the strength of the VR gene *I*. Race 2 was considered less fit to survive saprophytically as a result of having acquired a virulence gene to match the strong VR gene in tomato. P. Crill (*personal communication*, March, 1971) stated that, by 1968, race 2 had spread throughout Florida and was the limiting factor in tomato production on sandy soils, and that "There is absolutely no doubt that race 2 is very widespread in Florida."

It seems inconsistent to conclude at this time that the *I* gene in tomato or the *Sr* 6 gene in wheat are weak VR genes, when data to support stabilizing selection was based on the conclusion that these genes were strong. There is an alternative explanation; namely, that some complex races are equally or better able than simple races to survive parasitically or nonparasitically. Simple, fit races may be able to acquire extra genes for virulence while retaining their fitness. Complex races gradually or eventually may acquire survival attributes while retaining their genes for virulence, which may be the most logical explanation for the current prevalence of race 2 of the tomato pathogen after years of existence at low levels.

Recent studies by Nelson, Ayers, Cole, and Petersen (1970) on southern corn (*Zea mays* L.) leaf blight, incited by *Helminthosporium maydis* Nisik. & Miyake, suggest that virulent races can acquire certain parasitic fitness attributes over time. The 1970 epidemic of southern corn leaf blight in the U.S.A. was incited by a race of the pathogen, designated race T, with unprecedented virulence for corn hybrids carrying the so-called Texas male-sterile cytoplasm. Studies with isolates of the pathogen collected in previous years demonstrated that race T has been present in the U.S.A. at least since 1955. Comparative studies with isolates of race T collected in 1955 (preserved in limbo in leaf tissue since that time) and in 1970 revealed that the 1970 isolate possesses up to 15 times greater sporulation capacities, colonizes susceptible tissue more rapidly, and can cause infection under wider climatic regimes, than the 1955 isolate (R. R. Nelson, *unpublished data*). Any one of these attributes could contribute significantly to the increased frequency of the race; conversely, inferior sporulation abilities or an increased incubation requirement could operate against the increased occurrence of the race.

While there is evidence that complexity is associated in some cases with poor fitness (the evidence used to document stabilizing selection), there is no evidence that the virulence genes that confer complexity to a race are the same that condition fitness. Such a statement assumes that fitness attributes, as characterized previously, are under genetic domain, an assumption that appears most reasonable.

The suggestion that poor survival ability results from the "need" for races with newly acquired virulence genes to dismantle their "normal" parasitic apparatus in favor of alternate pathways to match VR genes merits comment. This suggestion is difficult to accept, and not just because we know so little about metabolic pathways associated with parasitism. It seems equally feasible that a plant's defense apparatus conferred against an avirulent race by a VR gene breaks down when confronted by a virulent race with a matching gene. The matching gene may enable the virulent race to produce greater amounts of some substance involved in its parasitic behavior; perhaps it produces a necessary metabolite that is not produced by a race lacking the virulence gene. It is difficult to perceive why these newly acquired metabolic abilities also must render a race less fit to survive.

Plant pathogens can acquire virulence genes by mutation, genetic recombination, or heterocaryosis. They also can *lose* virulence genes by mutation, genetic recombination, or the dissociation of heterocaryons. Complex races of *P. infestans* have been observed to revert to less complex or simple races. As examples, race (1,4) has been observed to revert to race (4), and race (1,3,4) to revert to race (3,4). The potato VR gene $R$ 1, which *P. infestans* races possessing the virulence gene to match it (any race with the virulence gene 1 in its designation) can overcome, is considered a strong VR gene. What happens to race (4), which originally was race (1,4), and to race (3,4), which arose from race (1,3,4)? Are these races automatically more fit because they are less complex? Do these races again restructure their parasitic apparatus after having become less complex? Does the lack of fitness ascribed to race (1,3,4) by virtue of its complexity pass on to race (3,4)? Presumably it would if the reversion from race (1,3,4) to race (3,4) is accomplished by a single genetic change. In such a case, the simple race would be no more fit than the complex race from which it arose, and decreasing complexity could not be equated with increasing fitness.

Even if added virulence genes *did* dismantle normal parasitic pathways in favor of alternate paths, it is difficult to understand why such a dismantling would have a deleterious effect on obligate parasites in their parasitic phase and on nonobligate parasites in their nonparasitic phase. Why should complex races of nonobligate parasites be less fit to survive nonparasitically? The only explanation available is that they are less fit because the concept of stabilizing selection deems them so.

Stabilizing selection is said to be operational only when strong VR genes are involved. Vertical resistance genes are said to be weak (van der Plank, 1968) when races that can match them are not less fit. This rationale could be viewed as a means of avoiding the fact that complex races can be fit. Parasite fitness is an attribute of a parasite. Fitness is relative not only among races, but also to the host(s). A parasitic attribute is not acquired from its host but, rather, is gained or lost by virtue of its own genetic constitution; the host serves only to *identify* parasitic attributes for the onlooker.

Perhaps stabilizing selection would be a more consistent concept if the assumption were made that unnecessary virulence genes in races of nonobligate parasites rendered them less fit to survive on simple members of their host(s). The parasitic dismantling-loss of fitness process then would be similar for obligate and nonobligate parasites. Many crops that serve as hosts to countless nonobligate parasites are cultured on a year-round basis in many areas of the world. Many areas support continuous growth of host species that harbor races in their parasitic phase when primary hosts are absent. To put it another way, many nonobligate parasites need not enter a nonparasitic stage at all. The current rationale for stabilizing selection for nonobligate parasites would not apply in these instances.

Studies by Scheifele and Nelson (1970) with northern leaf blight of corn, incited by *Helminthosporium turcicum* Pass., have clarified some of the factors conditioning the relative survival abilities of isolates of a nonobligate parasite. Some studies were designed with the rationale that relative survival abilities of simple and complex races of a nonobligate parasite in the parasitic phase on simple host cultivars may be as significant a determinant of racial composition as capacities to exist in the nonparasitic phase. Mixtures of simple and complex races (in accordance with van der Plank's definition) were evaluated for their differential survival on a simple maize inbred (simple in the sense of van der Plank) by serial transfers for several generations and by natural spread in the field. Without exception, the isolate predominating in each of the racial mixtures possessed superior infection and disease efficiencies, had a reduced incubation period, and/or possessed increased sporulation capacities. There was no relationship between genetic simplicity and relative fitness. In some mixtures, the simple race predominated because of superior fitness attributes. The complex race predominated in other mixtures, again because of superior survival abilities. The relative fitness of the isolates studied obviously was not governed by genetic simplicity but, rather, by independent factors for survival and relative fitness.

Van der Plank (1968) states that genes for VR are strong if complex races are stabilized on simple cultivars, and weak if complex races are not stabilized on simple cultivars. The same simple corn line was used by Scheifele and Nelson (1970) in all mixture studies and the same complex maize line was used to identify the different complex races. It would seem totally unacceptable to designate the VR genes in the complex corn lines as "strong" as a result of racial mixtures in which the simple race predominated, and then designate the *same* VR genes as "weak" as a result of racial mixtures in which the complex race predominated. These results support the previous suggestion that the philosophy of "strong" and "weak" VR genes may be an attempt to avoid the now apparent fact that simple races are not always the most fit to survive because simple races do not always possess superior capacities to survive. Races that are most fit to survive are those that are most fit, unless avirulence is limiting.

In another study by Nelson and Scheifele (1970) involving simple and complex races of *H. turcicum* and corn, the simple race predominated at the conclusion of the parasitic phase on corn, but the complex race was recovered more frequently at the conclusion of the nonparasitic (overwintering) phase. In other mixtures involving different isolates of the same simple and complex races on the same simple host, the simple race predominated during the nonparasitic stage in some cases and the complex race predominated in others. If the results of parasitic and nonparasitic survival abilities were viewed separately, two opposing conclusions relative to fitness would have been drawn. Factors governing fitness to survive nonparasitically may not be independent of factors conferring fitness to survive parasitically. This can be determined only by evaluating a given racial complex of a nonobligate parasite in both phases.

## Stabilizing Selection and Disease Control

The results of the varied research designed to evaluate the concept of stabilizing selection appear to negate the universality of the concept. The questioned validity of the concept is pertinent to disease control. The axiom that simplicity equals fitness forms the basis of the suggestion that stabilizing selection could function significantly in controlling stem rust of wheat in the Wheat Belt of the U.S.A. By planting simple cultivars in the South, the racial composition of *P. graminis tritici* would be stabilized in favor of simple races in that region. Complex cultivars would be planted in the more northern areas, for example, in Minnesota. As inoculum of simple races moved northward, the complex cultivars would be resistant and, thus, stem rust of wheat would be curtailed. However, in those instances in which the complex races are equally or more fit to survive, as some recent evidence has demonstrated, the planting of simple cultivars in the South would not only be ineffective, but also disastrous to the southern and northern wheat producers.

The genetic complexity of pathogenicity and virulence mechanisms of isolates of plant pathogens and their relative fitness to persist in parasitic and nonparasitic phases no doubt are attributes controlled by different genetic systems inherited independently from one another. Different isolates or populations of isolates should exhibit a variety of combinations of genetic complexity and fitness attributes. A knowledge that certain complex isolates may be less fit to survive on simple hosts or on saprophytic substrates could be utilized effectively to control epidemics or in predicting their potential occurrence.

# Conclusion

Horizontal or nonspecific resistance should have a stabilizing effect on populations of plant pathogens. Its effect does not alter the frequency of different races in favor of simple components, but rather curbs population shifts toward increased complexity. Its polygenic defense apparatus against serious levels of disease is effective so long as the parasite does not acquire all of the genes necessary to overcome the host's resistance genes. This is a matter of genetic probabilities. When large numbers of genes are involved, the probability of a total overcoming is the product of the probabilities of all the individual events.

Can VR genes be used to curb population shifts? The comments on multilines, gene deployment, and pyramiding of VR genes in this chapter are straightforward and need not be summarized again. What remains is the issue of stabilizing selection as proposed by van der Plank. In my opinion, the concept is not valid. In a real sense, stabilizing selection qualifies as no more than a hypothesis, which is a tentative proposition to explain certain results. The idea of stabilizing selection by VR genes never passed beyond the hypothetical phase because it was never put to test before it was presented as a concept fortified with the self-evident axiom that simplicity equals fitness and buffered by the strong- and weak-gene rationale.

It has been suggested to me that perhaps the controversy regarding the concept of stabilizing selection could be calmed to some degree if the position were taken that stabilizing selection may be a tendency, but not a law or axiom that is irrevocable and universal, as van der Plank contends. I cannot subscribe to that suggestion because the concept will not permit me to. It is a "black or white" concept. When complex races are stabilized against, the VR genes are strong. Vertical resistance genes are weak when complex races are not stabilized against. Genes are weak or strong "by definition." There can be no tendencies in such a concept.

It has been suggested to me that the validity of the concept has not been tested adequately and that more time and research are needed to reach a reasonable verdict. I cannot subscribe to that suggestion, again because the concept will not permit me to do so. Any future data that negate the concept can be dismissed readily by invoking the "weak gene" philosophy. Data already exist to support and refute the concept, and yet there is still controversy. If one cannot test the validity of the concept, does a concept really exist?

# References

Brown, J. F., and E. L. Sharp. 1970. The relative survival ability of pathogenic types of *Puccinia striiformis* in mixtures. Phytopathology 60:529-533.

Browning, J. Artie, and K. J. Frey. 1969. Multiline cultivars as a means of disease control. Annu. Rev. Phytopathol. 7:355-382.

Graham, K. M. 1955. Distribution of physiological races of *Phytophthora infestans* (Mont.) de Bary in Canada. Amer. Potato J. 32:277-282.

Martens, J. W., R. I. H. McKenzie, and G. J. Green. 1970. Gene-for-gene relationships in the Avena: *Puccinia graminis* host-parasite system in Canada. Can. J. Bot. 48:969-975.

Nelson, R. R., and G. L. Scheifele. 1970. Factors affecting the overwintering of *Trichometasphaeria turcica* on maize. Phytopathology 60:369-370.

———, J. E. Ayers, H. Cole, and D. H. Petersen. 1970. Studies and observations on the past occurrence and geographical distribution of isolates of race T of *Helminthosporium maydis*. Plant Dis. Reporter 54:1123-1126.

———, D. R. MacKenzie, and G. L. Scheifele. 1970. Interaction of genes for pathogenicity and virulence in *Trichometasphaeria turcica* with different numbers of genes for vertical resistance in *Zea mays*. Phytopathology 60:1335-1337.

Ogle, Helen J., and J. F. Brown. 1970. Relative ability of two strains of *Puccinia graminis tritici* to survive when mixed. Ann. Appl. Biol. 66:273-279.

Scheifele, G. L., and R. R. Nelson. 1970. Factors affecting the survival of *Trichometasphaeria turcica* (*Helminthosporium turcicum*) on *Zea mays*. Can. J. Bot. 48:1603-1608.

Thurston, H. David. 1961. The relative survival ability of races of *Phytophthora infestans* in mixtures. Phytopathology 51:748-755.

Van der Plank, J. E. 1968. Disease resistance in plants. Academic Press, New York. 206 p.

Watson, I. A. 1958. The present status of breeding disease resistant wheats in Australia. Agr. Gaz. New South Wales 69:630-660.

———. 1970. Changes in virulence and population shifts in plant pathogens. Annu. Rev. Phytopathol. 8:209-230.

———, and N. H. Luig. 1968. The ecology and genetics of host-pathogen relationships in wheat rusts in Australia, p. 227-238. *In* K. W. Findlay and K. W. Shepherd [ ed.] Third international wheat genetics symposium proceedings, Canberra Butterworths, London.

# 6 The Limits of Disease Control by Genetic Means

## R. R. Nelson

The purposeful manipulation of hereditary units to control plant disease is not the ultimate solution to all disease problems. This brief chapter identifies some of the real and suspected limitations of the use of disease resistance. The specific examples used merely illustrate the principles involved, and are not meant to represent unique instances.

Hereditary control of plant diseases is not possible when a cultivated species and its botanical relatives apparently possess no genes governing resistance. There seems to be a reasonably general principle governing disease relationships: specifically, it is unusual to find genes conditioning resistance to a pathogen with a wide host range. The principle seems particularly appropriate to diseases incited by soil-borne pathogens with wide host ranges. Resistance to *Phymatotrichum omnivorum*, *Sclerotium rolfsii*, and many damping-off and seedling blight fungi is rare. There are exceptions to the rule: genes for resistance to *Thielavia* root rot exist in tobacco, although the causal agent attacks many species.

The plant parts affected by disease often delineate the usefulness of resistance by genetic means. Breeding for resistance to pathogens causing rots of storage organs is not a primary means of control. Such storage organs as fruits and tubers are essentially passive in their response to attacking organisms. Organisms inciting rots of storage organs generally are poorly specialized and unsophisticated parasites, albeit potent pathogens. They gain entrance through wounds, lenticels, and so forth, and are not known primarily for their ability to penetrate directly. Varieties of fruits differ in thickness of skin, and potato varieties differ in lenticel characteristics and in the degree to which lenticel proliferation occurs under adverse conditions. To what extent such characteristics reduce rots of storage organs is debatable. In any event, these characters are not true resistance mechanisms, because they do not participate actively in defense. It may be concluded from this discussion that breeding for resistance to poorly specialized parasites attacking plant parts which do not react actively in defense has genuine limitations.

It is likely that related or ancestral species in their epicenters have co-evolved with certain of the parasites attacking them and their cultivated relatives. Natural selection and survival of the most fit would seem to dictate that substantial numbers of the current population of ancestral species have evolved to a suitable degree of some kind of resistance towards some parasites. Man's efforts to utilize the resistance of ancestral species is clearly evident in breeding programs involving wheat, oats, potatoes, tomatoes, and countless other cultivated crop plants. Frequently, however, genes for resistance may be located in a species too remotely related to the cultivated species to permit successful interspecific hybridization and transfer of genetic material. Although disease resistance has been identified in the wild species of the genus *Arachis*, it has not been possible to transfer that resistance to the cultivated peanut, *A. hypogaea*, by interspecific crosses. The cultivated peanut is an allotetraploid species with $2n = 40$ chromosomes. Most of the wild species of *Arachis* are diploids with $2n = 20$ chromosomes. There are several polyploid species of *Malus* that are very difficult to cross with cultivated apple varieties which are diploid. Usually some viable hybrids are obtained from polyploid x diploid crosses of *Malus* species. While the polyploidy of the noncultivated *Malus* species does not pose insurmountable barriers to hybridization, its use as a potential source of resistance is minimized. Incompatibility between species may be complete or partial. Embryo retardation may prevent the development of the hybrid plant. Often the hybrid develops to a fully grown plant but is unable to produce viable pollen and ovules. Interspecific hybridization may result in cytoplasmic male-sterility. The consequences of partial or complete sterility are the same, regardless of the specific causes. Recent advances in tissue culturing of embryos and techniques now available to cytologists and cytogeneticists are surmounting some instances of presumed cross-sterility. The resulting $F_1$ hybrid from a cross between the cultivated tobacco *Nicotiana tabacum* and the species *N. glutinosa* is a sterile amphihaploid. When seeds of the hybrid are treated with colchicine, the chromosomes of some plants are doubled and fertility is restored. Bridging interspecific sterility by gene transfer through a mutually compatible species to the desired species can be an effective means of utilizing desired genes located in wild species.

Linkage between genes conditioning desirable and undesirable traits often has posed a problem to researchers concerned with plant improvement by genetic means. Factors for resistance are often linked with such agronomic traits as poor plant height, vigor, quality, and yield. Much useful resistance has been lost or neglected when selection for agronomically desirable traits has received first priority. In many cases, it is not difficult to "break" linkages between desirable resistance and poor agronomic traits. The success of such efforts is directly proportional to the importance ascribed to disease resistance.

The consequences of linkage between genes conditioning resistance to one parasite and susceptibility to another became patently clear to plant pathologists in the 1930s and 1940s in their efforts to combat *Puccinia coronata avenae,* incitant of crown rust of oats. The variety "Victoria" was highly resistant at the time to all prevalent races of crown rust. Varieties derived from crosses of Victoria and Richland, a variety with outstanding resistance to stem-rust, soon occupied most of the acreage in the northern United States. The Victoria-derived varieties soon were destroyed by a then-unknown pathogen now recognized as *Helminthosporium victoriae.* Susceptibility to Victoria blight was linked with factors controlling resistance to crown rust. That linkage put an abrupt end to the promising Victoria-derivative oats.

The number of genes conditioning resistance to disease has a direct bearing on the ease and effectiveness by which resistance is attained. A few genes which give discrete, easily recognized classes among segregating populations are relatively easy to exploit. A large number of genes functioning collectively in a quantitative manner to govern disease response are more difficult to manipulate. Undoubtedly, genes conditioning quantitative resistance most often are located on several different arms of chromosomes. Transferring all of the genes into an agronomically acceptable background is a formidable task. Segregating populations fall into indiscrete, continuous classes, many possessing some merit but lacking a similar degree of resistance characteristic of one of the parental types. The limitations in handling polygenetically controlled resistance often are more philosophical than scientific. When the rewards are sufficiently great, the task often is accomplished. Both the rewards and the tasks are discussed elsewhere in this book.

The development of agronomically desirable varieties and consumer-acceptable products is seemingly an endless task. Fashions in vegetables, fruits, ornamentals, and the like are constantly changing to meet the demands of nutritionists, economists, consumers, and the new machines designed for improved mechanical harvesting and processing. The rate at which fashions and trends change places a significant stress on those concerned with the task of keeping disease resistance abreast with the changes. The factor of time may place real limitations on the use of resistance to control disease. By the time the Dutch developed resistance in the elm to Dutch elm disease, the vase-shaped elm was no longer popular and had given way to more compact, columnar trees suitable for smaller planting areas. The trend towards mechanical harvesting of tomatoes has increased the prevalence of certain fruit rot diseases. Similar problems associated with mechanized agriculture plague growers and researchers concerned with the production of a number of fruit and vegetable crops.

The length of the life cycle of a plant species has a bearing on the usefulness or feasibility of disease control by hereditary means. Progress in developing trees with resistance to one or more diseases is slow. The time required for the initial selection of certain trees exhibiting an acceptable level of resistance and the subsequent evaluation of the resistance of their progeny often spans many years. The tremendous capacities of most plant pathogens to generate untold numbers of new races within a period of several years prompt some investigators to conclude that the progeny originally selected as resistant may not be resistant when they are ready for distribution. Should seedling reactions be shown to represent mature plant responses to plant pathogens, it is conceivable that considerable time could be saved in developing resistance in plant species with long life cycles.

Crop plants frequently are subject to many different plant pathogens in a given place and at a given time. Rarely can it be expected that resistance to all pathogens will be integrated into a single commercial variety or selection. Although there are many outstanding examples of varieties with multiple resistance to different pathogens, the limitations in incorporating resistance to all pathogens are real.

A widespread epidemic of southern leaf blight of corn in the United States in 1970 clearly delineated a genuine limitation to the genetic control of plant diseases. The disease, incited by *Cochliobolus heterostrophus* Drechs. (*Helminthosporium maydis* Nisik. & Miyake) first became of serious concern when corn hybrids were first introduced into the South. The disease reached such serious levels in the South that intensive screening and selection were undertaken to find a suitable resistance to the pathogen. When resistance was found and incorporated into adaptable hybrids, the disease became relatively innocuous in subsequent years. What little is known about the inheritance of resistance suggests that resistance is conditioned by several genes. The fact that hybrids exhibit field resistance, highlighted by a restricted lesion size, supports the assumption that resistance is quantitative. The nature of the resistance probably accounts for the relative unimportance of the disease in recent years. The sporadic and localized outbreaks of the disease in the last 20 years may well have occurred on hybrids lacking suitable resistance available in most hybrids.

The increasing use of corn hybrids produced in Texas male-sterile cytoplasm, a procedure which obviates the need manually to detassel female rows in seed production fields, brought an end to the status quo of southern corn leaf blight. The male-sterile hybrids were highly susceptible to certain strains of the fungus, which apparently increased gradually among the populations of the pathogen concurrently with the increasing use of male-sterile hybrids.

In 1970, the vast expanse of highly susceptible hybrids, a race of the parasite which obviously was epidemiologically able to incite an epidemic, and favorable climatic conditions combined to produce an unparalleled epidemic of the disease. The unusual virulence of the new race was remarkably specific to hybrids produced with Texas male-sterile cytoplasm. Hybrids produced with normal cytoplasm by detasseling female rows in production fields were equally or more resistant to the new race than they had been in recent years to the "old race(s)" of the pathogen.

Of special interest is that corn hybrids produced with Texas male-sterile cytoplasm also are highly susceptible to yellow leaf blight, incited by an as-yet-undescribed species of *Phyllosticta*. The disease has occurred in localized epidemic fashion in the northern areas of the United States since 1967, apparently being favored by cooler climatic conditions in the early stages of the growing season. Once again, hybrids produced with normal cytoplasm exhibit a sufficient degree of resistance to minimize the likelihood of serious damage from the disease.

Cytoplasmic conditioning of disease susceptibility is a rarity, as far as we know. That a single major crop can exhibit a high degree of susceptibility concurrently to two vastly different pathogens for the same reason serves as a vivid reminder that unsuspected factors can dramatically affect our "thin line of plenty."

Although male-sterile cytoplasm conditions the susceptibility of corn hybrids to southern leaf blight and yellow leaf blight, that factor per se could not have caused a national epidemic. The widespread use of male-sterile hybrids throughout all of the corn-producing areas of the United States was the factor that made male-sterile susceptibility an economic catastrophe. The extensive use of a narrow base of germ plasm, or cytoplasm in the present case, invites catastrophe, and the history of such usage stands in graphic testimony to the fact. Genes condition effective resistance. Gene diversity among varieties is essential in the interest of disease control, and the deployment of resistance genes is discussed elsewhere in this book. The indiscriminate use of certain kinds of resistance genes points to another limitation of resistance and disease control.

The ability of many plant pathogens to produce a seemingly endless number of new and highly virulent races tempts us to conclude that stable control of diseases by resistance may not be feasible for some crops against some pathogens. Adequate resistance to leaf rust of wheat is a tenuous proposition, presumably because of the continual appearance of new races. A new race of crown rust of oats currently exists against which no known resistance is available. A new race of the pathogen causing *Fusarium* wilt of peas recently was discovered to be pathogenic to the currently used and previously resistant pea varieties. The notorious variability of the rice blast fungus has given rice workers reason to question the future of resistance in controlling the disease. Similar examples of a breakdown in resistance by the occurrence or increase of a new race are virtually endless.

When numbers of genes for resistance are matched against the potential numbers of races, the outlook could be construed as glum. Part of the theory behind the gene-for-gene concept states that one variety (genotype) of the host is susceptible to all races of the pathogen, while another variety (genotype) is susceptible to only *one* race. Conversely, one race of the pathogen can attack only one variety, and one race can attack *all* varieties. Those would appear to be unfair odds against the host for maintaining resistance to a variable pathogen. Flax, the pioneer model for the gene-for-gene relationship, is currently known to have only five loci conditioning rust resistance. Multiple allelic systems at these loci increase the number of available alleles for resistance to 27. To overcome resistance, genes in the pathogen segregate independently with no evidence for allelism. Thus it is possible, at least theoretically, for the pathogen to call upon more genes than the host can handle if 27 alleles actually constitute the maximum number of alternatives available to flax.

The apparent futility and/or inability in coping with new races by genetic means is caused not by the copious formation of new races by the pathogen, but by the attempted use of race-specific resistance to accomplish the task. All-or-nothing resistance is akin to feast or famine as far as crop productivity is concerned. A greater use of forms of nonspecific resistance that avoid disease losses by reducing amounts of disease seems to be a plausible alternative for disease control by genetic means. Thus, the limitations alluded to in this discussion are not limitations per se but are limitations to certain uses of inherent abilities to control disease.

In summary, certain real or suspected limitations to the use of genetic means for disease control have been identified. These include: (1) the absence of genes that can effect disease control; (2) the difficulty in finding resistance to poorly specialized parasites that attack plant parts which do not react actively in their own defense; (3) the inability to transfer genes for resistance from donor species to agronomically acceptable varieties; (4) close linkage between genes controlling desirable and deleterious traits; (5) the number of genes necessary to confer an acceptable kind of resistance; (6) the rapidity of changes in varieties to meet demands for new fashions in crop products; (7) length of the life cycles of plant species; (8) cytoplasmic conditioning of host susceptibility; (9) the incredible production of new races by many plant pathogens; and (10) the number of potential genes for resistance available to plant species.

To what extent future advances in technology and research will minimize the limiting effects of these factors is not known. The advancement of science since 1900 suggests that the future should hold considerable promise.

No single means of disease control is universally effective, and this chapter acknowledges that fact for disease resistance. It is written in order to be realistic, not discouraging. The remainder of this book illustrates the positive and encouraging facets of the use of genetic means to control plant diseases.

# 7 Breeding Methods for Disease Resistance

## Walter I. Thomas[1]

## Introduction

One of the main forms of control of disease is the use of resistant varieties. It was just over one hundred years ago that the concept of diseases being incited by microorganisms was accepted. Subsequently, Vilmarin, a French breeder, outlined the principles of the "progeny test" as the best way to determine the breeding behavior of a plant. However, the discovery that had the greatest influence on plant breeding was by Gregor Mendel, an Austrian monk. Mendel published the results of his work on the inheritance of characters in peas late in 1865. However, his work did not come into its own until the beginning of the twentieth century, and it was at this time that early investigators had started breeding resistance to disease in some crops and studying the nature of resistance in some varieties.

Most plant breeders feel that the science of genetics is the most important tool they have for improving plant varieties or other characteristics, as well as for the area of plant diseases. However, Van der

1. Associate Dean for Research, College of Agriculture, The Pennsylvania State University.

Plank (1963) feels that the plant pathologist and plant breeders must be as thoroughly trained in epidemiology as in such other sciences as mycology, virology, and genetics. He points out that plant breeding and other factors are all relevant to the control of a disease only as they reduce the initial population of the pathogen or are able to retard its subsequent increase.

One of the most important considerations in a breeding program involving disease resistance is the need for selection pressure imposed by subjecting plant populations to severe levels of the disease in question. (Techniques to exert selection pressure are discussed in Chapter 3.) Breeding for disease resistance requires a knowledge of both the host and the pathogen. The heritable characters of the host plant interact to a high degree with those of the pathogen. The etiology, distribution, host range, and ecological relationships of the pathogen should be understood. If a plant breeder is not fully trained in this area, he must obtain the assistance of a plant pathologist.

## The Application of Genetics to Plant Breeding

Prior to the application of genetics to the breeding for disease resistance an investigator must have some knowledge of the reproduction and pollinating processes of the plant species in use. Progress in plant breeding is dependent upon the development or discovery of new genetic combinations that may prove to be superior to existing varieties or types under practical growing conditions. To illustrate, assume that a plant breeder has two varieties of barley, one with rough awns and the other with smooth awns, with the plant characters designated $RR$ in the rough awn parent and $rr$ in the smooth awn parent. He also has a second character, such as resistance and susceptibility to stem rust. This character is also determined by a single factor pair, which we will designate $AA$ and $aa$, with resistance dominant. Assuming that genes for rust reaction are independently inherited from genes for awn-type and the rough awn parent is rust resistant, its genotype would be $RRAA$, and the smooth awn rust-susceptible parent would have the genotype $rraa$. Progeny of the cross between these two varieties would produce an $F_1$ which would be $RrAa$. In the formation of gametes from these plants, four different kinds would be obtained: $RA$, $Ra$, $rA$, and $ra$. With random union of these gametes from self-pollination, 16 combinations are possible, as illustrated in Table 7-1. Examination of the

TABLE 7-1 GENOTYPES OF $F_2$ PROGENY OF A CROSS FROM A ROUGH AWNED, DISEASE RESISTANT (*RRAA*) AND A SMOOTH AWNED, DISEASE SUSCEPTIBLE (*rraa*) PARENT.

| FEMALE GAMETES | MALE GAMETES | | | |
|---|---|---|---|---|
| | *RA* | *Ra* | *rA* | *ra* |
| *RA* | *RRAA* | *RRAa* | *RrAA* | *RrAa* |
| *Ra* | *RRAa* | *RRaa* | *RrAa* | *Rraa* |
| *rA* | *RrAA* | *RrAa* | *rrAA* | *rrAa* |
| *ra* | *RrAa* | *Rraa* | *rrAa* | *rraa* |

table shows that one-fourth of the progeny is homozygous for one factor pair, one-half of the progeny is heterozygous for one or both factor pairs, and one-fourth of the progeny is homozygous for both factor pairs. In addition, if one were looking for a homozygous, smooth awn, disease resistant type, only one-sixteenth of this population would meet the prerequisite. However, if dominance was complete, that is, if the heterozygous type *Rr* was indistinguishable from the homozygous *RR*, this desired genotype *rrAA* would be indistinguishable from the genotype *rrAa*. The important fact to be gathered from this example is that the $F_3$ generation would have to be evaluated to determine which plants were true breeding or homozygous for the desired characteristics. In each successive generation starting with the $F_2$ or the $BC_1$ (first backcross), the proportion of individuals homozygous for the various genes differentiating the parents increases. There is a closer and closer approach to homozygosity with each succeeding generation. This can best be illustrated with a single pair using the rough *RR* vs. the smooth *rr* awns in barley as follows: it can be seen from this example that the percentage of homozygous individuals increases in each generation, while the percentage of heterozygous individuals decreases by one-half.

If the same $F_1$ hybrid were backcrossed to the smooth awned parent for several generations, the expected results from the standpoint of homozygosity would be the same. That is, the per cent homozygosity in the $BC_1$, $BC_2$, and $BC_3$ generations would be 50, 75, and 87.5 per cent respectively (Table 7-2). These data illustrate several important facts. They show that for any factor pair, the approach to homozygosity is the same whether backcrossing or selfing is carried out where the $BC_1$ is comparable to the first segregating generation, $F_2$, following selfing. However, with backcrossing the homozygous individuals in any generation are like the recurrent parent. With continuous selfing only half of the homozygous individuals are like one parent and half are like the other.

TABLE 7-2   RATE OF APPROACH TO HOMOZYGOSITY WITH COMPLETE
SELF-FERTILIZATION AND ONE GENE PAIR.

|  | GENOTYPES | PER CENT HOMOZYGOSITY |
|---|---|---|
| $F_1$ | $Rr$ | 0 |
| $F_2$ | $1/4\, RR : 1/2Rr : 1/4rr$ | 50 |
| $F_3$ | $3/8\, RR : 1/4Rr : 3/8rr$ | 75 |
| $F_4$ | $7/16\, RR : 1/8Rr : 7/16\, rr$ | 87.5 |

Similar calculations can be made regarding the approach to
homozygosity with selfing or backcrossing for any number of postulated
heterozygous factor pairs in an $F_1$ hybrid, if no linkage and no selection
are assumed. Estimates with differing numbers of heterozygous factor
pairs in the $F_1$ and different generations are shown in Table 7-3. The
percentage in the table can be calculated by using the following formula:

$$\% \text{ homozygous individuals} = \left[ \frac{2^r - 1}{2^r} \right]^n \times 100$$

where $r$ equals the number of segregating generations, and $n$ equals the
number of heterozygous factor pairs in the $F_1$. An evaluation of Table
7-3 shows that the percentage of homozygous individuals of all
genotypes is the same in any particular segregating generation whether
selfing or backcrossing is carried out. We can also point out that even
with a large number of heterozygous factors in the $F_1$ parent, a large
number of individuals will be homozygous by the $F_{10}$ or $BC_9$
generation. However, there is a distinction between these two in that
with backcrossing all the homozygous individuals in any generation will
be of the recurrent parent genotype. In the case of selfing only $1/2\, n$ of
the homozygous individuals will be like one of the two parent
genotypes. This difference has prompted many plant breeders to use the
backcross method for breeding for disease resistant genotypes when a
reasonably good recurrent parent was available with good agronomic
value.

# Mode of Pollination

A breeding plan for incorporating disease resistance or any other
characteristic into a variety requires some knowledge of the mode of
pollination. For the sake of convenience, most crops can be divided into

TABLE 7-3  PERCENTAGES OF HOMOZYGOUS INDIVIDUALS FOLLOWING SELFING OR BACKCROSSING OF AN $F_1$ WITH $n$ HETEROZYGOUS FACTOR PAIRS ASSUMING INDEPENDENT INHERITANCE, NO SELECTION, AND AN INFINITE POPULATION.

| NUMBER OF HETEROZYGOUS FACTOR PAIRS IN $F_1$ ($n$) | PERCENTAGES OF HOMOZYGOUS INDIVIDUALS OF ALL TYPES | | | | | PERCENTAGES OF HOMOZYGOUS INDIVIDUALS LIKE ONE PARENT | | | | |
|---|---|---|---|---|---|---|---|---|---|---|
| | $F_2$ or $BC_1$ | $F_4$ or $BC_3$ | $F_6$ or $BC_5$ | $F_8$ or $BC_7$ | $F_{10}$ or $BC_9$ | $F_2$ | $F_4$ | $F_6$ | $F_8$ | $F_{10}$ |
| 1 | 50.0 | 87.5 | 96.9 | 99.2 | 99.8 | 25.0 | 43.7 | 48.4 | 49.6 | 49.9 |
| 10 | 0.1 | 26.3 | 72.8 | 92.4 | 98.0 | — | 0.03 | 0.07 | 0.09 | 0.10 |
| 25 | — | 3.6 | 45.2 | 82.2 | 95.2 | — | — | — | — | — |
| 50 | — | 0.1 | 20.4 | 67.5 | 90.7 | — | — | — | — | — |
| 100 | — | — | 4.2 | 45.6 | 82.0 | — | — | — | — | — |

three groups: (1) those naturally self-pollinated, (2) those often cross-pollinated, and (3) those which are naturally cross-pollinated.

1. Examples of naturally self-pollinated crops are wheat, oats, barley, soybeans, peas, and beans. In these plants, both the staminate and pistillate parts are enclosed within the glumes or petals of the flower.

Even though pressure exerted at the base of the flower by the enlarging lodicules forces the glumes to open slightly during anthesis, the amount of pollen shed outside the glumes is relatively small compared to that which comes directly in contact with the stigma. A relatively small percentage of natural crossing (generally less than one per cent) occurs between adjacent plants.

2. Cotton, sorghum, and sudan grass are examples of crops that vary considerably in the amount of cross-pollination that occurs. This group of species has perfect flowers, but their mode of pollination is affected by the fact that the stigma may be exposed to both self- and cross-pollination by either wind-blown pollen or pollen carried by insects. Sorghum and cotton, respectively, are examples of this.

3. Examples of naturally cross-pollinated crops are corn, rye, smooth brome grass, alfalfa, red clover, and many other crops. Species of these crops may be divided into three additional groups: those with perfect flowers, those with monoecious flowers, and those which are dioecious.

a. Examples of perfect flowered species are rye, reed canary grass, and alfalfa, which are structurally capable of natural self-pollination, but in which self-pollination is greatly reduced because of self-sterility factors. In the grasses, the glumes are opened widely when pollination occurs, so that the stigma is exposed to a great extent to pollen carried by the wind. In addition, the anthers are often extruded outside the lemma and palea before anthesis occurs, allowing this pollen to be scattered by the wind. In the legumes and in other ornamental species, in addition to self-sterility factors, the flowers are often attractive to pollinating insects.

b. Corn is one of the best examples of a plant with monoecious flowers. In this case, staminate flowers are borne on the tassel and the pistillate flower is borne on the cob. Although both the male and female flower parts are borne on the same plant, there is limited opportunity for self-pollination because of the wind-blown pollen.

c. Hemp and buffalo grass are examples of dioecious crops. These cannot be self-pollinated, because an individual plant is either staminate or pistillate only. This complete separation of sexes demands a different approach in breeding which can be compared more closely to the improvement of animals.

## Breeding Self-Pollinated Crops

Major early work in the improvement of crops with respect to disease resistance was with small-grain varieties and later with soybeans. Extensive programs of plant introduction brought varieties developed in other parts of the world. Examples of this are soybean varieties introduced from the Orient and wheat from Turkey, Russia, and Poland. Specific varieties are Turkey Winter Wheat, Red Fife Spring Wheat, and Landhoffer Oats. In the latter case the variety was of little value from an agronomic viewpoint, but it was a source of genes for crown rust resistance. Most recently, plant introductions have not been used primarily as varieties, but have been used as genetic sources for specific characters which include disease resistance.

Prior to the rediscovery of Mendel's laws of heredity in 1900 the common practice for breeding in naturally self-pollinated crops was to select individual plants from existing varieties, which were a mixture of genotypes. A good example of this is the work of Bolley in North Dakota, in which flax was grown on wilt-infested soil, and seed was saved from the surviving plants and used in the development of many varieties. This pureline method used extensively in the small grains was carried out by selecting a large number of single heads and observing the progeny row for such characteristics as disease resistance, grain type, bushel weight, and other desirable attributes. Those progeny rows which had the desired characteristics were saved and evaluated in comparison with currently used varieties. Subsequent opportunities for improvement by hybridization and slower progress of the pureline selection method led to the discontinuation of improvement by pureline selection. One of the earliest reports of the use of hybridization in self-pollinated crops was that of Shirreff (1873). However, it was not until the third decade of the twentieth century that the use of hybridization in the improvement of self-pollinated crops came into extensive use. In general, cereal grain breeders who have been involved extensively in breeding for disease resistance have classified several systems of improvement by hybridization, and these are generally referred to as the pedigree method, the bulk method, and the backcross method.

In the pedigree method, individual plants are selected in the first segregating generation ($F_2$) and are harvested individually with seed of each individual plant space-planted the next year. Individual heads are again selected within the progeny rows, and the sequence is followed until lines are uniform enough and meet the criteria established by the plant breeder for extensive testing and potential use.

The bulk method of breeding is different from the pedigree method, in that single plants are not selected in each generation after the $F_1$ cross. Instead, the plants are harvested in bulk and planted each year with a fairly sizable population until the $F_5$ or $F_6$ generation, when most plants are relatively homozygous. Then selected plant lines are evaluated in rows or under appropriate conditions for the crop in question.

The system of backcross breeding in self-pollinated crops, as previously stated, is used primarily when a desirable variety is available that has a deficiency in only one or two simply inherited characteristics. The desirable variety, sometimes called the recurrent parent, is crossed to another variety which has the gene or character that the recurrent parent lacks. In each successive year, desirable plants in the segregating population are crossed back to the recurrent parent, and progenies from each backcross which have the desired genes are carried through the next generation. This is continued for an adequate number of generations until the new variety with the desired genotype and degree of homozygosity is produced. Necessarily at the end of the backcrossing program a generation of self-pollination is necessary to stabilize the character obtained from the nonrecurrent parent in a homozygous condition.

The details of these methods of breeding self-pollinated crops are described in textbooks on plant breeding. However, a discussion that includes the breeding techniques as well as the pathologic techniques can be found in a monograph edited by Coffman (1961).

Of the techniques used for breeding for disease resistance, the backcross method deserves further consideration. Although this technique has been widely used for developing varieties resistant to the various rusts in the cereals, the principle also applies to other crops. In this technique genes for resistance, generally only one gene, are transferred into a recurrent parent that is already acceptable except for its resistance to a specific race of disease. Upon satisfactory completion of evaluation of the new variety, it is generally released. In many cases the success of the variety, based solely on its resistance to the disease, is so notable that such varieties come into wide use. As a consequence the same genotype is used over a wide area, and the incidence of a new race which parasitizes that variety may cause widespread destruction again only a few years after a variety was in production. An example of this is the oat cultivar Clinton, which was first introduced in the mid-1940s, and which by 1950 occupied over 75% of the total oat acreage in the United States. After only a few years, the Clinton cultivar and other derivatives from the same parentage were seriously damaged by race 7 of crown rust in the major epiphytotic of 1953. This same pattern has occurred in other varieties of small grain. This type of resistance, "vertical" as described by Van der Plank (1963), offers only temporary respite from the ravages of a disease. Van der Plank (1963) suggests that rust in North America could be controlled with varieties that are disease tolerant or which have stable polygenic resistance. This is sometimes termed "horizontal" or field resistance. However, experience in the midwest has shown that such varieties can be taken out by new rust races to which they are not tolerant but completely susceptible. Browning (1965) suggested that use of a multiline variety made up of a blend of isogenic lines, each carrying resistance to a different race of rust, might offer the same advantage as a variety with horizontal resistance. Since that time, Frey, Browning, and Grindeland (1970) have reported the release of new multiline oats. Their data indicate that multiline oats inhibit local epidemics of rust in oats fields, and periodic upgrading of the rust resistance could keep a multiline series useful for many years, as compared to cultivars which lasted only a few years because of the periodic occurrence of epiphytotics in the 30 years prior to 1960. However, the backcrossing technique will serve as the means by which isogenic lines are developed for use in the multiline synthetic.

Hegal and Moore (1970) reported that several varieties of oats grown adjacent to buckthorn (*Rhamnus cathartica*) showed moderate resistance to crown rust (*Puccinia coronata*). They noted in the development of the rust that there were fewer infections, hyphal growth was retarded, the number of days to onset of sporulation was greater, pustules were smaller, and fewer spores were produced per pustule in one variety as compared to another one. This type of nonspecific resistance would be useful in developing horizontal resistance espoused by Van der Plank (1963). Such resistance would have the effect of retarding the subsequent increase of the pathogen after its initiation.

# Breeding of Cross-Pollinated Crops

If one were to pick out the single phenomenon which contributed most to agricultural advance in recent years, it might be the development of hybrid corn. Even though inbreeding to develop lines of corn was started in the early part of the twentieth century, and crosses between varieties had already been proven to be superior to some varieties, almost all of the corn planted in the United States was of open-pollinated varieties until 1932. At that time, less than one-half of one per cent of Iowa's corn acreage was planted to hybrid corn. By 1943, essentially the entire state of Iowa was planted to hybrid corn. Other states of the Corn Belt quickly followed this trend. Today almost 100% of the corn acreage in the United States is planted with hybrid seed.

Although the reason for this rapid change was the easily recognized superiority of the hybrid in grain production, perhaps a more important reason from today's viewpoint is that the resistance of hybrid corn to lodging has permitted the general use of mechanical corn pickers, which would have been impossible with serious lodging problems caused by susceptibility to diseases. A comprehensive review of corn breeding is available in the monograph edited by Sprague (1955). Although this book is not new, it is the most recent single publication that gives extensive details of corn development, and it covers many of the basic principles of corn breeding.

Inbreeding of most cross-pollinated crops results in drastic vigor decreases in most cases. In addition, unusual characteristics and abnormalities appear among the progenies of some of the inbreds. After continued inbreeding and selection distinct lines are isolated which breed true.

The amount of reduction in vigor varies greatly among lines. For example, some lines of corn and sorghum may become so weak that they are impossible to maintain. On the other hand, some lines may yield over 50% of the variety from which they were obtained. While the yield and vigor of inbred lines in commercial crops may not be the most important consideration, the ability of lines to produce adequate supplies of seed, if the commercial seed sold is a single cross, is very important. In the early days of hybrid corn production, most commercial hybrids were of the double cross variety, so that the seed sold to the farmer was produced on a single-cross plant, and therefore the seed producer could market seed at a reasonable price. In recent years, with hybrid corn and hybrid sorghum, a good many single crosses have been found to be superior. In this case it is important that the inbred be vigorous enough to produce a good seed yield as well as having seed characteristics which make it suitable for planting on a commercial basis.

During the inbreeding process, selection is generally carried on to obtain lines with the desirable characteristics. It is during this selection that progress can be made for lodging resistance and disease reaction as well as other characters. Of course, the selection is most effective when the heritability is high for the character being selected.

The recovery of the vigor lost during the inbreeding process and the utilization of the desirable characters obtained during selection are the basic reasons for making hybrids from the inbreds developed. These inbreds may be used in several kinds of crosses.

Single crosses are hybrids made between two inbred lines. A single cross is relatively more uniform than other hybrids and is of special importance in certain crops where uniformity is desirable. The disadvantages of a single cross are the high cost of seed production and the poor size and shape of the seed, since it is produced on an inbred plant. In corn and in some other crops, single crosses were used primarily as foundation hybrids for three-way and double crosses.

Three-way crosses are hybrids between a single cross used as the female parent and an inbred line used as the male. The distinct advantage of this cross over the single cross is that the first generation hybrid single cross is used as the female or seed-producing parent, in order to obtain hybrid yields and high quality seed of the desired size and shape. A disadvantage of this type of cross is that some inbreds do not produce enough pollen to make the necessary pollinations in the commercial seed fields.

The third type of cross, which was once used almost exclusively for commercial corn, and to a lesser extent for sorghum, is the double-cross hybrid. This hybrid is a first generation cross between single-cross parents. Extensive data supporting the use of the double-cross was reported by Jones (1922).

Another type of cross that may be used extensively in a breeding program is called a top-cross. In this case, an open-pollinated variety may be used as a parent to test a number of inbred lines for their yielding ability and resistance to a disease and other hazards. For instance, an open-pollinated variety that is susceptible to a given disease may be used as the parent of a top-cross, and then top-crosses involving many inbred lines may be evaluated under conditions of an artifically established epiphytotic. If an inbred line crossed on a synthetic is capable of reducing the loss caused by the disease, one has evidence that such an inbred line combined with other similar lines may be highly satisfactory in commercial crosses.

Many inbred lines have been developed in certain cross-pollinated crops, which at the time of their development were highly satisfactory. Subsequently, defective characteristics may be observed in these inbred lines, and the backcross method previously described for self-pollinated crops may be used as a technique to transfer desirable characters in what is otherwise an acceptable inbred line. An example of this is the use of the single gene conditioning resistance to leaf blight (*Helminthosporium turcicum*), first reported by Hooker (1963).

A more recent technique used as a method of crop improvement is called "recurrent selection," a term first used by Hull (1945), and elaborated by Sprague (1952). This is a system of breeding that involves successive cycles of recombination of individuals in a population chosen on the basis of their superior yield, combining ability, or resistance to some disease. Based on their performance test in top-crosses, the best isolates, which may be first year selfed individuals or established inbred lines, are intercrossed by hand or allowed to intercross in isolated plots. This furnishes a new population upon which further selection may be practiced. These repeated cycles of selection may be carried on as long as it appears that the frequency of desirable plants is increasing or until the intensity of the characteristic being selected reaches the established level that may be desired. This technique has been used to improve agronomic and other characteristics.

In cases where it is necessary to make an evaluation of the hybrid progeny, two or three years are required to complete a cycle of selection. However, if selection for some disease resistance characteristic can be accomplished prior to pollination, interpollination of those plants with the highest degree of resistance can be completed each year, and a cycle of improvement can be completed in a shorter period of time. This technique was used with corn by Jenkins, Robert, and Findley (1954). This technique does not provide inbred lines or varieties that are satisfactory as a substitute for the hybrid corn, but it serves as a method of concentrating desired characters in a population from which pure lines can be extracted and which have a high frequency of desirable types.

## Plant Breeding Goals

The ultimate in a breeding program for disease resistance is the production of varieties or hybrids resistant or tolerant to prevalent diseases and useful to agriculture and society. Other objectives are: (1) to develop techniques suitable for use in breeding other plant species and resistance to other diseases; (2) to assist pathologists in genetic characterization of resistance to specific pathogens; and (3) to develop plant populations with a broad genetic background which carry genes for resistance to diseases that periodically inflict losses in crop production.

In order to understand the nature of resistance, knowledge of the mechanism of resistance is also a highly desirable goal. A plant breeder with plant populations having a diverse reaction to a specific disease can contribute to this objective by collaborating with plant pathologists and other scientists. This latter goal is paramount in improving our ability to cope with and perhaps prevent disease epiphytotics in the future.

# Related Literature

In addition to the standard textbooks available on plant breeding, the reader may wish to study some of the attempts made to improve various crops by plant breeders. An easy reference to some of the early work is presented in the 1936 and 1937 Yearbooks of Agriculture published by the United States Department of Agriculture. Although copies of these are not readily available, most libraries carry these volumes as a part of their reference material. The books are divided into individual chapters generally covering single species of crops or animals. They include work on many kinds of ornamental, vegetable, fruit, nut, and grain crops. In these two volumes the reader will find many references that give him necessary background if he expects to work with a crop that has a similar floral and pollinating characteristic to one of those described. Other references that may be of general value to the plant breeder or plant pathologist are: *Heterosis,* edited by John W. Gowen (1952); *Corn and Corn Improvement,* edited by G. F. Sprague (1955); *Wheat and Wheat Improvement,* edited by Quisenberry and Reitz (1967); *Oats and Oat Improvement,* edited by Coffman (1961); and *Plant Breeding,* edited by Frey (1966). In addition, a bibliography of reviews was published in *Phytopathology* 50:396-398 (1960) which gives an extensive list of breeding methods and programs for horticulture and other varieties.

# References

Browning, J. A. 1965. *In* Plant breeding, p. 223-236. Iowa State University Press, Ames.

Coffman, F. A. 1961. Oats and oat improvement. American Society of Agronomy, Madison, Wisconsin. 650 p.

Frey, K. J. 1965. Plant breeding. Iowa State University Press, Ames. 430 p.

———, J. A. Browning, and R. L. Grindeland. 1970. New multiline oats. Iowa Farm Sci. 24:571-572.

Gowen, J. W. 1952. Heterosis. Iowa State College Press, Ames. 552 p.

Hegal, A. S., and M. B. Moore. 1970. Some effects of moderate adult resistance to crown rust of oats. Phytopathology 60:461-466.

Hooker, A. L. 1963. Inheritance of chloratic-lesion resistance to *Helminthosporium turcicum* in seedling corn. Phytopathology 53:660-662.

Hull, F. H. 1945. Recurrent selection for specific combining ability in corn. J. Amer. Soc. Agron. 37:134-145.

Jenkins, M. T., A. L. Robert, and W. R. Findley, Jr. 1954. Recurrent selection as a method for concentrating genes for resistance to *Helminthosporium turcicum leaf blight in corn. Agron. J. 46:89-94.*

Jones, D. F. 1922. The productiveness of single and double first generation corn hybrids. J. Amer. Soc. Agron. 14:241-252.

Quisenberry, K. S., and L. P. Reitz. 1967. Wheat and wheat improvement. American Society of Agronomy, Madison, Wisconsin. 560 p.

Shirreff, P. 1873. Improvement of cereals, p. 1-26. William Blackwood & Sons, Edinburgh.

Sprague, G. F. 1952. Early testing and recurrent selection, p. 400-417. *In* G. F. Sprague, Heterosis. Iowa State University Press, Ames.

———. 1955. Corn and corn improvement. Academic Press, Inc., New York. 699 p.

Van der Plank, J. E. 1963. Plant diseases: epidemics and control. Academic Press, Inc., New York. 349 p.

# Part II
## Specific Considerations for Selected Crop Species

# 8 Rice
## S. H. Ou[1]

## The Rice Plant and Its Culture

Rice is the staple food of more than one-half of the world's population. The widening gap between the slow increase in rice production and the fast increase in population in rice-eating countries is one of the most urgent food problems of today's world.

The rice plant originated in Asia and has been grown in Asiatic countries since ancient times. About 2800 B.C. the crop was being grown in China. Today, about 90 per cent of the world's rice is being grown in Asia.

Rice belongs to the genus *Oryza* of the tribe *Oryzeae*, subfamily *Pooideae* in the grass family *Gramineae*. The genus includes between 22 and 24 species, depending upon the opinions of various workers. Two species, *O. sativa* and *O. glaberrima*, are cultivated. All rice varieties in Asia, Europe, and America belong to *O. sativa*, while many cultivated varieties in West Africa belong to *O. glaberrima*.

The cultivated varieties of *O. sativa* are commonly grouped into subspecies *indica*, *japonica*, and *javanica (bulu)*. Their distinctions are

1. The International Rice Research Institute, Los Banos, Laguna, Philippines.

based on some morphological characters and on their adaptation to temperature, photoperiod, and other growing conditions. The range of genetic variation, including disease resistance, is much greater in *indica* rices than in the other two subspecies. The International Rice Research Institute has a collection of more than 10,000 varieties.

Botanically, the rice plant is a grass, characterized by round, hollow, jointed culms; rather flat, sessile leaf blades; and a terminal panicle. Its vegetative organs consist of the roots, culm, and leaves. Its reproductive organs consist of panicle, spikelet, and flower. The flower consists of a lemma and palea; six stamens, each having two-celled anthers on a slender filament; and the pistil, containing one ovule which bears the short style and the bifurcate, plumose stigma. Rice is normally self-pollinated; out-crossing rarely exceeds 1 per cent.

In crossing two varieties, emasculation and pollination may be done in various ways. A technique successfully used at IRRI is as follows: select 20-30 flowers, usually from the middle of a panicle; clip off about one-third of the lemma and palea before they open; remove the young anthers inside these flowers, and cover the panicle with a bag. The next morning, at the time of flowering, dust the pollens of another panicle on these cut flowers.

Matured seed of rice has a period of dormancy that varies from a few days to 2 or 3 months or longer in some varieties.

Cultivated varieties vary in sensitivity to photoperiod. Some are not sensitive or are only slightly so; others are strongly photoperiod sensitive and flower only during the short days of the year.

The chromosome number of cultivated and wild rices is 24 ($n = 12$). Crosses between cultivated varieties and wild species, or between *indica* and *japonica* varieties result in partial or complete sterility. Crosses between *indica* varieties sometimes are partly sterile.

Although most of the world's rice is grown in fields flooded with water (some in water from one to several meters deep), some is grown under dry or upland conditions as a rainfed crop. As a rainfed crop, rice culture has been relatively primitive and its yield has been low.

Rice-growing in the United States, Australia, and some countries in Latin America is highly mechanized. Most rice culture in Asia uses a great deal of human labor and animal power. In most areas of Asia, land is prepared with the water buffalo after rain has softened the soil. The seed is first sown on seedbeds for 20-30 days. The seedlings are then transplanted by hand to the main field at 3-4 seedlings per hill. About 150,000 hills are planted to each hectare.

The rice crop matures in 90-200 or more days, depending on the variety. Good commercial varieties used at present mature in about

120-130 days. High-yielding varieties recently have been developed for tropical countries. Examples are IR8, IR5, IR20, and IR22. These new varieties have several agronomic advantages over the old varieties.

Several problems may be encountered in rice growing, ranging from deficiency of nutritional elements in the soil to attacks of weeds, insects, and diseases. Each of these alone may render rice culture unprofitable or even cause complete crop failure.

Rice diseases are caused by fungi, bacteria, nematodes, viruses, and even parasitic green plants. Over 200 species of fungi have been reported on rice. Some 70 parasitic diseases have been known. At least a dozen of the known rice diseases are very important. The disease problem becomes increasingly important as close planting and more fertilizers are used, and as new high-yielding varieties, whose high tillering tends to increase the field's humidity, are planted in large areas. Although rice is a very important staple and disease problems are serious, only a few people are engaged in the study of rice diseases.

Three of the most important diseases in Asia (blast, bacterial blight, and tungro virus) are discussed briefly below. Other important diseases are sheath blight (*Corticium sasakii*), stem rot (*Leptosphaeria salvinii* and *Helminthosporium sigmoideum* var. *irregulare*), brown spot (*Cochliobolus miyabeanus*), bacterial streak (*Xanthomonas translucens* f. sp. *oryzicola*) in the tropics, stripe and dwarf viruses in Japan and Korea, hoja blanca in Latin America, white tip (*Aphelenchoides besseyi*), and so forth.

## Blast Disease

Blast occurs in almost all rice-growing areas of the world. The Commonwealth Mycological Institute has recorded its presence in some 70 countries. In Japan, the entire rice crop is treated with fungicides to protect it against the disease. Tens of thousands of tons of chemicals are used each year. In countries of Southeast Asia, such as Thailand, Philippines, and Malaysia, many improved varieties are rendered useless by their susceptibility to blast. It is a serious problem in Latin American countries and in Africa. At the farm level, the losses caused by the disease may range from slight to complete.

The blast fungus produces spots or lesions on leaves, nodes, and different parts of panicles and grains. The leaf spots are typically elliptical with pointed ends and grayish at the center and brown at the margin. Fully developed lesions are 1-1.5 cm long and 0.2-0.5 cm wide. The shape, size, and color of the spots vary, depending upon the environmental conditions, the age of the spots, and the degree of susceptibility of the rice variety. Only minute brown specks may be observed on highly resistant varieties. On moderately resistant ones the spots are small, round, or short and elliptical, and a few millimeters long with a brown margin and a small gray center. The size and color of spots usually indicate a specific interaction between the race of the fungus and the host variety. Leaves may be infected by numerous spots and die.

When the node is infected, the sheath pulvillus rots and turns black, and in drying it often breaks off. Any part of the panicle may be infected and develop brown lesions. Areas near the panicle base are often attacked. Such an infection is called "rotten neck" or "neck rot."

The disease is incited by airborne spores. In temperate regions, such as Japan, the fungus overwinters in straw piles or in diseased seed. The population builds up to a peak in August for about a month. In the tropics, the fungus and the host grow the year round. The airborne spore population is usually high for 5 or 6 months of the rainy season. Under favorable conditions, the cycle of reproduction from spore to spore is about 6-7 days.

## Varietal Resistance
### Method of Testing

Early workers noticed that varieties differ in their resistance to blast. For the last 30-40 years researchers in various countries have conducted empirical tests to identify resistant varieties. In testing resistance, the varieties are artificially inoculated or exposed to the disease in the field or blast nurseries.

Artificial Inoculation.   Methods of artificial inoculation include spraying spore suspensions on plants, injecting spore suspensions, sheath inoculation, puncture inoculation, and so forth. Plants are usually tested at the seedling stage. Spraying a spore suspension is the most commonly used inoculation method. Spraying plants inside a closed chamber results in a much more uniform deposit of the spores on leaves than spraying directly on the plants in the open. Lesions appear initially on the fourth day under proper temperature (24-28 C) and humidity (saturation or nearly so) conditions. Readings may be made on the sixth or seventh day. The spore suspension may also be injected into the pseudostem (leafsheath bundle). The young leaves which emerge from the seedlings bear lesions.

In the sheath inoculation method a section of the leafsheath, usually of older plants, is cut and a drop of the spore suspension is placed on the inside of the sheath. The inoculated sheath is incubated for about 40 hours in a petri dish at 28 C. The epidermis of the inner surface is then stripped off and examined under a microscope. The amount of mycelial development from single conidia in the epidermal cells indicates the degree of susceptibility.

In the puncture inoculation method a small drop of spore suspension is placed on a punctured leaf. A lesion develops if the spores are infective to the variety.

To test susceptibility to blast at the flowering stage, or to the panicle blast, a spore suspension may be injected into the uppermost leafsheath which encloses the panicle, when the panicle has halfway emerged. This method gives a high percentage of infection, often 100 per cent.

Blast Nurseries.    Artificial inoculations are useful for specific studies, but they have limited application for general screening of rice varieties for resistance, mainly because the fungus has many races. Each inoculation can test only one or a few races.

The International Uniform Blast Nurseries are conducted as follows: use upland plots 1.2 m wide and of convenient length (about 10 m). Break the soil into fine particles and add fertilizers (120 kg N, 50 kg $P_2O_5$, and K when necessary). Heavy nitrogen application must be used. Plant 5 g of seed of each test variety in a 50 cm row at about the middle of the 1.2 m width (leave 40 cm at one side and 30 cm at the other), perpendicular to the plot, keeping the rows 10 cm apart. Plant a row of a susceptible variety after every two testing rows and a resistant check after every 10 testing rows; plant three rows on one side (40 cm) and two on the other side (30 cm), parallel to the plot as border rows with a susceptible variety. Water the plants daily to maintain a high humidity. During the growing season in rice-growing areas, there are sufficient airborne conidia to start infection. Disease readings are made 30 days after sowing. The blast nurseries permit the exposure of rice varieties to all the races of the fungus in the area. The development of races differs between seasons and repeated tests are necessary in the same locality.

Although many tests were made in various countries in the past, a relatively small number of varieties were tested in limited geographical areas for only a few seasons. Consequently, the varieties have not been exposed to the many races of the fungus. The selected resistant varieties may not actually possess a high degree of resistance. The International Rice Commission of FAO, United Nations, began an international uniform blast nurseries program (IBN) in 1961. The procedures were somewhat modified and the program strengthened in 1963. IRRI was asked to assume the responsibility of selecting the varieties, multiplying and distributing the seed to participating countries for testing, and compiling the results. Some 50 test stations in 26 countries are now cooperating in the program. The first set of 258 varieties have now gone through more than 150 tests.

In recent years, IRRI has also tested 8,200 varieties from its world collection, and the 300 varieties found resistant after seven tests are now also included in the IBN.

### Grading of Resistance

Various scales have been used for classifying degree of resistance or susceptibility. The IBN uses a scale of 1 to 7. Units 1 to 4 are based upon the types of lesions, and 5 to 7 are based upon the number of lesions or percentage of leaf killing. Units 1 and 2 are resistant reactions in which only brown specks appear. Unit 3 is an intermediate reaction in which small lesions are produced. Unit 4 refers to typical, large, elliptical lesions.

### Varietal Reaction

The IBN results and many other artificial inoculation experiments show that a wide range of resistance exists among varieties. Some varieties showed a broad spectrum of resistance with only very few susceptible reactions in the 150 or more tests of IBN and in the hundreds of artificial inoculations. In contrast, certain varieties were susceptible to nearly all tests or inoculations. Most varieties have various levels of resistance between the two extreme groups. No variety is resistant at all testing localities or to all races. The most resistant varieties so far known are Tetep, Carreon, Tadukan, C45-16, Ram Tulasi (Sel), Pah Leuad 29-8-11, Murungakayan 302, Thavalakkannan, and so forth. The most susceptible ones are Arlesiene, Fanny, Khao-tah-Haeng 17, and so forth.

Resistance to leaf and panicle blast were thought to be separate genetically, but our experiments have shown that they are positively correlated.

A limited attempt has been made to induce blast resistance with radiation, X-rays, thermal neutrons, and gamma rays. Susceptible varieties gain some resistance, but not a high degree.

Many wild species of rice have been tested for blast resistance. The reaction in these species and strains is as variable as that in cultivated species. No better sources of resistance have been found in them.

## The Causal Organism and Its Pathogenic Variability

### The Organism

The causal organism of rice blast is *Pyricularia oryzae* Cav. Researchers disagree about its nomenclature, its host range, and even the spelling of its generic name (many spell it as *Piricularia*). No perfect stage has been found.

The fungus is easily cultured on various media and natural substrates. Some isolates produce abundant spores in most media, others very few. Spore production is affected by substrate, *p*H, light, and other environmental conditions. To induce abundant spore production various kinds of media have been used, such as a special rice polish produced in the U.S.A., barley seed, starch-yeast, V-8 juice, and so forth. We often use short sections of rice straw including the node, which is easily available.

For inoculation the conidia are washed from the media with water and adjusted to a proper density (about $10^4$ per ml).

### Pathogenic Variability

As early as 1922, Sasaki in Japan noted the existence of pathogenic races in *P. oryzae*. Unfortunately, little work was done immediately after his report. During the 1950s several countries, particularly Japan, the U.S.A., and Taiwan (China), studied the races more intensively. Each country used a different set of differential varieties. Many races were characterized in each country during the early 1960s. Races were reported later from the Philippines, India, Korea, and Colombia. No comparison of races in these countries was possible because the differentials used in each country were not the same.

A cooperative study between Japan and the United States led to the proposal of a set of eight differential varieties for international use. They are Raminad Str. 3, Zenith, NP-125, Usen, Dular, Kanto 51, Sha-tiao-tsao(S), and Caloro. The races thus characterized are called international races and are designated as IA, IB, IC, and so forth, followed by numbers. All races that infect variety Raminad Str. 3 belong to IA, those that infect Zenith, to IB, and so forth, according to the order above. A standard race number following the designations (IA, IB, and so forth) has been suggested to avoid confusion among workers in different countries. Each country may use supplementary varieties for further differentiation. The system is being adopted gradually by various countries but differentials peculiar to a country are still used. In the Philippines we have characterized 150 races with

Philippine differential varieties and 45 international race groups with the international differentials.

Pathogenic variability in *P. oryzae* is complicated. Pathogenicity changes continuously as the fungus grows. The conidia produced on single lesions, reproduced from a single conidial culture or from single hyphal tips grown from one of the three cells of the conidia, may differentiate into numerous races. Forty-six single conidial cultures obtained from one leaf lesion were separated into 14 Philippine races or eight international races when inoculated on the respective differential varieties. In another instance, 182 single conidial subcultures from one isolate were differentiated into 42 Philippine races. These new races differ widely in pathogenicity. Some races infect only one or two of the differential varieties, and others infect all 12 varieties.

A typical blast lesion produces 2,000-6,000 conidia each night for 1 or 2 weeks. A petri dish culture may contain a million conidia. If more monoconidial isolates were tested, most of the 150 races known in the Philippines might be identified from single lesions or single cultures. As it multiplies, each conidium again produces many races. The concept of race or pathogenicity of an isolate in this case is that it is only a temporary phenomenon; it changes consistently and is not at all stable.

Many races probably are present in a given field, since the fungus changes as it multiplies. Sixty races have been found in our blast nursery by testing 363 conidia from the nursery.

The mechanism for such great variability is not known. Recent studies show that most cells in the mycelium and conidia are uninucleate. Anastomosis of hyphae has been frequently observed. Some early workers believed that the cells were multinucleate and heterocaryotic. This is another highly interesting area of study.

## Horizontal Resistance

Studies on horizontal resistance to blast have just begun. Recently, Sakurai and Toriyama in Japan reported that varieties St. 1 and Chugoku 31, which were originally developed for stripe virus resistance, also possess field resistance to blast. Compared with other varieties, they produce relatively few lesions in the field, in an upland blast nursery, or in inoculation tests with seven isolates of the fungus. St. 1, however, was reported to be severely attacked in other localities.

Small number of lesions and small size lesions may indicate possible horizontal resistance to blast. Both types were observed in our nursery on many of the 8,200 varieties screened for blast resistance. The difference in numbers of lesions varies from 10- to 100-fold. Small lesions (intermediate type, or type 3 according to international standards) produce 50-250 conidia each night, while typical large blast lesions produce 2,000-6,000 conidia each night. This difference is also 10- to 100-fold. Both types greatly influence the rate of disease development, particularly since the reproduction cycle of the fungus is only 6-7 days.

Many varieties showing small size lesions do not have horizontal resistance because they break down (produce many large lesions) in subsequent tests. Only 200 of some 400 varieties with small lesions remained resistant after five repeated tests in our blast nursery. These 200 remaining varieties will be further tested in the IBN. The only way to determine which varieties possess horizontal resistance is to expose them to all possible races of the fungus.

A variety may have only a few lesions because of insufficient inoculum, or because the variety has horizontal resistance. To determine the reason, the fungus may be isolated and used to inoculate the varieties from which the fungus was isolated. Many of the varieties consistently produced only a few lesions, while the susceptible checks produced 10 to 100 times more.

We found further that, at least in a few cases studied, the reason for the small numbers of lesions on the original variety is the change of the fungus into many races (42 races from the 182 fungus isolates mentioned above), a varying number of which cannot infect the original varieties. Of the 100 single conidial subcultures of an isolate from Tetep, 83 cannot infect Tetep; of the 85 subcultures of an isolate from Carreon, 80 cannot infect Carreon.

Because pathogenicity tends to change constantly, the fungus has the ability to adapt itself to new host varieties. On the other hand, it has a built-in weakness toward varieties with a broad spectrum of resistance, because the originally pathogenic race changes into new races and most of the new races cannot infect the original varieties. Since only a small portion of the population of each generation of the pathogenic race is pathogenic, the varieties are therefore considerably stable in their partial resistance to the fungus. They appear to be horizontally resistant.

## Breeding and Genetics

### Breeding for Blast-Resistant Varieties

Breeding for blast resistance started in the 1920s. Several varieties have been used as sources of resistance in Japan. Variety Sensho was used in the early years. From it, Futaba and other varieties were developed, which in turn were used as parents. Sensho was followed by resistant varieties Toto and Reishiki, from which the Kanto varieties (Kanto 51 to 55) were developed. The Kanto varieties were again used as parents. Variety Tadukan was used in developing the Pi varieties (Pi 1 to Pi 4), and Zenith and other varieties were also used. Many new varieties were used in succession. They were adopted for different lengths of time but were discarded, either because they became susceptible, or because they did not have enough desirable agronomic characters. During the last 20 years farmers in Japan have relied more on chemicals to control blast while planting high-yielding, but rather susceptible varieties.

In India, in the Madras State, variety Co 4 was used as a source of resistance many years ago. When it was crossed with Adt 10, two new resistant varieties, Co 25 and Co 26, resulted. These two new varieties rapidly replaced the popular Adt 10. Co 4 and its progeny have been used in various parts of India for breeding against blast.

During the last 20 years, Taiwan, the U.S.A., and other countries in Asia have conducted breeding programs for blast resistance with only limited success. Only in recent years have serious efforts been made by many countries in Asia and America.

*Genetics*

It is generally believed that resistance to blast is simply inherited. Resistance usually is dominant over susceptibility, and it is reportedly controlled mainly by one pair of genes. But control by two and three gene pairs has also been reported. Gene effects are said to be cumulative.

Before 1960, the idea of pathogenic races was not recognized in this type of work. The fungus used for inoculation consisted of several races. Recent work has used specific races. Most test results again show a simple inheritance: one pair of dominant genes and sometimes some minor modifier genes. Different genes control resistance to different races of the fungus. Most of these genes are independent. Kiyosawa and his associates in Japan have now identified at least 11 resistance genes, designated as *Pi-a, Pi-i, Pi-k, Pi-z, Pi-ta*, and so forth, according to the variety from which the gene was derived. In Taiwan and the U.S.A. the genes were designated as *Pi-1* and *Pi-6* (U.S.A.) or as *Pi-4, Pi-13, Pi-22, Pi-25* (Taiwan), and so forth, apparently according to the fungus races to which the gene is resistant.

Much less is known about the genetics of host resistance and about fungus pathogenicity than is known about some other major plant diseases.

## Nature of Resistance

The number of silicated cells of the leaves reportedly is closely correlated with resistance. Many studies have shown that nitrogen content, soluble nitrogen, or amino acids and amines are correlated with susceptibility. Resistance to mechanical puncture was also found to be correlated with resistance to the disease. Such correlations may not explain the real mechanism of resistance. For instance, if silica content causes resistance, why are varieties growing under identical conditions resistant to some races but susceptible to others?

Two toxins, Piricularin and $\alpha$-picolinic acid, have been isolated from the fungus and the diseased leaves. The chlorogenic acid produced in rice detoxifies the two toxins. The amount of the toxins produced by the fungus and the amount of chlorogenic acid produced by the host plants have been used to explain resistance and susceptibility.

Blast resistance has been attributed to phenolic compounds formed in the brownish margins of the blast lesions. Phytoalexin has also been detected from the interaction between the fungus and the host leaves.

# Bacterial Blight

Bacterial blight is one of the three major diseases of rice in Asia, particularly since wide areas have been planted to the high-yielding variety IR8, which is very susceptible to the disease. It has been reported in Malagasy in Africa, but it is not known in the Americas.

In temperate regions, the major symptoms of the disease are the blighted leaves, which turn white. The bacterium enters the leaves through water pores along the margin of the leaf blade. Elongated lesions with wavy margins form at one or both sides of the leaf edges. In a few days they enlarge both in length and width to cover all or a large portion of the leaves, turning the leaves first yellow, then white. Saprophytic molds often grow on these dead leaves, giving them a grayish-white color.

In the tropics there are two additional symptoms of leaf blight. One is "kresek." Two to four weeks after transplanting, seedlings wilt and die. "Kresek" was described as a bacterial disease in Indonesia in 1950. The causal organism was called *X. kresek* Schure. In 1964 it was found to be a severe form of bacterial blight. The second symptom is the yellowing of leaves. A young, pale yellow leaf emerges from a more or less normal green plant, and from then on the plant's growth becomes restricted. Since the disease in the tropics is not confined to the leaves, the disease has been called bacterial blight.

Heavily infected fields in Japan lose about 30 per cent in yields. In the tropics, "kresek" may cause complete loss in a field plot. In general, the disease causes more damage in the tropics because of the higher temperatures and more frequent rainstorms or typhoons, which quickly spread the disease, and also because the bacterial strains are more virulent in the tropics.

## Evaluation of Varietal Resistance

### Method of Testing

The usual method of inoculation is by spraying a bacterial suspension on leaves. Reitsma and Schure in Indonesia inoculated the leaves by needle pricking and by dipping the seedlings in a bacterial suspension in their study of the kresek. Muko and Yoshida found that the results of needle-pricking inoculation and of natural infection are similar. Inoculation by spraying separates the resistant and the susceptible varieties easily, but it does not distinguish various degrees of infection in the intermediate groups. Needle inoculation gives a more precise classification of different degrees of resistance. The immersion method is useful when a large number of plants are to be inoculated.

Various scales have been used for classifying the degree of resistance or susceptibility. Since there is a wider range of symptoms in the tropics, the scales used by various Japanese workers are not suitable for the tropics. We use a scale of 10 units, 0 to 9, to indicate degrees of resistance, ranging from immune to susceptible. Since the kresek and pale yellow symptoms are not found in adult plants, the description of symptoms in each unit at the two stages of growth are different, but a unit in seedling stage corresponds to the same unit in another. The scales are described in Table 8-1. Resistance at the seedling stage and at the flowering stage (flag leaf test) are positively correlated in general.

### Screening for Varietal Resistance

Varietal resistance has been tested in Japan for many years. Such varieties as Shigasekitori 11, Shobei, and so forth, have been used to develop resistant varieties. Recently, Sakaguchi and his associates tested 863 cultivated and wild varieties and found that varieties Lead, TKM-6, Nigeria 5, and two lines of wild rice were resistant.

TABLE 8-1 STANDARD SCALES FOR INDICATING DEGREE OF
RESISTANCE TO BACTERIAL BLIGHT (IRRI, 1965).

| SCALE | SEEDLING STAGE | FLOWERING STAGE— FLAG LEAF |
|---|---|---|
| 0 | No lesion observable (immune). | Same |
| 1 | Lesions restricted to 1-2 mm around points of inoculation. | Same |
| 2 | Lesions more or less elliptical; not more than 2-3 cm long. | Same |
| 3 | Lesions elongated, extend up to about 1/2 length of leaf blade or less. | Lesions elongated, less than 1/2 of leaf blade. |
| 4 | Lesions increase both in length and width, destroying 3/4 of leaf blade. | Lesions broad and coalescent; upper portions of leaves often dead; lesions extend to about 1/4 of lower half of leaf surface below points of inoculation. |
| 5 | Lesions extend to and destroy entire leaf blade. | Lesions coalescent and upper portion of leaves dead; lesions extend to about 1/2 of lower half of leaf surface. |
| 6 | Lesions extend to and destroy enitre leaf blade and less than 1/2 of leaf sheath. | Lesions extend to about 3/4 of the lower half of leaf blade. |
| 7 | Entire leaf blade and more than 1/2 of leaf sheath destroyed, and a few pale yellow symptoms appear. | Lesions extend to base and destroy the entire leaf blade. |
| 8 | Leaf blade and leaf sheath destroyed, many pale yellow plants, and less than 25 seedlings killed (kresck). | Lesions destroy entire leaf blade and extend to about 1/2 of leaf sheath. |
| 9 | About 50% or more seedlings killed (kresek). | Lesions completely destroy leaf blade and sheath. |

More than 8,000 varieties have been screened at IRRI during the last
few years. These varieties were needle-inoculated in the field with a
virulent strain on the flag leaves. Resistant selections were tested again
at the seedling stage in the greenhouse, then they were further tested
with several virulent strains because the strains were found to differ in
pathogenicity. Through repeated tests, about 30 varieties were selected.
When these 30 varieties were tested in India, however, many became
moderately susceptible. This again shows that bacterial strains differ in
their virulence. Varieties such as TKM-6, BJ 1, and so forth, have

broader bases of resistance than others. Scientists hope that resistance to specific virulent strains in different varieties can be combined by hybridization.

## Variability and Classification of the Bacterial Strains

The physiological properties of strains of *Xanthomonas oryzae* differ particularly in pathogenicity and in sensitivity to phages. They also differ somewhat in such physiological characters as gelatin liquification, $H_2S$ production, and acid production from sugar.

In Japan, four phages and five lysotypes have generally been recognized. Seven phages have been isolated in the Philippines, designated as BP1 to BP7, and 18 or more lysotypes (strains) have been found.

In Japan, 14-19 rice varieties, ranging from most resistant to most susceptible, have been used for testing virulence of strains. The strains were separated into two groups. Group I (or A) strains attack all varieties, while Group II (or B) strains attack only the intermediate and susceptible varieties. They are further separated into subgroups by the intensity of the disease on the varieties. We studied 50 strains on 24 varieties from the Philippines. Strains differ in virulence and in their pathogenic patterns. For instance, strains in Pattern I are highly virulent, causing an intermediate reaction on the resistant varieties, a susceptible reaction on the intermediate group of varieties, and death in susceptible varieties. Strains in Pattern II are weakly virulent, causing an immune reaction on resistant varieties, and only a moderately susceptible reaction on the susceptible varieties. Strains in Pattern III show great differences in reaction among varieties, from almost immune on the resistant, to death on the susceptible ones. Resistant and susceptible varieties do not show much difference to strains in Pattern IV. Weak strains never attack resistant varieties severely. Varieties resistant to most of the virulent strains may be attacked by a few virulent strains. Pathogenic races with distinct opposite reactions, such as those observed in *Pyriculariaoryzae,* have not been observed in *X. oryzae*.

Most strains from tropical countries are more virulent than Japanese strains when compared on many varieties. Some strains from India and Indonesia are highly virulent. A cooperative project between the University of Hawaii and IRRI on comparative studies of the bacterial strains should provide more information on the relative virulence of strains from various countries.

Variation in virulence seems to exist also within the single isolates. In isolation from the field the virulence of the culture also much depends upon which colony is selected during isolation.

## Breeding and Genetics

Breeding for bacterial blight resistance in Japan has been conducted for many years. The resistant variety Zensho 26 was developed from Shigasekitori. In turn, Zensho 26 was used as a resistant parent to develop the varieties Hoyoku, Kokumasari, and Shiranui. The variety Kogyoku was developed from Shobei. Kogyoku was further used for developing varieties Sachikaze and Nihonbare. Other varieties developed in various stations are: Oku 244, Fujiminari, Honen-wase, Shin 2, Nakashin 120, Koshihikare, Hogareshirazu, and Shirogane. None of these varieties is resistant to the current virulent strains of the bacterium.

Varieties are developed by IRRI, IR20 and IR22 resistant to the disease in the Philippines. Highly resistant varieties may be developed by combining several sources of resistance.

Little is known about the inheritance of resistance. Early studies indicated that genes for resistance may be dominant or recessive, monogenic or polygenic. More recently, Washio and his associates in Japan reported two complementary dominant genes, $X_1$ and $X_2$ in the variety Kidama, while there is only one dominant gene in Norin 274. Sakaguchi reported two genes for resistance, $Xa_1$ and $Xa_2$. The Kidama group of varieties has $Xa_1$, and Rantaj-emas carries $Xa_1$ and $Xa_2$.

Many crosses between resistant and susceptible varieties made in the Philippines show that the resistance may be recessive or dominant, depending upon the variety combinations and the degree of resistance in the resistant parents.

## The Tungro Virus Disease

Although tungro disease had occurred in Southeast Asia for a long time, the causal agent was learned only recently. It was found that a so-called physiological disease of Malaysia, "penyakit-merah" (red disease), known since 1938, is tungro disease.

"Mentek" disease of Indonesia, reported since 1859 and also considered a physiological disease, is now known to be tungro disease, although some other maladies may have been classified as "mentek." It was first reported from the Philippines in 1964. During the last 6 years, severe outbreaks of tungro occurred in Thailand. It is a major concern in East Pakistan and occurs in some parts of India.

The primary symptoms are stunting and leaf discoloration in shades of yellow to orange, both of which vary with the degree of susceptibility.

Both symptoms may gradually disappear on older plants of more resistant varieties. In susceptible varieties symptoms persist, and plants may be killed if infected while still young. Symptoms may not show when infection occurs in adult plants.

The virus is transmitted mainly by the rice leafhopper, *Nephotettix impicticeps* Ishihara. It may also be transmitted by *N. apicalis* (Motsch.) and *Recilia (Inazuma) dorsalis* (Motsch.). The virus may be transmitted within a few hours after the insect's acquisition feeding. Most *N. impicticeps* transmit the virus in a day. The virus does not persist in the vector. Viruliferous insects lose the ability to transmit the disease in 4 or 5 days unless they reacquire the virus.

## Screening for Varietal Resistance to the Virus

Varietal reaction to tungro virus may be tested in fields where the disease is epiphytotic each year. However, artificial inoculations usually are necessary. A mass screening system used at IRRI is essentially as follows: (1) about 3,000 adult insects are produced every other day by rearing them from eggs in a series of large cages. (2) They become highly viruliferous after feeding 2-4 days on diseased plants. Then they are used for inoculation for about 8-9 hours a day, are again fed on diseased plants for 15-16 hours to reacquire the virus, and are used again for inoculation the next day. This is repeated as long as they survive. (3) Seeds are soaked in water and transplanted at 29 seedlings per pot as soon as they germinate. (4) Plants are inoculated at the two- to three-leaf stage (11-13 days) with 2-3 insects per seedling inside two large cages, each containing 16 small pots. (5) The reactions of varieties are determined 12 days after inoculation. Sixteen varieties can be tested in duplicate each day, or 32 varieties without replication.

Since the reaction of varieties to seedling inoculation varies from about 0 to 100 per cent, the percentage of infection is used directly to indicate degree of susceptibility. Varieties are classified further into three arbitrary groups: (1) resistant, those with 30 per cent or less infected seedlings; (2) intermediate, 30 to 60 per cent infected seedlings; and (3) susceptible, more than 60 per cent infected seedlings.

The second criterion for classifying resistance is disease tolerance. Some plants "recover" from the disease at later stages of growth. For easy assessment, the degree of reduction in height of diseased plants, as compared with that of healthy plants, is used as the index. The reading is made about 6 weeks after inoculation. Scale unit $S_0$ indicates no reduction in height; $S_1$ indicates 25 per cent reduction; $S_2$ indicates 50 per cent, and $S_3$, 75 per cent or more.

Thousands of varieties and hybrid lines have been tested, and many resistant varieties have been identified. Pankhari 203 is the most resistant. Many commercial varieties in Indonesia and the Philippines are resistant: examples are Tjeremas, Bengawan, Peta, Intan, and Sigadis. These varieties were resistant to "mentek" of Indonesia before the tungro virus was identified, and they are also resistant to "penyakit-merah" of Malaysia. This is another indication that these diseases are similar or identical.

## Varietal Resistance to Insect Vector

When the insects were reared for virus transmission studies at IRRI, they survived and reproduced better on variety Taichung (Native)1 than on variety Peta. Further investigation on varietal resistance to the insect showed that several rice varieties were highly resistant. It was also found that varieties resistant to the virus and to the insect react independently. Pankhari 203 is resistant not only to the virus, but also to the insect vector. The variety IR8 is resistant to the insect but susceptible to the virus. Taichung (Native)1 is susceptible to both the virus and the insect, while Kai Lianh Hsung Ting is susceptible to the insect but resistant to the virus. Insect resistance provides an additional means of controlling the disease through hereditary characters of the plants.

## Breeding and Genetics

Varieties resistant to the virus and to the insect have both been used in breeding programs at IRRI and in Thailand. It is hoped that the new varieties resistant to the disease will be developed in the near future.

Only two strains of the virus have been detected thus far. They differ in symptoms on certain varieties, but not in varietal reaction. Breeding for tungro resistance is, therefore, not expected to be complicated. According to a recent brief report from East Pakistan, some seedlings of Pankhari 203 on which insects were feeding in an artificial inoculation experiment were infected by the virus. If this is confirmed, either the virus has other strains, or the insect has different ecotypes. Both possibilities need further study.

No detailed genetic studies have been made. Crosses between Pankhari 203 and Taichung (Native)1 or IR8 have shown that the $F_1$ is resistant and that the $F_2$ segregates in approximately a 9:7 ratio.

# Conclusion

Too little effort has been made to control rice diseases with resistant varieties. Consequently, there has been limited success in achieving effective rice disease control, particularly in the tropics. Much attention has been given recently to the study of rice diseases, but the number of scientists involved is still limited. Many problems await solution, both in practical and in academic areas.

Great genetic variability exists in cultivated rice varieties and in wild species. Resistance to any given rice pathogen may be found. The variability of the pathogen requires much more work than anticipated by earlier workers. For instance, considerable work has been done on the blast disease, but failure to recognize the great variability in the fungus has limited past successes. An international approach to identifying sources of resistance should be encouraged.

The success of using horizontal resistance to late blight of potato has stimulated the study of such resistance in diseases of rice, but the work has just begun in very few institutions.

Genetic information on the inheritance of resistance, on the pathogenicity of the fungus, and on the resistance of the host plant to disease is inadequate. In short, many important and challenging problems in the area of rice diseases remain which depend for their solution on greater research effort.

# References

Chang, T. T., and E. A. Bardenas. 1965. The morphology and varietal characteristics of the rice plant. IRRI, Los Banos, Laguna, Philippines. Tech. Bull. No. 4.

Ghose, R. L. M., M. B. Chatge, and V. Subrahmanyan. 1960. Rice in India, Revised Ed. Indian Council Agr. Res., New Delhi.

Ou, S. H. Rice diseases, in press. Commonw. Mycol. Inst., Kew, Surrey, England.

——— and P. R. Jennings. 1969. Progress in the development of disease-resistant rice. Ann. Rev. Phytopathol. Vol. 7.

Rice diseases and their control by growing resistant varieties and other measures. 1967. Proc. Symp. Agr. Forest & Fish. Res. Council, Min. Agr. & Forest., Tokyo, Japan.

Rice genetics and cytogenetics. 1964. Proc. Symp. at IRRI, February, 1963. Elsevier Publishing Co., Amsterdam.

The rice blast disease. 1965. Proc. Symp. at IRRI, July, 1963. The Johns Hopkins Press, Baltimore.

The virus diseases of the rice plant. 1969. Proc. Symp. at IRRI, April, 1967. The Johns Hopkins Press, Baltimore.

Trist, D. H. 1965. Rice, 4th Ed. Longmans, London.

# 9 Wheat
## E. L. Sharp[1]

Wheat is the world's most widely cultivated crop. The epicenter is the so-called fertile crescent area of the Middle East and includes Turkey, Iraq, Iran, Jordan, and Israel. Some of the oldest evidences of wheat culture have been found near Jarmo, in Iraq. The most primitive forms of wheat, wild einkorn and einkorn, appeared about 8000 B.C. The basic chromosome number of wheat (genome) is 7 but the various groups form a polyploid series of 7, 14, and 21 pairs of chromosomes. Table 9-1, which follows the classification of Morris and Sears (1967), lists the species and varietal groups of wheat.

Einkorn is a diploid containing only the A genome. All varietal groups under *Triticum turgidum* are tetraploids that contain the A and B genomes. *T. timopheevii* has the A and G genomes, while *T. timopheevii* var. *Zhukovskyi* is a hexaploid containing 4 doeses of the A and 2 doses of the G genome. All varieties of *T. aestivum* are hexaploids containing the ABD genomes, the main differences between varieties of *T. aestivum* being the result of single genes. Of the different wheats listed in Table 9-1, common wheat, club wheat, and durum are now by far the most widely grown.

1. Professor of Plant Pathology, Montana State University.

TABLE 9-1    THE GROUP OF THE VARIETIES (CULTIVARS) OF THE
GENUS *TRITICUM*.

| SPECIES | VARIETAL GROUP | BASED ON | COMMON NAME |
|---|---|---|---|
| *T. monococcum* L.[a] | | *T. monococcum* L. | Einkorn |
| *T. turgidum* L. | dicoccon | *T. dicoccon* Shrank (*T. dicoccum* Shrank) | Emmer |
| | durum | *T. durum* Desf. | Durum |
| | turgidum | *T. turgidum* L. | Poulard Wheat, Branched Wheat |
| | polonicum | *T. polonicum* L. | Polish Wheat |
| | carthlicum | *T. carthlicum* Nevski (*T. persicum* Vav.) | Persian Wheat |
| *T. timopheevii* (Zhuk) Zhuk. var. *timopheevii* | | *T. timopheevii* Zhuk. | None |
| var. *Zhukovskyi* (Men. & Er.) Morris & Sears, Comb. nov. | | *T. zhukovskyi* Men. & Er. | None |
| *T. aestivum* L. em Thell | spelta | *T. spelta* L. & *T. macha* Dek. & Men. | Spelt |
| | vavilovii | *T. vavilovii* Jakubz | None |
| | aestivum | *T. aestivum* L. (*T. vulgare* Host, *T. sativum* Lam.) | Common Wheat |
| | compactum | *T. compactum* Host | Club Wheat |
| | sphaero-coccum | *T. sphaerococcum* Perc. | Shot Wheat |

[a] The cultivated varieties of this entity have not yet been classified into groups.

Hybridization between various grasses and wheats followed by
chromosome doubling resulted in the various wheat groups. Evidence
indicates that *T. monococcum, T. speltoides,* and *T. tauschii* contributed the A, B, and D genomes respectively, now found in *T.
aestivum* (Morris and Sears, 1967). Emmer appeared about 8000 B.C.
and was replaced by durum about 300 B.C. following a series of
mutations resulting in a free-threshing wheat. Spelt is the most primitive
of the hexaploid wheats. After originating in the Middle East the various
wheats became distributed in various areas of the world. They were
transported to the New World, where they moved west with settlers.

Wheat is the major crop of the U.S. and Canada and is grown in
almost every country of Latin America, Europe, and Asia, with a
definite concentration in the Northern Hemisphere. Each year about
575 million acres are planted to wheat, and production is estimated at
about 8 billion bushels (Reitz, 1967). The four main wheat-producing

states in the United States are Kansas, North Dakota, Oklahoma, and Montana. In the United States per capita consumption of wheat exceeds that of any other single food, and it constitutes the main food staple in 35% of the world population (Reitz, 1967).

Genes for resistance to various diseases of wheat have been found in several species of *Triticum* as well as in related genera. Among species of *Triticum*, *T. dicoccon*, *T. durum*, and *T. timopheevii* have been valuable sources for disease resistance. The diploid relatives of wheat that have furnished genes for disease resistance belong mainly to the genera *Secale, Agropyron,* and *Aegilops.*

Breeding methods applicable to any other character may be used to utilize resistance, but it must be borne in mind that the genetic system of the pathogen is also involved. Techniques for breeding wheat have been published (Allard, 1966). Often the backcross method has been used for transferring resistance genes into an otherwise acceptable variety. The use of the bulk method has the advantage of allowing the incorporation of more diverse germ plasm. Transgressive breeding has, in many cases, resulted in desirable disease resistance not observable in either parent.

Different types of plant disease protection by genetic means have been reviewed by Caldwell (1968). A large share of the disease resistance incorporated into wheat has been *specific resistance*. This usually involves single genes, and the variety is highly resistant to certain clones of the pathogen but susceptible to others. Generally speaking, this type of resistance has been of short duration, but there are notable exceptions. Specific resistance expresses hypersensitivity, and it is easy to work with and follow in a backcrossing program. *General resistance* usually does not show hypersensitivity but presents barriers to penetration, development, and spread of a pathogen. It should show a stable and lasting protection against a disease. Any resistance is general until proven to be otherwise. It is questionable whether any resistance lasts indefinitely. There are some cases where plants that were believed to have a general type of resistance are slowly becoming more susceptible. *Tolerance* is a form of plant disease protection wherein the plants do not present any barriers to the pathogen. Symptoms and signs are not noticeably inhibited, but the effect on yield is much less than for completely susceptible varieties. The bulk of the work with plant disease protection has involved specific resistance, while tolerance has not been thoroughly investigated.

The use of multiline varieties, as advocated by Borlaug (1965), is a form of specific resistance that has not yet received adequate evaluation to determine its potential.

# Rust

Some of the most devastating losses to wheat have been caused by the rusts. Prevalence varies from year to year and between locations, but almost all areas where wheat is grown are subject to these diseases. Three conditions must be met for a rust epidemic to occur. There must be susceptible varieties, adequate inoculum, and a favorable environment. The rusts of wheat-stem rust, leaf rust, and stripe rust will be discussed separately in detail, but certain characteristics hold for all three.

The rusts, as obligate parasites, show a high degree of specialization, and resistance to them is usually simply inherited. Using differential hosts, a large number of physiologic races have been determined for each of the three rusts of wheat. The use of differential host varieties furnishes a gross classification of the pathogens and some tie-in with the past, but supplemental varieties of known resistance genes and those of greater relevance to present breeding programs are added for better characterization of the pathogen. The rust pathogens also have a genetic system involving virulence. Assuming a gene-for-gene relationship between host and pathogen (Flor, 1956), near isogenic host lines with specific single genes for resistance furnish precise materials for obtaining information on the pathogen population. The nuclear condition of the pathogen is an important consideration. In host resistance studies with the wheat rusts, the urediospores, which are dicaryotic, are routinely used in pathogenicity tests. One specific culture may be heterozygous for the virulence loci corresponding to certain resistance loci in standard differential varieties. Another culture may be identified as the same physiologic race but may differ considerably in genotype in relation to the standard differentials. This attribute, even barring sexual recombination, has great implications in the possible formation of new virulence genes by way of mutation or somatic recombination. Thus the culture, rather than the physiologic race, is the important entity.

The wheat rusts all require somewhat different environments for initial development of infection, and many host-pathogen combinations are particularly affected by temperature in the postpenetration phase. For prepenetration development free water is required on the host leaves for a minimum period of about 4 hours, but a 16-24 hour period is generally used. With pathogenicity tests with seedlings, inoculations are often made in settling towers to obtain uniform spore distribution on the leaves. In field trials, spreader rows are usually inoculated in the 4-

leaf stage. Often the spores are diluted with talc or mixed in non-phytotoxic oil. To assure a favorable environment for infection of spreader rows, the ground may be moistened and the inoculated plants may be covered overnight with clear plastic. Spreader rows should be interspersed throughout the trial plots for uniform development of rust. Resistance is detected by a scale of infection types which considers pustule size and/or relation of chlorotic or necrotic tissue area to the area with pustules. The basic methods used with the wheat rusts are modifications of those described by Stakman et al. (1962). Severity is a measure of the plant area covered with lesions, and the coefficient of infection combines both infection types and severity in a single per cent value.

TABLE 9-2    GENES FOR RESISTANCE TO STEM RUST OF WHEAT.[a]

| SYMBOL | REPRESENTED IN | CHROMOSOME |
|---|---|---|
| Sr1 | Hope | 2B |
| Sr2[b] | Hope | — |
| Sr3[b] | Marquillo | — |
| Sr4[b] | Marquillo | — |
| Sr5 | Kanred | 6D |
| Sr6 | Red Egyptian | 2D |
| Sr7a | Khapstein | 4B |
| Sr7b | Hope | 4B |
| Sr8 | Red Egyptian | 6A |
| Sr9a | Red Egyptian | 2B |
| Sr9c | *Triticum timopheevii* | 2B |
| Sr10 | No. 466 | — |
| Sr11 | Gaza | 6B |
| Sr12 | Marquis | 3B |
| Sr13 | Khapstein | — |
| Sr14 | Khapstein | — |
| Sr15 | Norka | 7A |
| Sr16 | Thatcher | 2B |
| Sr17 | Yaroslav Emmer | 7B |

[a] Varieties listed may contain more than one resistance gene.
[b] Mature plant resistance.

## Stem Rust

There are several special forms or varieties of *Puccinia graminis*, but the one causing stem rust of wheat, barley, and some related grasses is

*P. graminis* Pers. f. sp. *tritici* Eriks. and E. Henn. This variety has been divided into more than 300 physiologic races on the basis of the reaction on 12 standard wheat differential varieties (Stakman et al., 1962). The early work indicated no problem in developing wheat varieties resistant to stem rust, but soon after release of resistant varieties new races appeared which could attack them. Intensified studies on genes for virulence in the pathogen and the number and location of resistance genes in the host have led to a much better understanding of host-pathogen interactions.

Table 9-2 lists a number of stem rust (*Sr*) genes that have been located in various varieties. A number of resistance alleles have been determined for some loci, but genes for virulence occur at separate loci in the pathogen. Several workers have developed near isogenic lines in common backgrounds (Knott, 1963; Loegering and Harmon, 1969). The genes *Sr*2, *Sr*3, and *Sr*4 condition only mature plant resistance, while the others can also be effective in the seedling stage. The wheat variety Selkirk has resisted rust for more than 15 years. It contains the *Sr*6 gene plus mature plant resistance of the Hope type. Caldwell (1968) believes the latter to be an effective type of general resistance. Some *Sr* genes behave as recessives, and some as dominant genes. In fact, *Sr*6 is reported completely dominant with race 56, but recessive with 15B (Knott, 1963). Other genes have been found to be incompletely dominant with some races, and the host-pathogen interaction may be greatly influenced by temperature. The *Sr*6 gene, for example, ranges from a 0 to a 4 infection type as temperature is increased from 18 to 30 C (Loegering and Harman, 1969).

In cases where a wheat variety has several genes for rust resistance, the gene conditioning the lowest infection type is often epistatic to others and is the one expressed. There are cases, however, of additive gene action as described by Watson (1970).

In disease resistance studies, the author used all possible combinations of *Sr*6, *Sr*8, and *Sr*9 in evaluating genes for virulence in race 15B - 1L and found additive gene action of *Sr*6 with either *Sr*8 or *Sr*9. A combination of all three genes produced the lowest infection type. Resistance genes *Sr*8 and *Sr*9 both conditioned intermediate infection types but showed no additive action with each other. Knott (1963) found a combination of these 3 genes gave effective resistance against 29 races of stem rust. In Table 9-2, *Sr*9c and *Sr*17 are examples of resistance genes occurring in different species of *Triticum*. Resistance genes have also been transferred to wheat from the genera *Secale* and *Agropyron* (Knott, 1963; Stewart et al., 1968).

## Leaf Rust

The causal organism of leaf rust, *Puccinia recondita* Rob. ex Desm., attacks wheat, rye, and some grasses. A total of 183 physiologic races have been identified using 8 differential varieties of wheat (Johnston, 1967). The pycnial and aecial alternate host is of little importance in the Western Hemisphere, and physiologic races probably develop mainly as a result of mutation and somatic recombination. Supplemental and highly resistant varieties have also been used to obtain more relevant information about the leaf rust population. Wheat varieties from Kenya and South America, as well as interspecific and intergeneric crosses, have furnished sources of resistance.

Studies on the inheritance of resistance have been reviewed by Anderson (1963). As with the other cereal rusts, much emphasis has been placed on isolation of specific genes for resistance and on determination of the interaction of these *Lr* genes with the virulence pool in the pathogen. A number of different *Lr* genes have been transferred to the susceptible varieties Thatcher and Prelude. Multiple alleles were found at the *Lr2* locus in Webster, Carina, Brevit, and Loros (Dyck and Samborski, 1968; Samborski and Dyck, 1968). Some other *Lr* genes reported were: *Lr1* in Malakoff, *Lr3* in Mediterranean, and *Lr11* in Hussar. These genes behave as recessives, incomplete dominants, or dominants depending on the specific host-pathogen combination. Temperature may also have a marked effect on host-pathogen interaction.

The variety Transfer was developed from a cross of Chinese Spring x *Aegilops umbellulata.* Samborski (1963) selfed a culture of *P. recondita* giving a 1+ infection type on this variety and obtained cultures which resulted in 0, 1+, and 4 infection types on Transfer. This work showed that resistance genes from other species are not necessarily superior to those within *Triticum.* The work also showed that changes from avirulence to virulence can occur by several steps.

Growth stage resistance may become evident in some varieties with the development of the flag leaf, while in other varieties it may appear at the 3-leaf stage (Samborski, 1963). Caldwell et al. (1957) reported a mature plant resistance in the wheat variety Dual which was superior to any of its parent varieties. The parent varieties had some seedling resistance, but Dual did not. Reaction to 8 physiologic races was evaluated in artificially induced epidemics in the field. This mature plant resistance was believed to be caused either by an interaction of combined parental resistances, or by an additive effect of the parental resistances. Caldwell (1968) reported that a mature plant resistance in

Chinese Spring conditioned by 4 genes finally succumbed to leaf rust after 10 years of protection.

The genetic background of a host variety is often important in genetic studies. Modifier genes may either increase or decrease the level of resistance of a major gene. In breeding programs, very large populations may be required to obtain plants with the original degree of resistance. Law and Johnson (1967) used intervarietal chromosome substitutions of Hope into Chinese Spring in determination of the leaf rust resistance genes in Hope. Several resistance genes were found, two on chromosome 7B, and at least one each on chromosomes 7A and 7D. One gene on 7B and the genes on the 7A and 7D all acted as modifiers of a major gene on 7B. Chromosome 7A of Hope increased the mean level of resistance in the $F_2$, while chromosome 7D of Hope decreased the mean.

General resistance and tolerance to leaf rust have also been reported (Caldwell, 1968).

## Stripe Rust

This disease, caused by *Puccinia striiformis* West. is important in many areas of the world because it attacks wheat, barley, and many related grasses. In the United States it is confined mainly to the cooler and higher elevations in the West. It is considered the most important rust of wheat in Europe.

Of all the cereal rusts, it is most influenced by environment, particularly temperature. Its particular adaptation to low temperatures probably restricts its presence in many areas. Studies have shown that stripe rust can endure high temperatures during the light phase, but resistance is exhibited in all varieties if the temperature exceeds 15 C in the dark phase of the postpenetration period (Sharp, 1965). High temperatures in the preinoculation period may also result in lowered infection type for some host-pathogen combinations.

Stripe rust may be evident on the leaves, leaf sheaths, and heads. The peculiar growth habit results in long stripes of pustules or lesions, following a course of least resistance between veins of the leaves. On susceptible seedling leaves, one successful penetration may result in coverage of the entire leaf because of the growth of runner hyphae from the initial point of infection.

Stripe rust was an important disease problem of wheat in the Pacific Northwest in the 1930s, then remained largely of minor importance until about 1958. At that time extensive acreages of three very susceptible winter wheats, Omar, Itana, and Westmont, were grown in the area. Prevalence has varied since 1958, but continuous outbreaks have occurred on susceptible wheat varieties.

The apparent absence of an alternate host for the pycnial and aecial stages precludes sexual recombination, but a number of races do occur, probably arising through mutation, heterocaryosis, and the parasexual cycle. Work on physiologic specialization of *P. striiformis* was initiated as early as 1905 when Biffen (1905) demonstrated that resistance to stripe rust was inherited in a Mendelian fashion. In this classical work the very susceptible Red King (*Triticum aestivum*) was crossed with resistant Rivet (*T. turgidum*). The segregating progeny of the $F_2$ indicated that resistance was controlled by a single recessive gene. Dominant genes were later also shown to condition resistance. One dominant gene in Chinese 166 conditioned immunity to 18 physiological races in Europe. Race identification of *P. striiformis* in Europe is centered at Braunschweig, Germany. Using the 13 standard European host differentials, 57 races of *P. striiformis* have been reported (Fuchs, 1960).

The differential host varieties used to determine physiologic races of stripe rust in Europe are unsuitable for the use of workers in the United States. Researchers in the U.S. have established a tentative set of host differentials that will be used in determination of the virulence gene pool and will be more relevant to resistance sources presently used in wheat breeding programs of the area (Line et al., 1970). Using these differentials, plus some supplements, 11 distinct pathogenic types have been determined (Volin, 1971). The wheat variety P.I. 178383 has been extensively used in the northwestern United States as a source of resistance to stripe rust. However, a new race soon appeared which could overcome the specific resistance conditioned by a major dominant gene in P.I. 178383 (Beaver and Powelson, 1969). A diallel cross analysis study to determine the resistance genes in P.I. 178383 showed that in addition to the major dominant gene, three minor recessive additive genes were also involved (Sharp, 1968). Monosomic analysis showed the major gene to be on chromosome 1B, and the minor genes are probably on chromosomes 5A, 5B, and 5D. Minor gene lines were developed containing all possible combinations of the 3 genes, and to date they have retained the same level of resistance to all known races in the area. All of these lines were temperature-sensitive and expressed highest levels of resistance at relatively high temperature profiles. Similar infection types developed on both seedling and mature plants, but other combinations indicate greater resistance in mature plants than in seedlings. The wheat variety Crest, a recent release from Montana, contains the major gene from P.I. 178383 and is heterogeneous for the minor genes.

A number of stripe rust workers have detected specific types of resistance in different wheat selections. Additionally, transgressive segregation indicates a large number of additive genes among varieties evaluated (Lewellen et al., 1967; Pope, 1968). In fact it has been difficult to find a wheat variety or selection which does not contribute some resistance to its progeny. Studies indicate that the genes involved are recessive in action, may be readily accumulated, and may present a long-term resistance to a wide array of physiologic races of *P. striiformis*.

Macer (1963) reported on 7 loci which contain genes for stripe rust resistance (*Yr* genes). Resistance was mainly dominant, but some loci behaved as recessives. Multiple alleles were found to occur at the *Yr3* and *Yr4* loci.

## Smut

The important smut diseases of wheat are closely associated with the historical development of plant pathology. They all belong to the Ustilaginales order of the Basidiomycetes. The two smut families, Ustilaginaceae and Tilletiaceae, are differentiated on the basis of teliospore germination. In the former the basidiospores are produced laterally on the basidium or are lacking, while in the latter the basidiospores are produced terminally. The Tilletiaceae include the common bunt species *Tilletia caries* (D.C.) Tul., *T. foetida* (Wallr.) Liro, and dwarf bunt *T. controversa* Kuhn. Flag smut, *Urocystis tritici* Koern, is also a member of the Tilletiaceae and is characterized by sterile cells that surround fertile teliospores. Loose smut, *Ustilago tritici* Pers. (Rostr.) is the lone member of the Ustilaginaceae attacking wheat.

There is a basic difference between the common infectious units of the smuts and rusts. The smut teliospores (chlamydospores) cannot be propagated independently, as with urediospores of rust. With each inoculation, smut teliospores undergo reduction division, and various recombinations for virulence may occur if the virulence loci are heterozygous. Before infection can occur with the smut fungi, there must be fusion of compatible lines to reform the dicaryotic phase. Smut isolates or races therefore include collections of teliospores having the same virulence on the selected test varieties. Where adequate evaluation has been performed, a gene-for-gene relation between virulence in the pathogen and resistance in the host appears to be present.

With the exception of *U. tritici,* which is a floral-infecting smut, the smuts of wheat are initiated as seedling infections. With artificial inoculations, teliospores are thus applied to the seed or to the young ovaries in the case of loose smut. In the latter case, the fungus lives within the scutellum of the seed as dormant mycelium and becomes active when the seed is planted to form the next crop. Resistance or susceptibility is determined on the basis of the percentage of plants that develop smut. In field trials, a universally susceptible wheat is used which serves as a check on inoculation procedures. Wheat varieties or lines showing 0-10% infection are usually considered resistant, while those showing more than 10% infection are considered susceptible. In some cases an intermediate category is used for plants showing 11-40% infection. In view of the gene-for-gene hypothesis, recent emphasis has been placed on the pathogenic race, and races are grouped and characterized according to the number of resistance genes they can attack. A smut percentage of 0-10 indicates avirulence, while a percentage of 11 or greater indicates virulence (Hoffman et al., 1967). A number of excellent literature reviews on smut are available, including those of Holton (1967) and Halisky (1965).

## Common Bunt

The two causal agents of common bunt, *T. caries* and *T. foetida,* are similar in many respects. Morphologically, the former has a spiny epispore, while the epispore of the latter is smooth.

The main sources of resistance to common bunt were reviewed by Kendrick (1961). These, with their respective symbols, are Martin (M and M2), Hussar (H), Turkey (T), Hohenheimer (Ho) and Ridit (rd). The latter two contained resistance factors not analyzed for specific genes. The wheat variety Omar contains both the *T* and *M* genes for resistance. Using differential varieties containing these different factors for resistance, Kendrick reclassified the "T" races (*T. caries*) and "L" races (*T. foetida*) into a total of 17 pathogenic types. Many races previously considered separate on the basis of minor differences were combined. Furthermore, most of the "L" races had nearly identical counterparts among the "T" races. The 17 pathogenic types were further condensed into 6 groups corresponding to the main genes or factors for resistance represented in the differential varieties. This reclassification greatly facilitated screening of varietal material for resistance. The club wheat Omar, which contains both the *T* and *M* genes for resistance, was grown extensively in the Pacific Northwest, but its resistance was overcome by race T-18. In recent years several new genes have been found conditioning resistance to common bunt, and the symbol *Bt* has been adopted by all concerned investigators for designating genes for bunt resistance. Metzger and Silbaugh (1971) have presented the most recent information on sources of resistance. Table 9-3 summarizes data on bunt resistance. The genes *Bt*4 and *Bt*6 are closely linked with a gene for red glume color on chromosome 1B. Only one gene, Bt8, conditions resistance to all known races, but *Bt*9 and *Bt*10 are each attacked by only one race. The wheat variety P.I. 178383 has been used extensively in breeding programs because of its high resistance to smut. Three genes appear to be involved, and *Bt*9 and *Bt*10 have been identified (Metzger, *personal communication*).

TABLE 9-3    GENES FOR RESISTANCE TO COMMON AND DWARF BUNT.

| GENES | | |
|---|---|---|
| New Symbol | Old Symbol | ORIGINAL SOURCE |
| *Bt*1 | *M* | Martin |
| *Bt*2 | *H* | Hussar |
| *Bt*3 | *rd* | Ridit |
| *Bt*4 | *T* | Turkey |
| *Bt*5 | *Ho* | Hohenheimer |
| *Bt*6 | *R* | Rio |
| *Bt*7 | *M*2 | Martin |
| *Bt*8 | — | P. I. 178210 |
| *Bt*9 | — | C. I. 7090[a] |
| *Bt*10 | — | P. I. 116306 |

[a] Personal communication from R. J. Metzger.

## Dwarf Bunt

Dwarf bunt caused by *T. controversa* was once considered as a special form of *T. caries,* and it is similar in many respects to the common bunt species. There are minor differences in morphology of the teliospores and considerable differences in the response to environment. As the common name indicates, diseased plants are usually noticeably stunted, but the amount of dwarfing may vary with different race-host variety combinations. Interspecific hybridization between smuts may also give rise to intermediate types. Since dwarf bunt cannot be economically controlled by seed treatment chemicals, disease resistance assumes a major role.

The same major sources of resistance used for common bunt are also used in determination of races of dwarf bunt. Hoffman et al. (1967)

described 10 pathogenic (D) races of *T. controversa* and noted that many races had counterparts within the common bunt species. Race D-9 was virulent on all of the 7 differential varieties. P.I. 178383 was immune to all races.

## Flag Smut

Flag smut has been most destructive on winter wheat in areas having mild winters. In the U.S. it is most important in localized areas of Washington and Oregon. A systemic infection is initiated in the seedlings, but smut sori may appear on all above-ground parts of the plant. In many cases, spikes fail to develop on infected plants.

In comparison to other cereal smuts, flag smut shows less tendency for pathogenic specialization. Only four races have been determined (Halisky, 1965). Purdy and Allan (1967) studied heritability of flag smut resistance in several wheats and concluded that resistance could be selected for. Norin 10 x Brevor 14 was highly resistant, but many varieties of semidwarfs originating from this variety that had not been screened for flag smut resistance were susceptible. Watson (1970) reported that *Urocystic tritici* was easily controlled in Australia by cultivars having a polygenic system. The variety Nabawa, released in 1922, has retained its resistance for almost 50 years.

## Loose Smut

This floral-infecting smut caused by *U. tritici* causes most damage with relatively high temperatures and humid conditions during the blossom period. The teliospores are short-lived, and those in the soil are of no consequence. The mycelium within the seed, however, can apparently persist just about as long as the seed remains viable. The soft winter wheat areas of the U.S. may be particularly vulnerable to this disease.

At least 20 physiologic races of this pathogen have been identified on the basis of reaction on differential varieties (Haliskey, 1965). The resistance of Kawvale was found to be dominant and caused by at least 2 genes (Heyne and Hansing, 1955). Additive gene action was involved, in that a major gene conditioned 0-10% smut, a minor gene conditioned 11-45% smut, and the combined genes resulted in an immune reaction to race 11. It was found that the embryonic constitution of the $F_1$ plant controlled resistance. In other studies, an effect of the maternal tissue on infection has been shown (Gaskin and Schafer, 1962), and both one and two recessive genes were involved in conditioning resistance to loose smut. It is interesting that the specific resistance of Kawvale to loose smut has been effective since 1932 (Caldwell, 1968).

# Powdery Mildew

The causal agent of powdery mildew, *Erysiphe graminis* DC f. sp. *tritici* E. Marchal, forms powdery mycelia and conidia on the leaf surface on infected plants. In later stages of development, the mycelium gives rise to the dark colored cleistothecia-containing the ascospores. Primary infection usually results from ascospores that have overwintered in cleistothecia, but the fungus may survive the winter in some cases as vegetative mycelium on living leaves. The short-lived conidia are the repeating spore stage of the fungus and are produced in copious amounts on susceptible varieties. They may germinate over a wide range of humidities, but infection is best with a high humidity and the absence of free moisture. Following germination, appressoria are formed from which penetration pegs enter the epidermal cells, giving rise to digitate haustoria. The haustoria are the only fungus structures produced internally, and they derive nourishment for the other, superficial fungus structures.

The fungus is heterothallic, and because the infectious units are haploid, new recombinations for virulence occurring through the sexual stage are immediately expressed. Expression of host-pathogen interaction (infection type) is basically the same as that described for the rusts. A scale from 0 to 4 is based on the relative amounts of external mycelium, sporulation, chlorosis, and necrosis. Pugsley (1963) has reviewed recent investigations on the genetics of resistance to powdery mildew of wheat. There appears to be a gene-for-gene relationship between genes for resistance in the host and genes conditioning pathogenicity in the pathogen. A total of 30 physiologic races have been identified using 8 host differentials, and 9 different genes for resistance (*M*1 genes) have been identified among different wheat varieties. Race 10 from Germany has the widest host range and can overcome the resistance of four different *M*1 genes. The resistance genes usually have been dominant in action, but resistance in the differential variety St. 14/44, derived from *Secale,* is recessive. A number of the *M*1 genes have been transferred into the susceptible variety Federation for use in specific genetic studies.

Roberts and Caldwell (1970) found evidence of a general resistance to powdery mildew in the wheat variety Knox. With heavy mildew in the field, Knox never showed more than 30% infection. Pustule number and size were both relatively less on Knox than on highly susceptible varieties. The general resistance of Knox was transmitted to many of its progeny.

## Septoria Diseases

There are three species of the genus *Septoria* which may attack wheat: *S. tritici* Rob ex. Desm., *S. nodorum* Berk., and *S. avenae* Frank F. sp. *triticea* T. Johns. All of these species may cause leaf blotches, but *S. nodorum* also attacks the heads, causing a glume blotch. *S. avenae* f. sp. *triticea* is the least important of the three species.

*Septoria* may be noted in the early part of the growing season as yellow areas on the lower leaves which later change to irregular blotches. The blotches assume a speckled appearance as the fruiting bodies or pycnidia are formed. The pycnidia contain long, septate, hyaline spores. Spore size and number of septa are used in species determination. *Septoria* can be particularly damaging if moisture is abundant late in the growing season. On highly susceptible varieties the flag leaves and areas of the leaf sheaths may die within a few days after infection, indicating the presence of a toxin.

Studies indicate that resistance to *S. tritici* is simply inherited (Narvaez and Caldwell, 1957). Resistance of some varieties was controlled by single dominant genes, while in other cases resistance was controlled by two partially dominant genes of additive effect.

Brönnimann (1970) has done considerable work in evaluating reaction of wheat to the glume blotch fungus *S. nodorum*. Inoculum for field trials was increased on autoclaved wheat kernels. No physiological races are known, but there are differences in aggressiveness between isolates due mainly to sporulation ability. There appears to be no true resistance, but some wheat selections show tolerance to the disease. The thousand kernel weight in relation to the noninoculated control gave the best measurement of tolerance. The genetic basis for the tolerance is largely unknown, but preliminary trials indicate additive gene action. The recently licensed spring wheat variety, Fortuna, is extremely susceptible to *S. nodorum*.

# Common Root and Foot Rot

A complex of fungi may be involved in this disease, but the most important member is *Helminthosporium sativum* PKB, the imperfect stage of *Cochliobolus sativus* (Ito & Kurib.) Drechsl. ex Dast. The ascomycetous perfect stage is probably of little consequence in nature. In addition to root and foot rot, *H. sativum* may also cause spot blotch of the leaves and be a member of the black point disease appearing at the heading phase. The fungus produces dark, oblong, multicelled spores which may persist on plant residues or in the soil for several years.

In many trials on disease resistance naturally infested soils have been used, but inoculum potentials may be adjusted by growing the fungus on a sand-cornmeal mixture and then adding the inoculum to natural soils in known amounts. Root rot ratings are usually based on the amount of discoloration and lesions occurring in the crown area or on the sub-crown internodes.

The early work with *H. sativum* showed that the fungus was extremely variable in culture. Even though the spores usually form germ tubes from the two terminal cells only, any one hypha can contain several genetically different nuclei. Heterocaryosis and the parasexual cycle contribute to great genetic diversity in the fungus. Some wheat cultivars were obviously more damaged by the pathogen than others, but control through resistant varieties alone did not appear promising.

The more recent work on resistance of wheat to common root and foot rot has been reviewed by Larson and Atkinson (1970). Substitution lines and monosomics were used for determining the locations of factors contributing to root rot resistance in Apex and Cadet Spring wheats. Chromosome 5B had a major gene conditioning a recessive type of resistance. Minor genes conditioning resistance to *H. sativum* were found on chromosomes 2B and 2D. In other work, hybridization of moderately resistant parents led to transgressive segregation for highly resistant selections. In this latter case, a large number of genes appeared to be involved.

# Cercosporella Foot Rot

Winter wheat may be heavily attacked by *Cercosporella herpotrichoides* Fron. The fungus invades near the soil line, and eye-spot lesions are formed on the leaf sheaths and culms. In later stages the stems may fall in a criss-cross manner; thus the name "strawbreaker." A dark fungus stroma usually develops in the center of the lesions. Heads of affected plants may be sterile or partially sterile, and the tangled nature of the fallen stems impedes the harvest operations. The long, needle-like, multiseptate spores are produced during moist periods in the stem lesions and may be disseminated by splashing rain.

Bruehl et al. (1968) used dried-oat inoculum in evaluating the resistance of about 2,000 winter wheats from the United States Department of Agriculture World Wheat Collection. All hard red wheats were found to be susceptible, but a few soft red and white wheats showed some resistance. The resistant types were from northern Europe and had relatively thick stems and broad leaves. Most resistant wheats were late maturity types. The wheat variety Odin was moderately resistant, as indicated by reduced vertical spread, and it contributed resistance to some of its progeny. There were cases of transgressive segregation for resistance as crosses between susceptible parents often led to selections showing moderate resistance to foot rot.

## Black Chaff

Black chaff incited by *Xanthomonas translucens* f. sp. *undulosa* (Smith, Jones and Reddy) Hagb. is distributed widely on wheat but is usually of minor importance. The bacteria invade through natural openings, and wounds and lesions first appear on leaves, floral bracts, and culms as small water-soaked areas. Further development gives a translucent appearance to the leaves, and with moist conditions a yellow exudate forms on the surface. Bacteria within the exudate may be spread by splashing rains. In later stages brown to black areas of discolored tissue appear on the infected plant parts. The disease may be confused with pseudo black chaff, which occurs on Hope, H44, and derivatives and is the result of the formation of melanin pigments under high humidity and temperature.

Woo and Smith (1962) reported that resistance to black chaff was controlled by a single dominant gene. The pseudo black chaff reaction is also simply inherited.

## Soil-Borne Wheat Mosaic

This virus disease primarily affects winter wheat and is apparently vectored by *Polymyxa graminis,* a slime mold. There are a number of different strains, which result in symptoms ranging from a light green to a yellow mosaic and from moderate to severe dwarfing.

Genetics of disease resistance to soil-borne mosaic has been reviewed (Sebesta and Bellingham, 1963). Resistance has been reported to be controlled by a single dominant gene, and by two dominant genes. Additionally, there have been reports of transgressive segregation for resistance. Wheat varieties which have been reported to be resistant include Egypt 95, Gabo, Kenya Standard, Kenya 338.2C2.E.2, and Supremo 51A.

## Wheat Streak Mosaic

This virus disease is prevalent throughout the western hard red winter and hard red spring wheat areas of the United States and Canada. The virus is transmitted by the eriophyid mite, *Aceria tulipae.* The mites live over the summer on volunteer wheat in fallow land and live over winter in a semi-dormant state deep in the crown of the plants or in the curl of the leaves. They may be wind-disseminated for distances up to at least one mile. Wheat streak mosaic virus causes yellow streaks, stripes, yellow-green mottling, and stunting of susceptible plants. In severe cases wheat plants may be killed by this disease.

None of the wheat species have a high degree of resistance to the virus, but tolerance has been reported in some varieties of common wheat. Intergeneric hybrids between wheat and *Agropyron* show the most promise for control (Sebesta and Bellingham, 1963). *Agropyron elongatum* has been extensively investigated as a possible source of resistance. One derivative of wheat x *A. elongatum,* P₃-19, was highly resistant to wheat streak mosaic, and resistance was conditioned by a single pair of chromosomes from *A. elongatum.* By use of irradiation techniques it should be possible to transfer the genes for resistance to a wheat chromosome.

## Conclusion

The use of specific resistance, usually conditioned by a single gene, has served a useful purpose in the past and has led to enormous economic benefits. However, this resistance has usually been of short duration, and its selection by plant breeders often resulted in the loss of other valuable but less obvious resistance genes. In the future more emphasis will need to be placed on other types of resistance. Combinations of specific genes with a broader resistance base offer some promise. The pathogen is less likely to gain the necessary virulence genes to overcome the resistance, and interallelic interaction could further extend the resistance over that attainable by the collective effects of the individual genes. New sources of resistance can probably best be sought in centers of wheat origin where maximum opportunity has existed for different host-pathogen combinations to evolve.

Both general resistance and tolerance offer promise for plant disease control, and there is considerable evidence that both participate against a number of plant diseases in different wheat cultivars. A more complete understanding of the factors contributing to these types of disease control is needed. Their incorporation and use in commercially acceptable varieties appears desirable. It should also be possible to further advance the benefits of general resistance and tolerance by judicious hybridization and selection.

Evidence is accumulating that transgressive segregation for resistance occurs with many wheat-pathogen combinations. Often the results obtained were not directly sought, but were an added bonus from some other study. Earlier studies on incorporating high types of resistance in wheat varieties have usually precluded the incorporation of the hypostatic genes, which become obvious in transgressive segregation. The research with stripe rust further indicates that these additive genes may furnish a source of long-lasting disease resistance. Recent studies also suggest that resistance to a number of facultative pathogens of wheat may be most readily detected and developed through transgressive segregation.

# References

Allard, R. W. 1966. Breeding methods with self-pollinated crops, p.109-165. *In* R. W. Allard, Principles of plant breeding. John Wiley and Sons, Inc., New York.

Anderson, R. G. 1963. Studies on the inheritance of resistance to leaf rust of wheat. Proceedings of the Second International Wheat Genetics Symposium, Lund, Sweden, August 19-24. Heriditas Supplementary Volume 2:144-155.

Beaver, R. G., and R. L. Powelson. 1969. A new race of stripe rust on the wheat variety Moro, C.I. 13740. Plant Dis. Reporter 53:91-93.

Biffen, R. H. 1905. Mendel's laws of inheritance and wheat breeding. J. Agr. Sci. 1:4-48.

Borlaug, N. E. 1965. Wheat, rust and people. Phytopathology 55:1088-1098.

Brönnimann, A. 1970. Zur Vererbung der Toleranz des Weizens gegenüber Befall durch *Septoria nodorum* Berk. Z. für Pflanzenzuchtung 63:333-340.

Bruehl, G. W., W. L. Nelson, F. Koehler, and O. A. Vogel. 1968. Experiments with Cercosporella foot rot (strawbreaker) disease of winter wheat. Washington Agr. Exp. Sta. Bull. 694, p. 1-14.

Caldwell, R. M. 1968. Breeding for general and/or specific plant disease resistance. Third International Wheat Genetics Symposium, Canberra, August 5-9. Australian Academy of Science. p. 263-292.

————, J. F. Schafer, L. E. Compton, and F. L. Patterson. 1957. A mature plant type of wheat leaf rust resistance of composite origin. Phytopathology 47: 691-692.

Dyck, P. L., and D. J. Samborski, 1968. Genetics of resistance to leaf rust in the common wheat varieties Webster, Loros, Brevit, Carina, Malakof and Centenario. Can. J. Genet. Cytol. 10:7-17.

Flor, H. H. 1956. The complementary genic systems in flax and flax rust. Adv. Genet. 8:29-54.

Fuchs, E. 1960. Physiologische Rassen bei Gelbrost (*Puccinia glumarum* (Schm.) Erikss. et Henn.) sur Weizen. Nachrichtenbl. Deutsch. Pflanzenschutz. (Braunschweig) 12:49-63.

Gaskin, T. A., and J. F. Schafer. 1962. Some histological and genetic relationships of resistance to loose smut. Phytopathology 52:602-607.

Halisky, P. M. 1965. Physiologic specialization and genetics of the smut fungi. III. Bot. Rev. 31:114-150.

Heyne, E. G., and E. D. Hansing. 1955. Inheritance of resistance to loose smut of wheat in the crosses of Kawvale x Clarkan. Phytopathology 45:8-10.

Hoffman, J. A., E. L. Kendrick, and R. J. Metzger. 1967. A revised classification of pathogenic races of *Tilletia controversa*. Phytopathology 57:279-281.

Holton, C. S. 1967. Smuts. Wheat and wheat improvement. Agronomy No. 13:337-353. American Society of Agronomy, Inc., Madison, Wisc.

Johnston, C. O. 1967. Leaf rust of wheat. Wheat and wheat improvement. Agronomy No. 13:317-325. American Society of Agronomy, Inc., Madison, Wisc.

Kendrick, E. L. 1961. Race groups of *Tilletia caries* and *Tilletia foetida* for varietal resistance testing. Phytopathology 51:537-540.

Knott, D. R. 1963. The inheritance of stem rust resistance in wheat. Proceedings of the Second International Wheat Genetics Symposium, Lund, Sweden, August 19-24. Heriditas Supplementary Volume 2:156-166.

Larson, R. I., and T. G. Atkinson. 1970. A cytogenetic analysis of reaction to common root rot in some hard red spring wheats. Can. J. Bot. 48:2059-2067.

Law, C. N., and R. Johnson. 1967. A genetic study of leaf rust resistance in wheat. Can. J. Genet. Cytol. 9:805-822.

Lewellen, R. T., E. L. Sharp, and E. R. Hehn. 1967. Major and minor genes in wheat for resistance to *Puccinia striiformis* and their responses to temperature changes. Can. J. Bot. 45:2155-2172.

Line, R. F., E. L. Sharp, and R. L. Powelson. 1970. A system for differentiating races of *Puccinia striiformis* in the United States. Plant Dis. Reporter 54:992-994.

Loegering, W. Q., and D. L. Harmon. 1969. Wheat lines near-isogenic for reaction to *Puccinia graminis tritici.* Phytopathology 59:456-459.

Macer, R. C. F. 1963. The formal and monosomic genetic analysis of stripe rust (*Puccinia striiformis*) resistance in wheat. Proceedings of the Second International Wheat Genetics Symposium, Lund, Sweden, August 19-24. Heriditas Supplementary Volume 2:127-142.

Metzger, R. J., and B. A. Silbaugh. 1971. A new factor for resistance to common bunt in hexaploid wheats. Crop Sci. 11:66-69.

Morris, R., and E. R. Sears. 1967. The cytogenetics of wheat and its relatives. Wheat and Wheat Improvement. Agronomy No. 13:19-87. American Society of Agronomy, Inc., Madison, Wisc.

Narvaez, I., and R. M. Caldwell. 1957. Inheritance of resistance to leaf blotch of wheat caused by *Septoria tritici.* Phytopathology 47:529. (Abstr.)

Pope, W. K. 1968. Interaction of minor genes for resistance to stripe rust in wheat. Third International Wheat Genetics Symposium, Canberra, August 5-9.    Australian Academy of Science. p. 251-257.

Pugsley, A. T. 1963. Genetics and exploitation of resistance to powdery mildew in wheat. Proceedings of the Second International Wheat Genetics Symposium, Lund, Sweden, August 19-24. Heriditas Supplementary Volume 2:178-182.

Purdy, L. H., and R. E. Allan. 1967. Heritability of flag smut resistance in three wheat crosses. Phytopathology 57:324-325.

Reitz, L. P. 1967. World distribution and importance of wheat. Wheat and Wheat Improvement. Agronomy No. 13:1-18. American Society of Agronomy, Inc., Madison,

Roberts, J. J., and R. M. Caldwell. 1970. General resistance (slow mildewing) to *Erysiphe graminis* f. sp. *tritici* in Knox wheat. Phytopathology 60:1310. (Abstr.)

Samborski, D. J. 1963. A mutation in *Puccinia recondita* Rob. ex Desm f. sp. *tritici* to virulence in Transfer, Chinese Spring x *Aegilops umbellulata* Zhuk. Can. J. Bot. 41:475-479.

———, and P. L. Dyck. 1968. Inheritance of virulence in wheat leaf rust on the standard differential varieties. Can. J. Genet. Cytol. 10:24-32.

Sebesta, E. E., and R. C. Bellingham. 1963. Wheat viruses and their genetic control. Proceedings of the Second International Wheat Genetics Symposium, Lund, Sweden, August 19-24. Heriditas Supplementary Volume 2:184-201.

Sharp, E. L. 1965. Prepenetration and postpenetration environment and development of *Puccinia striiformis* on wheat. Phytopathology 55:198-203.

———. 1968. Major and minor genes for resistance to stripe rust in P.I. 178383 wheat. Proceedings of the Cereal Rusts Conference, Oeiras, Portugal, August 4-14.

Stakman, E. C., D. M. Stewart, and W. Q. Loegering. 1962. Identification of physiologic races of *Puccinia graminis* var. *tritici.* ARS, USDA, E617 (Revised).

Stewart, D. M., E. C. Gilmore Jr., and E. R. Ausemus. 1968. Resistance to *Puccinia graminis* derived from *Secale cereale* incorporated into *Triticum aestivum.* Phytopathology 58:508-511.

Volin, R. B. 1971. Physiologic race determination and environmental factors affecting development of infection type in stripe rust (*Puccinia striiformis* West.,). Ph.D. thesis. Montana State Univ.

Watson, I. A. 1970. Changes in virulence and population shifts in plant pathogens. Ann. Rev. Phytopathol. 8:209-230.

Woo, S. C., and G. S. Smith. 1962. A genetic study of leaf sheath barbs, auricle hairs, and reaction to stem rust and "black chaff" in crosses of (ND 105 x ND 1) and ND 113 with Conley wheat. Acad. Sinica Bot. B. 3:195-203.

# 10 Maize

## A. L. Hooker[1]

## *The Plant*

*Origin.*　　Maize is native to the Americas, and among cultivated crops it is one of the oldest. The origin of maize is still a matter of speculation. A current theory suggests it had its origin from a primitive type of corn, remains of which have been found in cave dwellings of early man about 7,000 years old.

*History.*　　Maize has been cultivated for many centuries by the Indians in Central America and in the adjacent areas to the north in southern Mexico and to the south in the adjacent highlands of South America (Sprague, 1955). Here, many races and varieties of maize were selected. It was without question the most important cultivated food plant of the time, and the great Aztec, Maya, and Inca civilizations in this area were based on agriculture with maize cultivation as their principal enterprise. Maize culture in the Western Hemisphere subsequently

1. Professor of Plant Pathology and Genetics, Departments of Agronomy and Plant Pathology, University of Illinois.

extended northward and southward from the area of domestication.

Following the discovery of maize in Cuba by Columbus in 1492, maize was introduced into Spain. While first regarded only as a garden curiosity, its value as a food crop was soon recognized. In a few years the crop spread to France, northern Africa, and into Italy and southeastern Europe. Early in the sixteenth century the Portuguese introduced maize along the west coast of Africa. At about the same time it was introduced into India, the Philippines, and the East Indies. From India the plant spread into China. In most of these regions maize played a secondary role to other cereal or starch crops that had become dominant in local agriculture.

It was in America, however, that maize (Indian corn, or simply corn, as it is now known in the U.S.A.) played such an important part in white man's history. It was of major significance to the explorations of the Spanish conquerors in Mexico and Central and South America, and to the settlement of the English colonists on the eastern seaboard of North America. Maize culture in North America received its greatest advance when the American colonists moved into the grasslands and woodland clearings of the Middle West. Although, for various reasons, maize had not been cultivated extensively by the Indians in this area, the area was to become the most important maize-growing region of the world. In this area numerous productive dent varieties were selected, largely from the blending of Iroquois flint varieties and Mexican soft dent varieties that had migrated into the region. Elsewhere in the world other successful varieties were also selected. In many instances these were derived from an ancestry of mixed types. From these varieties, either directly or indirectly, inbred lines that are now used to produce modern hybrids were developed (Sprague, 1955).

*Economic Importance.*    Among the cereal grains, maize ranks third behind wheat and rice in world production. Over half of this production is grown in the U.S.A. Maize is used as a feed grain and as a silage or green fodder crop for livestock, directly as human food, and in numerous industrial products.

## Botany

Maize, unlike other cereals, has separate male and female flowers. A tassel bearing staminate spikelets is produced at the apex of the main stem. Ear shoots bearing pistillate flowers are produced on lateral branches from the mid-portion of the stalk.

Pollen production is abundant and is usually initiated before silk emergence. Under warm, dry field conditions pollen retains its viability for only a few hours. Silks are receptive for several days, and fertilization occurs within 12-28 hours after pollination.

## Breeding

Several methods of breeding are used in maize improvement (Jugenheimer, 1958). Each can be applied to disease resistance. The main types are (1) selection within and among inbred lines; (2) cumulative selection; (3) backcrossing; (4) convergent improvement; (5) mass selection; and (6) recurrent selection. Some methods result in the development of new inbred lines, others in the improvement of existing inbred lines, and still others in the improvement of varieties or the development of source populations from which superior lines can be selected.

## Agronomic Characters Desired in Addition to Disease Resistance

Many features are needed in a desirable maize variety or hybrid. These usually include satisfactory yield, maturity, standability, tolerance to crowding, ability to respond to soil fertility, plant type, suitability for mechanical harvesting, resistance to ear dropping, grain and forage quality, insect resistance and tolerance, and tolerance to heat, drought, and cold. Attributes of plant type include plant height; ear height, placement, and covering; leaf angle; suitability as a seed parent; and suitability as a pollen parent. Most of these desirable attributes are fitness characters which influence the development of diseases, and in turn their attainment is affected by the degree to which it is possible to achieve disease resistance.

# Breeding for Resistance to Specific Diseases

More than 25 diseases occur on maize; the importance of each varies in different parts of the world. Several diseases have the potential of causing nearly complete destruction of the crop, others are less devastating, but can cause appreciable yield losses, and still others, for one reason or another, never have become destructive over large areas. Although some endosperm mutants, such as sugary-1, opaque-2, and so forth, are commonly more susceptible to certain seedling and other diseases, maize disease problems and methods of solution do not differ greatly among the dent, flint, pop, sweet, and other endosperm types.

## Northern Leaf Blight

Northern leaf blight is one of the more destructive diseases of maize because the pathogen causing it can develop to epiphytotic proportions. Young seedlings, when repeatedly infected, can be killed. When the pathogen attacks older plants under favorable conditions it can reduce yields by two-thirds or more. The disease is representative of several leaf blight diseases.

*Nature of the Pathogen.*    Northern leaf blight is caused by the fungus *Helminthosporium turcicum* Pass. In the laboratory the ascomycete stage *Trichometasphaeria turcica* Luttrel is formed.

Apparently the ascospore stage is unimportant in the life cycle of the pathogen in nature. The fungus passes from one growing season to the next as mycelium or as chlamydospores formed in dormant conidia. During the following growing season the mycelium and chlamydospores form conidia, which are then airborne to infect leaves of the host plant. Lesions form in 8-14 days after infection. Under cool, moist conditions abundant reproduction by means of conidia occurs on the lesion. Secondary spread is from airborne conidia.

Pathogen variation has not complicated breeding for resistance to the disease. The fungus shows some specificity of virulence to such host genera as maize, sorghum, johnsongrass, and so forth. Strains differing in their ability quickly to produce lesions which enlarge and sporulate rapidly can be distinguished (Nelson, Robert, and Sprague, 1965; Nelson, MacKenzie, and Scheifele, 1970). Strains apparently differ in survival ability in nature. Chlorotic-lesion resistant corn having gene *Ht* has been resistant in the greenhouse to many isolates of the pathogen from numerous parts of the world and has exhibited resistance in the field to isolates of the fungus occurring in North and South America, Europe, Africa, and parts of Asia (Hooker, Nelson, and Hilu, 1965).

The fungus loses pathogenicity and ability to sporulate when cultured for long periods on artificial media. A solution to this problem is to maintain the fungus in infected leaf tissue kept cool and dry or as a water suspension of spores kept at 8-10 C.

*Nature and Expression of Resistance.*    Several types of resistance to *H. turcicum* exist in maize. Two of the main types affect lesion number (Hughes and Hooker, 1970) and lesion type (Hooker, 1963). In the first type, resistance of plants in the field is expressed as a low number of lesions that form on the plant. Lesions are usually most numerous on the lower leaves. Individual lesions differ only slightly in size and shape on different plants. In the second type, some form of lesion difference is seen between resistant and susceptible plants. On plants having gene $Ht_1$ chlorotic lesions or lesions surrounded by a yellow to light brown border form on seedlings as well as on older plants. Fungus sporulation in the lesion is inhibited, resulting in a reduction of reinfection.

Spores of *H. turcicum* germinate on the surface of corn leaves and penetrate them directly. In susceptible corn leaves hyphae penetrate the xylem, grow vigorously there, and spread some distance in the leaf. About six days after infection, hyphae leave the xylem and penetrate the surrounding bundle sheath and chlorenchyma cells. This results in wilt-type lesions. Under humid conditions, numerous conidiophores and conidia are produced in the lesion. In leaves of lesion-number resistant plants, hyphae grow poorly in the xylem, and lesion development is delayed. However, sporulation occurs. In chlorotic-lesion resistant leaves, hyphae rarely penetrate the xylem, and lesion enlargement is limited to the slow growth of hyphae in the mesophyll tissue. Sporulation is suppressed; the degree of sporulation suppression is usually greater in genetic backgrounds expressing few lesions than in backgrounds having a large number of lesions.

Resistant plants produce toxic chemicals called phytoalexins that inhibit fungal growth (Lim, Hooker, and Paxton, 1970). These phytoalexins are produced in resistant plants when host and pathogen interact. They are first detected about 3-4 days after inoculation, the same time interval after inoculation when histological studies have shown a slowing of hyphal growth in the leaf. The speed of formation and amount of phytoalexin produced vary with different maize genotypes. The phytoalexins have not been found in uninfected plants, nor in infected susceptible plants. No morphological barriers that limit hyphal growth have been seen.

*Inoculation Methods.*    To inoculate a large number of plants in the field, simple procedures are needed. A workable one is to collect infected leaves from susceptible plants 4-6 weeks after anthesis. In order to have inoculum not contaminated with other corn leaf pathogens a special planting of a corn hybrid susceptible only to *H. turcicum* may be needed. The leaves are air-dried without heat or in a seed corn dryer and then ground in a feed mill. The dried leaf material is kept in cool, dry storage. When plants are about 18-24 inches high, a small amount of the ground leaf material is placed in the whorl of each plant or dusted over the entire plant. Infection is best if the inoculations can be made during or just prior to light rains, as dew forms on the plant, or preceding irrigation. Plants may need to be inoculated two or more times during the season. Unless the weather is unusually hot and dry, infection takes place and secondary spread of the fungus occurs. Spreader rows of susceptible inbreds or hybrids are often desirable to provide inoculum for secondary spread.

To inoculate seedling plants in the greenhouse (glasshouse) or to inoculate fewer plants in the field, spore suspensions can be made from fungus cultures growing on nutrient media. Fungus isolates that sporulate well and are pathogenic should be selected. Spore suspensions may also be obtained by placing leaf pieces with lesions on moist absorbant material, such as filter paper, in covered dishes for 24 hours. These sporulating leaf pieces are then agitated in water and strained. Spore suspensions in water or 2% sucrose solution are sprayed on seedlings in the greenhouse or placed in the leaf whorl of plants in the field. Seedlings are incubated at 100% humidity for 12-16 hours. Lesions develop in 7-14 days.

*Recording Difference in Disease Reaction.*   In the lesion-number type of resistance, plants are rated in the field after silking. Plants severely infected soon after silking are damaged more by the disease than are plants heavily infected late in the season. Therefore, stage of plant development relative to maturity at the time notes are taken is important in measuring resistance relative to disease losses.

The following scale, developed by workers in the United States Department of Agriculture, is widely used to record the intensity of infection:

| RATING | DESCRIPTION |
| --- | --- |
| 0.5 | Very slight infection; one or two restricted lesions on lower leaves. |
| 1.0 | Slight infection; a few scattered lesions on lower leaves. |
| 2.0 | Light infection; moderate number of lesions on lower leaves. |
| 3.0 | Moderate infection; abundant lesions on lower leaves, few on middle leaves. |
| 4.0 | Heavy infection; lesions abundant on lower and middle leaves extending to upper leaves. |

| RATING | DESCRIPTION |
|--------|-------------|
| 5.0 | Very heavy infection; lesions abundant on all leaves; plants may be prematurely killed. |

In addition, intermediate classes of 1.5, 2.5, 3.5, and 4.5 can be used. For ease in statistical manipulations some workers use a 1-9 scale but in a different order, so that 1 equals the least and 9 the most resistance. This has advantages in calculating selection indexes and in other calculations where low numbers indicate the least amount of character expression, such as yield, standability, and so forth. Lesion-number type of resistance cannot be measured readily in the seedling stage.

In the lesion-type form of resistance, ratings are based on the character of individual lesions. This form of resistance is expressed by seedlings and by older plants. Susceptible plants usually have wilt-type lesions without distinct yellow margins. In most genetic backgrounds, these plants are in marked contrast to resistant plants with chlorotic lesions. In certain genotypes (usually lesion-number susceptible) this contrast is greatest soon after infection, while in other genotypes (usually lesion-number resistant) the contrast is greatest late in the infection period. Thus, it is desirable to have suitable susceptible check plants in all test plots. In a backcross program, where a dominant gene for chlorotic-lesion resistance is being transferred, it is usually desirable to have the recurrent parent inoculated so that the lesion type typical of the susceptible parent can be compared with the resistant lesion type in segregating populations. Young lesions can be checked also for sporulation suppression. Sections of susceptible leaves with lesions floating on water or placed on a moist substrate in covered glass dishes usually sporulate abundantly in a few days. Lesions from resistant plants remain relatively clean.

*Sources of Resistance.*   Numerous sources of the lesion-number type of resistance are known. From most genetically variable varieties and synthetics, resistant lines can easily be selected with a good inoculation program. Certain varieties and synthetics have a higher frequency of resistant plants than others. Several inbreds outstanding in resistance have been developed.

Two of the best known sources of the chlorotic-lesion resistance are inbred GE440 and the popcorn variety Ladyfinger (PI217407). Other sources of this type of resistance include white and yellow dent, flint, sweet, and popcorn endosperm types from many areas of the world (Hooker et al., 1964).

*Inheritance of Resistance.*    Resistance based on lesion number is under multiple-gene control. At least 12 chromosome arms carry genes for resistance (Jenkins and Robert, 1961). Most of the gene action is additive (Hughes and Hooker, 1970). Other forms of gene action also exist. Two resistant inbred lines with identical ratings as inbreds may transmit different levels of resistance in hybrid combinations. In a similar manner, susceptible inbreds may contribute different degrees of susceptibility to their hybrids. This indicates that part of the gene action is caused by dominance or epistasis. Thus, for a more complete evaluation of the resistance of several inbreds with multigenic resistance, they may need to be studied in hybrid combinations.

Chlorotic-lesion resistance is conditioned by the single dominant gene $Ht_1$ in the GE440 and Ladyfinger popcorn sources (Hooker, 1963). The gene is located on chromosome 2. Other sources of resistance also carry single dominant genes at this locus. Genetic studies indicate the existence of at least one other dominant gene locus that segregates independently of $Ht_1$. With a number of sources of monogenic resistance, homozygous resistant plants seem to be more resistant than heterozygous plants. Plants with dominant genes at two loci appear to be more resistant than plants with dominant genes at a single locus. Genes for lesion-number resistance enhance the expression of chlorotic-lesion resistance.

*Breeding Methods.*    Since most of the gene action is additive in lesion-number resistance, simple selection procedures should be effective in isolating inbred lines with multigenic resistance. Recurrent selection has been shown to be an effective means of concentrating genes for resistance. Jenkins, Robert, and Findley (1954) found that the greatest progress was made during the first two or three cycles of recurrent selection.

With chlorotic-lesion type of resistance, homozygous resistant plants are readily selected in a homozygous condition following selfing and selection.

Susceptible lines can be improved by the backcrossing technique. In the case of lesion-number resistance large populations need to be grown, and linkage to other genes, such as those affecting maturity, must be kept in mind. Since the chlorotic-lesion type of resistance is simply inherited, linkage to other genes is not much of a problem in the backcross program, but 6-10 backcrosses are needed to fully recover the character of the recurrent parent.

*Use of Resistance Types in Maize Production.*   Both the lesion-number and lesion-type forms of resistance can be used in the breeding program. Each seems to enhance the effectiveness of the other. Being simply inherited, the lesion-type form of resistance is the easiest to use in a backcross program. Combinations of several genes for lesion-type resistance are possible in the same line. Homozygosity of genes in the final hybrid seems to be desirable. Combinations of different genes for lesion-type resistance are possible in the form of multiline hybrids; these hybrids would be blends of several seed lots, each resulting from crosses of backcross-derived inbred lines having different genes for lesion-type resistance, but similar in other respects.

## Leaf Rust

Common maize rust, caused by *Puccinia sorghi* Schw., is distributed throughout the world in many areas where maize is grown. It is an endemic disease in the Western Hemisphere. Another rust, *P. polysora* Underw., is widely distributed in the warmer, more tropical areas of the world. When *P. polysora* was first introduced into Africa it developed in epidemic proportions and caused substantial yield losses. Leaf rust is representative of several obligate parasite diseases of maize.

*Nature of the Pathogen.*   *P. sorghi* is a macrocyclic heteroecious fungus. The aecial and pycnial stages develop on several species of *Oxalis.* In most parts of the world aecial infection is limited, and the teliospores play a limited role in the life cycle of the fungus. In North America the fungus overwinters in the uredial stage in southern regions and spreads northward each spring and summer. The fungus is favored by humid weather, and secondary spread by means of urediospores is rapid. The fungus has a high reproductive rate and no exacting environmental requirements. Like other cereal rusts, *P. sorghi* has the potential of developing in epiphytotic proportions.

There are numerous races of *P. sorghi* that differ in virulence to maize lines carrying single genes or a combination of a few genes for the hypersensitivity form of resistance. No attempt has been made to establish a standard set of differential varieties, although lines with monogenic differences for the 30 or more resistance genes and in isogenic backgrounds have been developed. The classification of physiologic races of *P. sorghi,* commonly done in other cereal rust fungi, seems academic and of little practical value. Naturally occurring biotypes of *P. sorghi* vary in virulence, ranging from those that are virulent to many genes for resistance, to those that are virulent to only a few. This variation in virulence accounts for the fact that certain maize genes condition good resistance in some parts of the world but are completely ineffective elsewhere.

*Nature and Expression of Resistance.*    Resistance in maize to *P. sorghi* is clearly of two types: (1) specific resistance; and (2) nonspecific resistance (Hooker, 1969). The latter has also been called generalized resistance, or simply general resistance.

Specific resistance is expressed by young plants in the form of chlorotic or necrotic flecks, or small pustules surrounded by chlorotic or necrotic tissue. Older plants simply may not show evidence of infection. Specific resistance is based upon host plant hypersensitivity, is qualitative in expression, and is clearly specific to certain biotypes of the fungus. A maize genotype can be highly resistant to certain rust strains and fully susceptible to others. Resistance to a specific rust biotype apparently persists throughout all plant stages from seedling to adult. Some maize genotypes express resistance to a large number of rust strains, while others express resistance to only a few. A strain of maize may express good resistance to rust on one continent but none at all on another continent. This has been observed with several maize genotypes when tested against rust cultures found in Australia and in the U.S.A.

Nonspecific resistance is usually expressed by the fully grown plant in the field. It is expressed in the form of a low number of rust pustules on the plant. It is quantitative in expression and ranges in a continuous series from highly resistant plants, which have only a few pustules, to highly susceptible plants, which have most of the leaf surface covered with pustules.

Specific resistance has been studied histologically (Hilu, 1965). In resistant plants, evidence of incompatibility is seen soon after the cytoplasm of host and fungus have made intimate contact. This can take various forms, all of which result in subnormal development of the fungus.

The nature of nonspecific resistance has not been determined.

*Inoculation Methods.*    *P. sorghi* is an obligate parasite, and inoculum must be produced on living plants. Various ways are used to collect spores. Spores retain their viability for 6-12 months or more if stored dry at 2-10 C. Special storage methods are available for longer periods.

Individual plants may be inoculated in the greenhouse or field by means of a hypodermic needle and syringe. With larger plants in the field, pouring a spore suspension into the whorl late in the day or when the weather is cool and cloudy is satisfactory. Very young tissue does not always express resistance.

In the field, spreader rows of susceptible maize are usually planted at the ends of the test plots. The plants in the spreader rows are inoculated as described above, and the rust is allowed to spread into and within the test plots throughout the growing season. The method works well when environmental conditions are favorable for rust development and when purity of individual rust biotypes is not a factor in the testing program.

In the greenhouse, spores are usually diluted with some diluent such as inert talc, water, or light mineral oil and dusted or sprayed over the plants. The inoculated plants are then incubated for 8-14 hours at 100% humidity, allowed to dry slowly, and placed on the greenhouse bench. Disease reaction can be determined in 8 to 14 days after inoculation.

*Recording Difference in Disease Reaction.*   Specific resistance is recorded on the basis of infection type. These differ only slightly from those described for other cereals. The types most commonly seen are as follows:

| INFECTION TYPE | DESCRIPTION |
|---|---|
| 0 | Small chlorotic flecks. |
| 1− | Small necrotic spots. |
| 1 | Small pustules surrounded by necrotic tissue. |
| 2 | Small pustules surrounded by a chlorotic area. |
| 3 | Medium-sized sporulating pustules without chlorosis. |
| 4 | Large sporulating pustules. |
| X | Mixture of resistant- and susceptible-type pustules interspersed over the leaf. |
| Z | Resistant-type pustules on the older leaf tissue inoculated and susceptible-type pustules on the younger leaf tissue such as that in the leaf whorl at the time of inoculation. |

Only a few plants are needed to measure reaction type if the plants are genetically uniform.

Nonspecific resistance is recorded on the basis of the percentage of leaf area covered with pustules. Scales developed for use with other cereal rust fungi are satisfactory. When rust develops late in the growing season, differences in amount of infection on different leaves of the same plant are not as great as those seen for *H. turcicum* leaf blight. Records of disease severity are usually taken 2-4 weeks after anthesis. Resistance may not persist later in the growing season, and ratings become more difficult to determine.

*Sources of Resistance.*    A large number of American inbred lines have the nonspecific form of resistance (Hooker, 1969). Most American open-pollinated varieties have a high frequency of resistant plants, and rust-resistant inbred lines can easily be developed by selection. Relatively few highly susceptible inbreds are used in the production of American hybrids. However, few commercial inbreds have specific resistance. Specific resistance is also rare in American open-pollinated varieties. Nevertheless, over 100 sources of specific resistance have been identified by the author and associates by empirically testing inbred lines and open-pollinated varieties from many parts of the world for rust reaction in the seedling stage.

*Inheritance of Resistance.*    Specific resistance is usually simply inherited. Dominant genes at six or more loci on chromosomes 3, 4, and 10 have been identified (Hooker, 1969; Wilkinson and Hooker, 1968). Several distinguishable alleles are distinguished on the basis of differential reactions to various biotypes of *P. sorghi*. Three loci are tightly linked on the short arm of chromosome 10. At the $Rp_1$ locus in this series, several "alleles" have recently been shown to be pseudoalleles or very closely linked genes (Saxena and Hooker, 1968). When the recombinants were tested against a series of diverse and differentially virulent *P. sorghi* cultures and compared with the reaction of the heterozygote of the respective alleles, the disease reaction of each recombinant was the same as the heterozygote. The recombinants had resistance to groups of rust biotypes equal to the additive effects of the two parents. Structurally, $Rp_1$ is a complex region with several genes located within a minute chromosome segment. Each "allele" may be viewed as one variation of this segment carrying at least one dominant gene for rust resistance and one or more recessive genes for rust susceptibility, closely linked together. Chromosome segments conditioning widely effective ranges of resistance to rust can be produced by recombination and selection, and, once constructed, they can be easily maintained and used in conventional plant breeding. Specific resistance may also result from the action of 1-3 recessive genes (Hooker, 1969). Various forms of gene interaction occur.

Nonspecific resistance in maize to *P. sorghi* is inherited as a polygenic character (Hooker, 1969). Plants in segregating $F_2$ populations range in disease reaction from one parent to the other, sometimes exceeding one or both parents in disease reaction. The mean of the $F_2$ is near the average of the two parents. Hybrid reaction is usually between the average of the two parents and the most resistant parent. Nonspecific resistance is a highly heritable character and easy to select for in a breeding program.

*Breeding Methods.*    Selfing and selection should be an effective breeding procedure to isolate inbred lines with either specific resistance or nonspecific resistance. Since the number of genes conditioning nonspecific resistance is probably not large, the character should be fixed within a few generations of selfing. Backcrossing can be used to add dominant genes for specific resistance to susceptible inbred lines or to inbred lines with nonspecific rust resistance. Simple mass selection or some form of recurrent selection should be useful in concentrating genes for resistance. Hybrids with adequate resistance to rust should be produced when inbred lines with a moderate to high degree of rust resistance are combined.

*Use of Resistance Types in Maize Production.*    Specific resistance, as mentioned previously, functions against certain biotypes of *P. sorghi,* and therefore its utility in disease control is limited. When rust biotypes appear that are virulent to a widely cultivated maize variety with only specific resistance, those rust biotypes are favored over others in terms of survival and tend to become the dominant biotypes in a region. Under these conditions, for all practical purposes the usefulness of the specific resistance is lost, and the variety appears as if it were fully susceptible. In the U.S.A., where maize rust is not an important disease, specific resistance is rare in the commercial crop. It is doubtful if specific resistance alone will give lasting protection against rust. Judging from the situation prevailing in the American Corn Belt, specific resistance should be used sparingly, if at all.

Nonspecific resistance is believed to function against all prevailing strains of the rust fungus. No evidence for specific host-variety rust-biotype interactions has been seen. The naturally occurring inoculum is usually comprised of many biotypes. If a given maize genotype has the nonspecific type of resistance it always has a low number of pustules when exposed to inoculum in the field. Under comparable favorable conditions for disease development, susceptible maize has a large number of pustules. This form of resistance is common in American maize and is believed to be the major reason why *P. sorghi* fails to develop in destructive proportions in the U.S.A. (Hooker, 1969).

For pathogens like *P. sorghi,* nonspecific resistance should be used in the commercial crop whenever possible. Specific resistance can be used if it is recognized that its value is only temporary. In other cereal crops the effectiveness of specific resistance is presumably enhanced when used in multilineal varieties. The technique is applicable to maize.

## Stalk Rots

Root and stalk rots are caused by several fungal and bacterial organisms. Some affect the plant early in its development, but the most common form of stalk rot occurs after anthesis when the plant is maturing. These organisms cause direct yield losses in the form of premature plant death and are important factors contributing to lodged plants and harvesting difficulties. In the U.S.A. *Diplodia maydis* (Berk.) Sacc. and *Gibberella zeae* (Schw.) Petch are two of the most common and destructive stalk rot pathogens. Other organisms are also important in the U.S.A. and elsewhere in the world. Regardless of the pathogen, most root and stalk rot diseases have common features, and similar problems are encountered in breeding for resistance to them.

*Nature of the Pathogens.*    Most of the root and stalk rot pathogens persist from year to year in the soil or in crop refuse. They commonly invade the plant by way of the root system. Stalks are usually not killed until several weeks after anthesis.

Stalk rots are more prevalent in plants subjected to stresses. Stresses cause an early senescence and a drop in the sugar content of roots and stalks. Such senescent tissue is unable to prevent late season invasion by stalk-rotting fungi. Stalk rot tends to be more prevalent in plants grown under favorable conditions for growth early in the season but subjected to unfavorable conditions after silking. Virus infections and leaf injuries caused by disease, insects, and other factors make the plant more susceptible. Within a given area, early maturing hybrids tend to be more susceptible to stalk rots than full-season hybrids. There is also evidence to indicate that high yielding plants tend to be more susceptible to stalk rots and lodging than low yielding plants. In certain areas, root-feeding insects contribute to root rots, stalk rots, and lodging. Root and stalk rot diseases are favored by high soil fertility and high plant populations.

Thus, most of the cultural practices aimed at higher plant yields favor this type of disease. Without disease control, the full benefits from these cultural practices are not achieved.

Although several pathogens comprised of many strains are involved in causing stalk rot, pathogen variability is believed to be of little importance in breeding for root and stalk rot resistance. Strains of *D. maydis* vary in pathogenicity (aggressiveness) but have shown little evidence of selective pathogenicity (virulence) toward corn inbreds or hybrids. As a general rule, corn hybrids resistant to one stalk rot pathogen tend also to be resistant to other pathogens that attack the plant at a similar time and in a similar manner (Hooker, 1956). Some strains of stalk rot fungi are weakly pathogenic and should be avoided in any inoculation program to measure plant reaction.

Cultures of *D. maydis* lose pathogenicity and ability to sporulate when maintained for long periods on artificial culture media. Cultures maintained in infected plant tissue, however, retain pathogenicity. In the case of *Diplodia,* cultures can be inoculated into ears 2-4 weeks after pollination and the resulting infected kernels stored. Pathogenic cultures can be reisolated when needed.

*Nature and Expression of Resistance.*   Resistance to rot spread is related to tissue senescence, which is preceded by a loss of cell vigor in the pith (Pappelis, 1965; Wysong and Hooker, 1966). Living cells seem to be capable of suppressing fungal invasion and growth, and the suppression appears to be chemical in nature. Suppression deteriorates with stalk maturity and occurs more rapidly in certain genotypes than in others. Resistant hybrids are characterized by a continued increase in vegetative dry matter for several weeks after pollination, whereas susceptible hybrids are characterized by a cessation of vegetative growth at pollination, followed by a rapid senescence of the plant (Mortimore and Ward, 1964; Wall and Mortimore, 1965).

Resistance to late season stalk rots seems to be of a nonspecific type. Hybrid reaction is similar to a range of strains of the commonly experienced stalk rot fungi.

*Inoculation Methods.*   Spores of *D. maydis* are readily obtained from fungus cultures grown on cooked grain seed. Light colored sorghum seed or oat seed is steeped in water for about 24 hours with several changes of water. The water is drained away. Stoppered containers with the steeped grain are autoclaved at 20 psi for 30 minutes on two successive days. A pure culture of *Diplodia* spores in water or a mycelial block from a pathogenic sporulating isolate is introduced into the sterile grain using aseptic techniques. After several days growth, the inoculated grain is shaken to distribute the fungus uniformly throughout the container. After 3-5 weeks in diffuse light at room temperature, spore suspensions are readily obtained by gently crushing or kneading the grain seed covered with pycnidia under water. The resulting spore suspension is then filtered through cloth or a fine screen. Spore concentration apparently is not critical, but each ml of inoculum should contain several hundred spores.

The spore suspension is introduced into the stalk by means of a specially made inoculator equipped with a sturdy, hollow needle (Jugenheimer, 1958). Another method is to use a hypodermic needle on a syringe. Holes may first need to be punctured in the stalk by means of a sharp instrument. The above method should be suitable for any pathogen where water suspensions of spores or bacterial cells can be made.

Another method for *Diplodia,* and commonly used for *G. zeae* and other fungi that do not sporulate readily in culture, utilizes round toothpicks. They are boiled repeatedly in water to remove the resins, stacked on end in a suitable container, soaked with nutrient broth, and sterilized. Fungus cultures are grown directly on the toothpicks. Holes are punched or drilled into the stalk, and the toothpicks are inserted in the holes and left in this position.

Inoculations are usually best made in the first elongated stalk internode above the brace roots. Time of inoculation will vary with maturity of the plant. Early maturing maize should be inoculated near anthesis, while later maturing maize in inoculated progressively later, up to 3-4 weeks after anthesis.

*Recording Differences in Disease Reaction.*   Depending upon maturity and season, at least 4-6 weeks should elapse between inoculation and data collection. The reactions of check strains of known reaction help in determining the proper time to take notes. For critical data stalks are split lengthwise, and the degree of rot spread is recorded according to a scale such as the following:

| RATING | DESCRIPTION |
| --- | --- |
| 0.1 | 0-3% of inoculated internode rotted and discolored. |
| 0.3 | 3.1-6% of inoculated internode rotted and discolored. |

| RATING | DESCRIPTION |
|--------|-------------|
| 0.5 | 6.1-12.5% of inoculated internode rotted and discolored. |
| 1.0 | 12.6-25% of inoculated internode rotted and discolored. |
| 2.0 | 26-50% of inoculated internode rotted and discolored. |
| 3.0 | 51-75% of inoculated internode rotted and discolored. |
| 4.0 | 76-100% of inoculated internode rotted and discolored. |
| 4.5 | Discoloration of less than 50% of adjacent internode. |
| 5.0 | Discoloration of more than 50% of adjacent internode. |
| 5.3 | Discoloration of 3 internodes, including inoculated internode. |
| 5.4 | Discoloration of 4 internodes, including inoculated internode. |
| 5.5 | Discoloration of 5 internodes, including inoculated internode. |
| 6.0 | Plant prematurely killed. |

In ratings 4.5-5.5 only internodes above the inoculated internode are considered. Some workers combine ratings 0.1, 0.3, and 0.5 with rating 1.0 and combine ratings 4.5, 5.3, 5.4, and 5.5 with rating 5. As with leaf blight, the numerical ratings can be in reverse order for statistical reasons.

The stalk rot reactions of genetically uniform inbreds or single crosses can be determined reliably from 10 inoculated plants in each of 3 or 4 replications. A larger number of plants are needed for genetically heterogeneous populations.

For less critical work in a breeding program, stalks are pushed over at harvest to determine their relative stalk strength. The inoculations serve primarily to increase the rate of stalk deterioration. Since inoculations are usually made during several successive inbreeding generations, protection is achieved against escapes in infection.

Stalk rot ratings may be based upon natural infection. Plants are grown in fields where stalk rot was severe in previous crops. The incidence of stalk rot in the test plot may also be increased by using a high fertility program with ample nitrogen, high plant population, drought stress late in the season, and partial defoliation of the plants. The percentage of prematurely dead plants is determined at appropriate intervals late in the season. Firmness of the stalk, surface discoloration, and presence of fungus fruiting bodies on the stalk are sometimes used for rating plants for degree of rotting. Late in the season the percentage of broken or lodged stalks is determined. Since the severity of natural infection varies, replication and a large number of plants are helpful in determining stalk rot reaction. Data from several locations and from several years is more valuable than data from a single test.

Although very useful in identifying susceptible material, none of the evaluation methods is completely satisfactory in the identification of hybrids that will be relatively rot free under all conditions. This may be because of the numerous organisms involved in causing stalk rot and the numerous plant, climatic, and edaphic factors that influence disease development (Cloninger et al., 1970). Nevertheless, germ plasm resistant to stalk rot and lodging in one location tends to be resistant in other locations of similar temperature and day length. The reaction of maize inbreds and hybrids in one year tends to be highly correlated with their reaction in other years. There is also a good agreement between stalk rot reaction as determined by artificial inoculation and the reaction resulting from natural infection. Resistance to stalk rot is usually positively associated with resistance to lodging, even though not all stalk lodging is pathological in origin. There is a significant and positive correlation between the stalk rot reaction of inbreds and stalk rot reaction of their hybrids.

*Sources of Resistance.*    Selection for stalk rot resistance has taken place during the adaptation of maize varieties to certain localities. Selection during inbreeding within these varieties has resulted in the identification and stabilization of resistant segregates and inbred lines. Some open-pollinated varieties have been good sources of resistant lines, while other varieties have been poor sources. Early maturing varieties frequently have little stalk rot resistance in comparison to later maturing varieties. Production of synthetics by combining ten or more of the best inbred lines has given source populations from which a number of excellent stalk rot and lodging resistant inbreds have been obtained. In contrast to other maize diseases, exotic maize varieties have been explored very little for sources of resistance.

*Inheritance of Resistance.*    Inheritance data suggest that resistance to *D. maydis* is a quantitative character (Kappelman and Thompson, 1966). Hybrid reaction tends to follow the calculated average of parent reactions. Stalk rot reaction seems to be highly heritable with most of the gene action additive. The $F_1$, however, tends to approach the most resistant parent in reaction, and in crosses some inbreds are more effective in transmitting resistance to their hybrids than are other inbreds with similar resistance. This suggests that some gene action is nonadditive in effect.

Evidence exists that resistance in some sources may depend upon a few genes or blocks of genes (El-Rouby and Russell, 1966). In certain populations, stalk rot reaction is fixed relatively soon during inbreeding. It has also been possible to transfer susceptibility to a resistant line and resistance to a susceptible line by means of a modified backcrossing procedure.

*Breeding Methods.*    Corn breeders have been selecting for stalk rot resistance for over 30 years. Marked progress has been made in the identification of stalk rot and lodging resistant germ plasm. Inoculations and other evaluations of stalk rot reaction have proved useful in the selection of resistant inbred lines. Data indicate that most of the genetic advance is achieved during the first four or five selfing generations.

Recurrent selection has been a useful breeding procedure for concentrating genes for resistance to stalk-rotting fungi (Jinahyon and Russell, 1969). By outcrossing to a superior source of resistance and then backcrossing to the line, it has been possible to improve the stalk rot resistance of a line. Alternate selfing and backcrossing can also be used.

*Use of Resistance Types in Maize Production.*    The various components of resistance to stalk rot diseases have not been fully identified. Hence, it is not possible to predict on the basis of stalk rot reaction alone which breeding lines will be the most useful in commercial production. It follows, then, that regardless of the method of development, resistant inbred lines should be further evaluated in some type of test-cross, utilizing a susceptible inbred or single-cross as a tester (Russell, 1961). Finally, to identify the most useful lines, they should be put in performance tests where maturity, drying rate, standability, yield, and other characters are measured. This seems necessary, because some forms of stalk rot resistance may be negatively associated with yield. In addition, higher stalk rot resistance than necessary may contribute to slow grain drying, and some lodging susceptibility is caused by factors other than disease.

## Other Diseases

Primary emphasis has been placed on three diseases and on conditions and research prevailing in the U.S.A. in the major portion of this chapter. This should not be taken as an indication that only the diseases discussed are of importance to maize production, or that research on breeding for disease resistance has been done only in the U.S.A. Because of space limitations, the author elected to limit the discussion to diseases where he has had first-hand experience and to those diseases that seem best to illustrate the general principles and problems of breeding for disease resistance in maize. Great progress has been made in disease control through resistance to numerous other diseases and in many parts of the world (Bojanowski, 1969; Boling and Grogan, 1965; Lockwood and Williams, 1957; Loesch and Zuber, 1967; Nelson and Ullstrup, 1964; Storey and Howland, 1967a, 1967b; Thompson, Rawlings, and Moll, 1963; Wernham and MacKenzie, 1968; and Wiser, Kramer, and Ullstrup, 1960).

## Goals Still to be Achieved

Maize hybrids completely resistant to all diseases and superior in other respects have yet to be produced.

Inbred lines, varieties, and corn hybrids have resistance to one or more diseases. These have been located largely through empirical screening of available germ plasm. Little is known about sources of superior resistance, or in what races and varieties of maize a high frequency of genes for resistance can be found. Little effort has been put forth to explore the wild relatives of maize and exotic germ plasms as sources of resistance.

Methods of inoculation have been developed to produce field-scale epiphytotics of some major diseases and to eliminate disease escapes. Improvements in inoculation techniques are needed for other diseases. Techniques that permit identification of resistant genotypes prior to pollination would be useful.

Methods by which resistance is expressed, the components of resistance, and the nature of disease resistance need to be determined. If possible, the nature of resistance should be equated to the expression of specific genes for resistance, so that when these genes are used in breeding an accurate prediction of the resistance expression can be made. The expression of resistance in relation to the various strains or biotypes of the pathogen must be elucidated. Some forms of resistance are specific for certain biotypes of the pathogen, others seem to be non-specific, and others are someplace in between.

A more complete interpretation of the genetic basis of resistance to disease is needed. The importance of major genes, their modifiers, and the action of polygene systems need to be resolved. In some instances, the work on host resistance must be related to corresponding genetic studies with the pathogen concerned. The role of various types of cytoplasm in affecting disease reaction needs to be studied. Reaction to at least two maize leaf diseases is influenced to a considerable extent by the cytoplasm type (Hooker et al., 1970; Scheifele, Nelson, and Koons, 1969; Villareal and Lantican, 1965).

Studies on the genetics of virulence and of aggressiveness in maize pathogens are particularly limited. Little information is available on the influence of host varieties grown or the relative prevalence of different strains and biotypes of pathogens.

Various procedures have been used or suggested in breeding maize for yielding ability; however, data are not available concerning the relative efficiency of these procedures in breeding for disease resistance. Also, as disease resistance involves the genetic interaction of two living organisms, breeding procedures other than those already suggested may need to be developed. More effort is needed to achieve genetic diversity for disease resistance in the commercial crop. The hybrid nature of the crop and the method of seed production permit the exploration of multilineal hybrids or blends of hybrids.

# References

Bojanowski, J. 1969. Studies of inheritance of reaction to common smut in corn. Theor. Appl. Genet. 39:32-42.

Boling, M. B., and C. O. Grogan. 1965. Gene action affecting host resistance to *Fusarium* ear rot of maize. Crop Sci. 5:305-307.

Cloninger, F. D., M. S. Zuber, O. H. Calvert, and P. J. Loesch, Jr. 1970. Methods of evaluating stalk quality in corn. Phytopathology 60:295-300.

El-Rouby, M. M., and W. A. Russell. 1966. Locating genes determining resistance to *Diplodia maydis* in maize by using chromosomal translocations. Can. J. Genet. Cytol. 8:233-240.

Hilu, H. M. 1965. Host-pathogen relationships of *Puccinia sorghi* in nearly isogenic resistant and susceptible seedling corn. Phytopathology 55:563-569.

Hooker, A. L. 1956. Association of resistance to several seedling, root, stalk, and ear diseases in corn. Phytopathology 46:379-384.

———. 1963. Monogenic resistance in *Zea mays* L. to *Helminthosporium turcicum*. Crop Sci. 3:381-383.

———. 1969. Widely based resistance to rust in corn, p. 28-34. *In* J. A. Browning [ed.] Disease consequences of intensive and extensive culture of field crops. Iowa Agr. and Home Economics Exp. Sta. Sp. Rept. No. 64. 56 p.

———, H. M. Hilu, D. R. Wilkinson, and C. G. Van Dyke. 1964. Additional sources of chlorotic-lesion resistance to *Helminthosporium turcicum* in corn. Plant Dis. Reporter 48:777-780.

———, R. R. Nelson, and H. M. Hilu. 1965. Avirulence of *Helminthosporium turcicum* on monogenic resistant corn. Phytopathology 55:462-463.

———, D. R. Smith, S. M. Lim, and J. B. Beckett. 1970. Reaction of corn seedlings with male sterile cytoplasm to *Helminthosporium maydis*. Plant Dis. Reporter 54:708-712.

Hughes, G. R., and A. L. Hooker. 1970. Gene action conditioning resistance to northern leaf blight in maize. Crop Sci. 11:180-184.

Jenkins, M. T., A. L. Robert, and W. R. Findley, Jr. 1954. Recurrent selection as a method for concentrating genes for resistance to *Helminthosporium turcicum* leaf blight in corn. Agron. J. 46:89-94.

———, ———. 1961. Further genetic studies of resistance to *Helminthosporium turcicum* Pass. in maize by means of chromosomal translocations. Crop Sci. 1:450-455.

Jinahyon, S., and W. A. Russell. 1969. Evaluation of recurrent selection for stalk-rot resistance in an open-pollinated variety of maize. Iowa St. J. Sci. 43:229-237.

Jugenheimer, R. W. 1958. Hybrid maize breeding and seed production. FAO Agr. Dev. Paper No. 62. Rome. 369 p.

Kappelman, A. J., Jr., and D. L. Thompson. 1966. Inheritance of resistance to *Diplodia* stalk-rot in corn. Crop Sci. 6:288-290.

Lim, S. M., A. L. Hooker, and J. D. Paxton. 1970. Isolation of phytoalexins from corn with monogenic resistance to *Helminthosporium turcicum*. Phytopathology 60:1071-1075.

Lockwood, J. L., and L. E. Williams. 1957. Inoculation and rating methods for bacterial wilt of sweet corn. Phytopathology 47:83-87.

Loesch, P. J., Jr., and M. S. Zuber. 1967. An inheritance study of resistance to maize dwarf virus in corn (*Zea mays* L.). Agron. J. 59:423-426.

Mortimore, C. G., and G. M. Ward. 1964. Root and stalk rot of corn in southwestern Ontario. III. Sugar levels as a measure of plant vigor and resistance. Can. J. Plant Sci. 44:451-457.

Nelson, O. E., and A. J. Ullstrup. 1964. Resistance to leaf spot in maize. Genetic control of resistance to race I of *Helminthosporium carbonum* Ull. J. Hered. 55:195-199.

Nelson, R. R., A. L. Robert, and G. F. Sprague. 1965. Evaluating genetic potentials in *Helminthosporium turcicum*. Phytopathology 55:418-420.

———, D. R. MacKenzie, and G. L. Scheifele. 1970. Interaction of genes for pathogenicity and virulence in *Trichometasphaeria turcica* with different numbers of genes for vertical resistance in *Zea mays*. Phytopathology 60:1250-1254.

Pappelis, A. J. 1965. Relationship of seasonal changes in pith condition ratings and density to *Gibberella* stalk rot of corn. Phytopathology 55:623-626.

Russell, W. A. 1961. A comparison of five types of testers in evaluating the relationship of stalk rot resistance in corn inbred lines and stalk strength of the lines in hybrid combinations. Crop Sci. 1:393-397.

Saxena, K. M. S., and A. L. Hooker. 1968. On the structure of a gene for disease resistance in maize. Proc. Nat. Acad. Sci. 61:1300-1305.

Scheifele, G. L., R. R. Nelson, and C. Koons. 1969. Male sterility cytoplasm conditioning susceptibility of resistant inbred lines of maize to yellow leaf blight caused by *Phyllosticta zeae.* Plant Dis. Reporter 53:656-659.

Sprague, G. F. [ed.] 1955. Corn and corn improvement. Academic Press, New York. 699 p.

Storey, H. H., and A. K. Howland. 1967a. Inheritance of resistance in maize to the virus of streak disease in East Africa. Ann. Appl. Biol. 59:429-436.

————— —————. 1967b. Resistance in maize to a third East African race of *Puccinia polysora* Underw. Ann. Appl. Biol. 60:297-303.

Thompson, D. L., J. O. Rawlings, and R. H. Moll. 1963. Inheritance and breeding information pertaining to brown spot resistance in corn. Crop Sci. 3:511-514.

Villareal, R. L., and R. M. Lantican. 1965. The cytoplasmic inheritance of susceptibility to *Helminthosporium* leaf spot in corn. Philipp. Agric. 49:294-300.

Wall, R. E., and C. G. Mortimore. 1965. The growth pattern of corn in relation to resistance to root and stalk rot. Can. J. Bot. 43:1277-1283.

Wernham, C. C., and D. R. MacKenzie. 1968. Plot techniques with maize dwarf mosaic virus: I. Field studies, 1966. Plant Dis. Reporter 52:24-28.

Wilkinson, D. R., and A. L. Hooker. 1968. Genetics of reaction to *Puccinia sorghi* in ten corn inbred lines from Africa and Europe. Phytopathology 58:605-608.

Wiser, W. J., H. H. Kramer, and A. J. Ullstrup. 1960. Evaluating inbred lines of corn for resistance to *Diplodia* ear rot. Agron. J. 52:624-626.

Wysong, D. S., and A. L. Hooker. 1966. Relation of soluble solids content and pith condition to *Diplodia* stalk rot in corn hybrids. Phytopathology 56:26-35.

# 11 OATS: A. Continental Control Program

## J. Artie Browning[1]

## *Introduction*

Oats have been notorious for disease problems, and continuing programs of breeding for resistance to major oat diseases have been accepted as inevitable. But oat diseases need not be a limiting factor in future oat production. With oats being grown more intensively, and with high-yielding, high-protein cultivars in the offing, oat workers must use newly developed resistance theory to initiate a control program to achieve nonephemeral resistance.

Crown rust, stem rust, and yellow dwarf are the most serious diseases of oats from the upper Mississippi River Valley to the Prairie Provinces of Canada where most North American oats are grown. Not coincidentally, these diseases are caused by obligately parasitic, continental pathogens with great epidemic potential. This chapter presents a continental control program based on the use of host resistance genes to

1. Professor of Plant Pathology, Iowa State University. Journal Paper No. J-6932 of the Iowa Agriculture and Home Economics Experiment Station, Ames, Iowa. Project 1752.

protect oats from these diseases. The ideas presented are directed toward the goal of obtaining resistance to today's pathogenic races, but the resistance obtained also should stand against the races of tomorrow. Principles elucidated from "the oat model" for disease control in mid-America may apply to major diseases of other crops.

## Importance and Distribution of Oats as a Crop

A major grain crop of the world, oats are grown primarily in the Northern Hemisphere and rank in world production after wheat, rice, and corn. Best adapted in cool, moist climates, oats are grown for grain primarily north of 40° latitude in North America, and north of 50° latitude in Europe and Asia. Leading oat-producing countries are the U.S.A., Russia, Canada, Germany, France, and the United Kingdom (Coffman, 1961). Oats are grown also at upper elevations in Mexico and South America and have the highest potential for forage production of any crop tested in the high, cool, tropical climate of Colombia (Crowder et al., 1967).

In the U.S.A., oats are third after corn and wheat in acreage planted to grain crops, and fourth after corn, wheat, and sorghum in value. About 18.6 million acres of oats were harvested for grain in the U.S.A. in 1970, yielding a crop valued at $577 million. Future cultivars could yield 150 bu/acre and contain 25% protein. Contemporary oat cultivars have groat protein contents ranging up to 19%. Oats provide the most and best balanced nutrition of any grain crop, and 6% of current production goes directly for human consumption. Additionally, in the southern U.S.A. millions of acres of fall-sown oats are grown for pasture. This acreage is of prime importance in epidemiology of the rust diseases.

## Species of Avena Important to Oat Cultivation and Improvement

Cultivated oats are annual members of the genus *Avena,* family Gramineae. *Avena* spp. important in cultivation or in oat improvement are (Coffman, 1961):

Diploid, $n = 7$
*A. strigosa* Schreb.　　　　　sand oat

Tetraploid, $n = 14$
*A. barbata* Brot.　　　　　slender oat

*A. magna* Murphy & Terrell[a]

Hexaploid, $n = 21$

| | |
|---|---|
| *A. byzantina* C. Koch | red oat |
| *A. fatua* L. | wild oat |
| *A. sativa* L. | common oat |
| *A. sterilis* L. | wild red oat |

It now is believed that *A. sterilis* is the primitive ancestor of *A. byzantina* that, in turn, gave rise to *A. sativa*. The center of origin of cultivated oats is in extreme southwestern Asia where, along with the progenitors of cultivated wheat and barley, large natural populations of *A. sterilis* still abound. Oats spread over Europe and Asia as weed-like mixtures in barley and wheat and were domesticated there. After selection by nature and man, *A. byzantina* types showed better adaptation in southern Europe (and later, in the southern U.S.A.), and *A. sativa* types became a dominant cereal in northern climes (Coffman, 1961). *A. sativa* makes up the bulk of the oat acreage in the northern U.S.A., Canada, and northern Europe. The remaining acreage is sown largely to cultivars of the red oat, *A. byzantina*. Most fall-sown and very early cultivars of spring-sown oats in the U.S.A. are red oats. Many cultivars are intermediate between *A. sativa* and *A. byzantina*. The other species serve as gene donors. Oats cross readily within (but not between) ploidy groups, so there is little difficulty in transferring genes among hexaploid species. *A. sterilis,* the progenitor of all hexaploid oats, has become the major donor species, contributing genes to cultivated oats for disease resistance, large seed, winter hardiness, high protein percentage, and so forth (Zillinsky and Murphy, 1967). There is great need to preserve indigenous populations of *A. sterilis* as a gene bank for future generations.

# Diseases of Oats

Oat diseases were described and illustrated recently by Simons and Murphy (1961, 1968). Certain oat diseases (for example, gray speck and soil-borne oat mosaic) occur only in limited and discrete areas; others, like root rots, probably occur in most fields every year and cause small annual losses characteristic of endemic diseases; still others,

[a]Murphy et al., 1968.

especially yellow dwarf, crown rust, and stem rust (although they may be absent some years), have the potential for causing severe epidemics in susceptible cultivars when environmental conditions are favorable. Thus, in Iowa, crown rust caused an estimated reduction in yield of 30% in 1953 and 12% in 1957; stem rust, a 10% reduction in 1953 and 11% in 1954; and yellow dwarf, 15% in 1949 and 12% in 1959. Because losses caused by such environment-sensitive diseases occur unpredictably, they are more consequential to farmers than endemic diseases such as root rot. These diseases, especially crown rust, frequently have been the limiting factor in oat production, and obtaining adequate resistance has been a major objective of most oat improvement programs.

The major effort in oat genetics also has been devoted to disease resistance, further attesting to the importance of diseases. The lion's share of inheritance work has been directed at crown rust, stem rust, the smuts, and Victoria blight, although resistance is used in controlling other diseases also. See Murphy and Coffman (1961) and Simons, Zillinsky, and Jensen (1966) for references on genetics of disease resistance.

## Crown Rust of Oats

Crown rust is the common name both of a serious leaf-rust disease of oats and of the causal agent, *Puccinia coronata* Cda. var. *avenae* Fraser & Led. Both "crown" and "coronata" refer to the crown-like digitate projections from the teliospore apices. Crown rust was monographed recently (Simons, 1970).

*P. coronata* is a heteroecious, heterothallic, macrocyclic rust fungus. The uredial and telial stages occur on oats and many other grasses; the spermagonial and aecial stages on buckthorn (*Rhamnus* spp.). The uredia are bright orange-yellow, round to oblong pustules surrounded even in susceptible cultivars by some light green, chlorotic host tissue. Telia are dark brown or black, and shiny from the intact host epidermal covering. Teliospores survive northern winters on host debris, and their germination in the spring may result in development of the sexual stage on *Rhamnus*. The aecial stage is conspicuous when the bright orange-yellow aecial cups are forming and discharging aeciospores; later, when rust is building up on oats, buckthorn leaves show only necrotic lesions or shot-holes. Aeciospore infection of oats results in reestablishment of the repeating or uredial stage. *P. coronata* also can overwinter in the South in the uredial stage and cycle between North and South independently of *Rhamnus*.

*P. coronata avenae* is a highly variable fungus, with new races arising from the sexual process on *Rhamnus,* from mutation, and from heterocaryosis. Between the 1930s and 1953, 112 pathogenic races were identified on a standard set of 13 differential cultivars. This set was replaced in 1953 by a set of 10 differentials on which races 201 to 462 have been identified (Simons, 1970). Some differential cultivars have been used as sources of resistance in developing commercial cultivars, which gives maximum but transitory utility to a differential. Thus, the Bond and Landhafer resistances, for example, have appeared extensively in commercial cultivars, and, for some time, most crown-rust isolates have been virulent on these differentials; that is, they have ceased to differentiate among clones of the crown-rust population. This problem is met in part by using supplementary differentials that represent new sources of resistance and/or lines important in breeding programs. As they become available, isogenic lines are being used as supplementary differentials.

The crown-rust fungus has changed continuously in response to the living substrate provided by man. Races identified since 1953 can be organized into groups, each represented by a "type race." The type race is based, not on priority, but on representativeness of the group for certain man-made purposes. Thus, since 1953, the important race groups in the U.S.A. have been represented by type races:

202—virulent on Bond, but not Landhafer. Includes races 203 and 205;

216—virulent on Bond and Victoria, but not Landhafer. Includes 213;

290—virulent on Bond and Landhafer, but not Victoria. Includes 294 and 295;

326—virulent on Bond, Landhafer, and Victoria. Includes 321;

325—virulent on most differentials. Includes race 264B.

These five race groups represent successive and overlapping stages of rust-race evolution of *P. coronata* on commercially grown cultivars. Whether *P. coronata* may be held in check by stabilizing selection (Van der Plank, 1968) is not certain. The most virulent race, 264, appeared in the U.S.A. in the mid-1950s. Even though it could parasitize all commercial oat cultivars and all hexaploid sources of resistance and, thus, was a "super race," its prevalence soon decreased (Michel and Simons, 1966). Browning and Frey (1969) interpreted this to mean that race 264, with several unnecessary genes for virulence, lacked fitness genes necessary for epidemic potential in competition with less virulent but more aggressive races 216 and 290.

## Stem Rust of Oats

Stem rust is the common name both of a serious disease of oats and of the causal agent, *Puccinia graminis* Pers. f. sp. *avenae* Erikss. & E. Henn. *P. graminis avenae* develops as well as crown rust on leaves in the greenhouse; in the field, however, stem rust is a disease of stems, leaf sheaths, and panicles. *P. graminis'* life cycle parallels completely that of *P. coronata,* except that the alternate host is barberry (*Berberis* spp.), different plant parts are attacked, and uredia and telia of the two fungi are markedly different in appearance. Uredia of *P. graminis* are brick red, much larger than those of *P. coronata.* Uredia and telia rupture the host epidermis, making the infected area very ragged in appearance and to the touch.

Fewer pathogenic races have been reported for the stem-rust fungus than for crown rust, which reflects that fewer resistance genes have been identified and used in differential and commercial cultivars. A new, modified international system of identifying and registering races of *P. graminis avenae* recognizes 97 races of this fungus as having been identified on a standard set of seven differential cultivars (Stewart and Roberts, 1970). Each differential, except *A. strigosa* 'Saia,' contains a single resistance gene. Most genes also have been incorporated into commercial cultivars. As with crown rust, this resulted in their transitory value as differentials. The important race groups over the last half-century have been represented by type races:

2—virulent on cultivars containing gene $Pg$-3;

8—virulent on cultivars containing genes $Pg$-2 and $Pg$-3;

7—virulent on cultivars containing genes $Pg$-1 and $Pg$-3;

21 (old 7A)—virulent on cultivars containing genes $Pg$-1, $Pg$-3, and $Pg$-4;

6—virulent on cultivars containing genes $Pg$-1, $Pg$-2, and $Pg$-3;

31 (old 6AF)—virulent on cultivars containing genes $Pg$-1, $Pg$-2, $Pg$-3, $Pg$-4, and $pg$-8.

The race groups represent successive and overlapping stages of rust-race evolution of *P. graminis avenae* on commercially grown cultivars. Race 31 now is the most common race in North America (Martens et al., 1970; Stewart and Rothman, 1971). Since race 31 (and other North American races) carry several unnecessary genes for virulence, Martens et al. (1970) concluded that "Unnecessary genes for virulence do not necessarily reduce competitive ability." Leonard (1969), on the other hand, also working with *P. graminis avenae,* presented experimental evidence to support the theory that simple races are more fit to survive.

## Yellow Dwarf of Oats

Although an oat disease commonly called red leaf had long been recognized as of major importance, it remained for Oswald and Houston (1953) to show that this disease was caused by an aphid-transmitted virus, the barley yellow dwarf virus (BYDV). Serious to some degree in most years, a pandemic of yellow dwarf (YD) occurred in the north-central states in 1959. This was described in Special Supplement 262 to the Plant Disease Reporter and followed by comprehensive reviews of the disease (Bruehl, 1961; Rochow, 1961).

Leaves of infected plants turn yellow-to-red hues that merge gradually into normal green. Infected plants may be dwarfed, mature early, suffer severe blast, and sustain marked reductions in yield and test weight. Clintland, a very susceptible cultivar, was reduced in yield by 75% in an artificially induced epidemic (Endo and Brown, 1963). YD causes a malformation of phloem, interfering with normal translocation of sugars that exude to the outside. In humid areas most YD-killed plants, late in the season, show a blackened appearance from the presence of fungi, especially *Alternaria* spp., living saprophytically on sugar exudates.

The BYDV is obligately transmitted by at least 11 aphid species to a wide range of gramineous hosts. Once it has acquired the circulative BYDV by feeding on a diseased plant, an aphid can transmit the virus for most of its remaining life. Considerable work has been done on variants of the BYDV based on the pattern of transmission by test clones of aphids, on incubation period, on the relative severity of symptoms in selected cultivars of oats and barley, and on properties of purified preparations. Most variants fall into one of four or five main groups, some of which may actually be distinct viruses. Some variant groups can be transmitted by a single species of aphid; others can be transmitted nonselectively by several species. As is the case with all biological entities, there can be much variation within a major group of variants. Variants differ markedly in the severity of disease they cause in "susceptible" cultivars. In general, variants of the BYDV transmitted nonselectively by several aphid species seem the most common in nature and the most destructive (Gill, 1969; Jedlinski and Brown, 1965; Rochow, 1969; Rochow and Jedlinski, 1970).

# Epidemiology

## Residual vs. Continental Pathogens

A basic dichotomy in epidemiology is whether a pathogen is *residual* or *continental*. A residual pathogen is one that can perpetuate itself locally. It can be influenced by the grower(s) affected. "A continental pathogen, on the other hand, like a continental climate, originates outside the area, covers large areas, and local growers are at the mercy of whatever blows their way" (Browning et al., 1969). Interfering with the capacity of a pathogen to remain residual in an area is the first goal of most control programs, and a successful program means the disease is controlled. But crown rust, stem rust, and yellow dwarf all are caused by pathogens that have both residual and continental capabilities. Severe disease outbreaks involving these pathogens in mid-America result from their continental capabilities.

## Epidemiological Unity of the Puccinia Path

"The South and the North are mutually supplementary in the annual development of wheat stem rust in a vast area of North America, extending from Mexico through the Mississippi basin of the U.S.A. and onward to the Prairie Provinces of Canada, a distance of some 2,500 miles" (Hamilton and Stakman, 1967).

This great area, the heartland of North America, called by Browning et al. (1969) the "Puccinia Path," is the epidemiological unit for control of the continental crown- and stem-rust fungi, and apparently also for the BYDV.

"As the uredial stage of the rust seldom survives the winter north of central Texas, the area northward is dependent on Texas or Mexico for most of the initial inoculum early in the growing season, especially since barberries (*Berberis vulgaris* L.) have been largely eradicated from the north-central states, where they once abounded. Conversely, the uredial stage of the rust seldom survives the long, hot summer in Texas and northern Mexico between the wheat harvest in the spring or very early summer and the autumn sowing of winter wheat. That area is then dependent on the northern areas for windblown inoculum in the fall" (Hamilton and Stakman, 1967).

Crown rust and buckthorn parallel stem rust and barberry closely. Although buckthorn eradication was not required by law in Iowa until 1955, eradication now has progressed to the point where buckthorn no longer is believed a factor in epidemics in Iowa (except possibly on a local scale).

Eradication of these alternate hosts did the job it was designed to do. Eradication removed sources of new races and sources of early spring inoculum, thereby thwarting the residual capacities of these fungi. "But possibly the most important contribution of these programs was to unify the Puccinia Path epidemiologically. *P. graminis* and *P. coronata* are no longer residual pathogens, able to survive and recycle in many farm or county units in the North; the entire Puccinia Path is the unit. The ability of these fungi to make the transition from residual to continental pathogens has enabled them to continue as threats to small-grain production in the Puccinia Path" (Browning et al., 1969).

## Epidemiology of Crown and Stem Rust

The effective overwintering of the crown- and stem-rust fungi in the Puccinia Path is along the Gulf Coast of the U.S.A. and Mexico, where large acreages of fall-sown, spring-type oats are grown for grazing. Oats grown far enough south that the fungi are not thrown into the telial stage by cold weather can be damaged severely by rust and supply urediospores to initiate rust development in the North. Convection currents take the inoculum from the upper portions of actively sporulating plants in an early dough stage of growth, and low-level jet winds transport it to young plants to the north. The northward movement of *Puccinia* urediospores has been well documented (Asai, 1960; Hamilton and Stakman, 1967; Rowell and Romig, 1966).

For an epidemic to develop in the North, inoculum either must arrive from the South in quantities that are directly damaging, or arrive early enough in the season to allow time for several cycles of increase. Since inoculum traversing great distances suffers tremendous dilution and attrition, the latter situation is far more common. Early in the season, low temperatures limit rust establishment in the North. At Ames, Iowa, oat crown- and stem-rust establishment is prevented on nights when the temperature drops below about 10 C, such as clear nights favorable for heavy dew deposition (Browning, *unpublished data*). Rainy nights, on the other hand, are likely to have temperatures above that level, even early in the season. Rain serves to remove urediospores from the air, convey them to host plants, and to supply moisture and moderate temperatures necessary for rust establishment (Rowell and Romig, 1966).

Once established, early rust increase is logarithmic. Plotting a developing crown-rust epidemic by trapping spores daily outside large plots, Cournoyer (1970) (1) showed that crown-rust spore release was maximal from 1000-1900 hr and peaked between 1100 and 1200 hr, when temperature and wind velocity are higher and relative humidity is lower; and (2) produced sigmoid growth curves (characteristic of population increase in a limiting environment) that are described by the logistic growth function.

Oat crown- and stem-rust development in the Puccinia Path is at least a host-host-pathogen interaction (Van der Plank, 1968) since it involves minimally a host genotype in the South, another in the North, and a continental pathogen limited by (and that limits) host genotypes in both the North and the South.

## Epidemiology of Yellow Dwarf of Oats

The BYDV is a systemic pathogen transmitted rapidly from plant to plant by efficient vectors. For YD to be serious, many of the approximately 1.25 million oat plants per acre must be inoculated with the BYDV within a few weeks of emergence. In a pandemic this must take place over extensive areas, and it requires astronomical numbers and high mobility of vectors moving from areas of great concentration (Bruehl, 1961). Both local and migratory aphids are involved.

The BYDV overwinters in perennial grasses and fall-sown small grains, predisposing them to winterkilling. It then is transmitted by aphids to spring-sown grains nearby or hundreds of miles away. The pattern of spread in a field may indicate the vector involved. Of the more common vectors on oats, the English grain aphid (*Macrosiphum avenae*) is very active and results in infected plants being scattered at random; the oat bird-cherry aphid (*Rhophalosiphum padi*) lands along field borders and results in YD-diseased plants being concentrated there and, later, in circular patches within the field; and the greenbug (*Schizaphis graminum*) behaves as a vector similar to the English grain aphid, but causes feeding injury to the leaf that many mistake for early season rust pustules. Barley is the preferred small-grain host of the corn leaf aphid (*R. maidis*), but in the field it will feed on oats long enough to transmit the BYDV.

BYDV variants prevalent in Illinois consistently have been in groups transmitted nonselectively by *R. padi* and *M. avenae* or selectively by *R. padi*. In New York, variants have cycled gradually between groups like those in Illinois and variants transmitted selectively by *M. avenae*, which usually have predominated (Rochow and Jedlinski, 1970). In contrast, Manitoba has abrupt changes in major variants from year to year (Gill, 1969). The relative stability of virus populations in Illinois and New York probably is a result of the local presence of winter grain hosts. Evidently the BYDV is a residual pathogen transmitted by residual vectors in Illinois and New York and a continental pathogen transmitted by continental vectors in Manitoba. Illinois, with an extensive winter grain acreage, usually shows the relative stability of a residual system but, as in 1959, also is part of a continental system. Probably the Puccinia Path also is the BYDV Path, and the BYDV variants present in Manitoba are those that overwinter in grains to the south (especially winter wheat in north Texas and Oklahoma) and are carried north by aphid vectors riding low-level jet winds (Bruehl, 1961; Browning et al., 1969; Wallin et al., 1967).

YD usually involves minimally a host-host-vector-pathogen system, with one host usually not being oats. The overwintering host (usually not an oat) will determine which variants survive winter. The first aphid to transmit virus in spring may be most important; an early inoculation by a mild variant of the BYDV may prevent subsequent, possibly more severe but closely related, variants from developing because of cross protection (Jedlinski and Brown, 1965). The degree of disease development may depend on whether the aphid is an efficient vector of the locally predominant virus strain, whether the variant is mild or severe, and on the activity of the aphid. Continental vectors may introduce new variants into an area. New variants of the virus may result from phenotypic mixing of serologically unrelated variants, especially in doubly infected winter grains (Rochow and Jedlinski, 1970). Also, such plants may be damaged severely from synergism following multiple infections by unrelated variants.

Whether the BYDV is subject to stabilizing selection has not been established. It seems, however, that the BYDV population is relatively stable and simple (compared, for example, with the rusts); most variants fall into one of four or five major groups. Oat breeding should be less likely to upset this complex, relatively well-buffered host-host-vector-pathogen system. The YD epidemic ends on oats (at least in the North); therefore, the overwintering host and the vector(s) should exert relatively more selection influence on BYDV variants than oats, making breeding of tolerant cultivars more promising. Encouraging, too, is the experience of Rochow and Jedlinski (1970) that it is more difficult to recover virus from tolerant than intolerant cultivars.

# Experiences From Breeding Oats for Rust Resistance

## Lessons from the Past—the Vicious Circle

Crown rust, stem rust, and YD are crowd diseases: they become more serious as susceptible hosts are brought closer together. Theoretically, they should be held in check by several mechanisms (Browning and Frey, 1969): (1) interspecific diversification, (2) intraspecific diversification, (3) host resistance, and (4) stabilizing tendencies in the pathogens. The last two are basic to the operation of the others, though all are interrelated. With oats, man decreased interspecific diversity by controlling weeds and intraspecific diversity by producing pure-line cultivars. At first, pure-line cultivars resulted in homogeneous genotypes only for a given agricultural area. But one of the characteristics of oats in comparison with, say, wheat is that oats are primarily a feed grain, and the area over which an adapted oat cultivar is cultured is not limited by market or other considerations. Thus, adapted oat cultivars spread over vast areas. Even when climate dictated that different agronomic types be grown in different parts of the U.S.A., the same resistance genes spread across agronomic types and across state and national borders. A single resistance gene was incorporated into cultivars sown across the continent.

Widespread use of the same resistance genotypes changed the host-host-pathogen system to a host-pathogen system in which the pathogen was able to recycle *within* the crop (that is, on the same resistance genotypes in the North and in the South) and which narrowed the pathogen population to clones virulent on the homogeneous host. Popular, widely planted cultivars gave oats and vertical resistance a notoriously poor record.

Repeatedly, the wide use of a single resistance genotype simultaneously excluded avirulent races and homogenized the pathogen population to strains virulent on that host genotype. This, in turn, necessitated a new genotype of the host that, in turn, selected a new virulence genotype of the pathogen. This is the "vicious circle" of oat improvement, and it resulted in "boom and bust" years for oat production. This unwitting guiding of the evolution of the rust fungi has occurred wherever specific wheat- and oat-rust resistance genes have been used—the U.S.A., Canada, Australia, and Kenya (Johnson, 1961). It possibly is best illustrated with oats because the appearance of a "new" pathogen, *Helminthosporium victoriae* M. & M., itself a product of a homogeneous oat culture, forced a dramatic change in oat cultivars polar in rust response. The resultant and sudden shift induced in the pathogen population by the host population was brought into clear focus. This was reviewed for oats by Murphy (1965), Browning and Frey (1969), and Browning et al. (1969). Johnson (1961), too, showed that the pathogenic response of the rust can be "causally related to the man-made modification of the hosts."

## The Current Situation

Early failures of single resistance genes before a highly variable pathogen did not lead to new theories of controlling diseases via resistance; they led only to more intensive searches for that pure line of "greatest value" that incorporated a gene or genes giving resistance of such a high type to such a broad spectrum of races that the pathogen would be unable to bridge the gap.

But pathogens have been able to bridge the gap. The production of new resistant pure-line oat cultivars has been paralleled by a gradual upgrading of the level of virulence in prevalent crown- and stem-rust races. Several oats formerly listed as crown-rust resistant now are susceptible; very few commercially available oats have satisfactory crown-rust resistance. Only one pure line, Portal, currently recommended for Iowa, offers adequate vertical resistance. Fortunately, recent efforts also have been directed toward increased horizontal resistance or tolerance in pure lines, and toward population resistance in multilines. Certain pure lines, such as Grundy, O'Brien, Orbit, Otter, and Pettis, have superior yields in spite of crown rust and must be considered as having some tolerance, even though they are rated susceptible. The oats currently most resistant, however, and which also offer the greatest promise of lasting protection from crown rust, are the multilines.

With stem rust, the level of virulence in the pathogen has not paralleled resistance in the host—it has outdistanced it. Genes $Pg$-1, $Pg$-2, and $Pg$-4 (only recently combined) are common in commercial cultivars, but the prevalent race 6AF has virulence for genes $Pg$-1, $Pg$-2, $Pg$-3, $Pg$-4, and $pg$-8. Thus, no commercially available cultivar has stem-rust resistance today, and there is scant promise of one in the near future.

Tolerance to YD has been increased gradually until it is quite high in Jaycee and Pettis. Most cultivars, however, remain fully susceptible and are protected only by escaping the disease.

# Breeding Pure-Line Oat Cultivars for Resistance or Tolerance to Major Diseases

## Crown Rust

*Use of Oligogenic Resistance.*   Vertical resistance (VR) is race specific. As such, lines with VR reduce incoming inoculum ($x^0$) but not the rate of increase ($r$) of virulent races (van der Plank, 1968). Promising genes for VR to crown rust are available from *Avena sterilis.* Incorporating them is straightforward breeding. They can be incorporated into commercial cultivars either singly or doubly, or they can be pyramided. Assume three VR genes, *A, B,* and *C,* and that virulence for these loci does not occur in the fungus population at the time the new cultivar is released. Virulence alleles could arise from mutations or recombinations in existing races. Obviously, the likelihood of virulence arising for *ABC* together in a single cultivar (if *A, B,* and *C* do not occur singly in commercial production) is far less than for *A, B,* and *C* incorporated separately (Johnson, 1961).

Considering the versatility of the crown-rust fungus, I see no reason for optimism that oligogenic resistance used singly, doubly, or pyramided will be successful if past management practices are used. Pyramiding VR genes may only use them faster, unless something keeps the pathogen population simple (Van der Plank, 1968). However, with new management systems, especially gene deployment, there is hope for oligogenic resistance.

*Horizontal Resistance.*   As used herein, horizontal, generalized, and field resistance are equated as race-nonspecific, moderate resistance (Heagle and Moore, 1970). Cultivars with horizontal (HR) and vertical (VR) resistance are similar in that both look resistant (though HR plants usually less so), but they affect the epidemic and the pathogen population differently (Van der Plank, 1968). Plants with VR reduce $x_0$ and therefore delay the onset of the epidemic, while plants with HR reduce *r* and therefore delay the epidemic progressively after its onset. Also, since cultivars with HR do not respond differentially to the pathogen population, unlike those with VR, they should not favor a more virulent race, and they should be more stable over time. Therefore, as defined, cultivars with HR are the indicated control measures for the cereal rusts. In practice, however, it is impossible to know that a given host genotype is resistant to all present—much less future—clones of a pathogen; therefore, HR must be regarded as a useful working concept (Browning and Frey, 1969).

Development of cultivars with HR will not be easy. Horizontal resistance may be rare and difficult to detect and recover from hybrid populations. Van der Plank (1968) has emphasized that, to breed for HR, effective VR genes must be excluded so they do not mask the effect of the HR genes. Parents in an HR breeding program may be chosen from (1) commercial cultivars with moderate resistance to a broad spectrum of races (Heagle and Moore, 1970), or which at least have yielded well over time despite rust; (2) entries in the USDA International Oat Rust Nurseries that show resistance of a nonspecific nature; (3) "slow rusting" cultivars (for example, Red Rustproof); and (4) susceptible *A. sterilis* and *A. fatua* plants that withstand rust in native populations. More crosses must be made, and larger populations must be grown, to combine HR and desirable agronomic traits, than is necessary for VR. Recovering good agronomic traits, especially maturity, may be necessary to test adequately for HR. Since HR is polygenic, HR x HR crosses may give still higher resistance. Unfortunately, candidate lines from such crosses cannot be evaluated adequately for rust response in ·small plots because heavy spore loads from nearby susceptible lines may mask their resistance. They should be grown in large plots or fields where rust that causes damage builds up and spreads on the cultivar under test. Paired fungicide-sprayed and rusted plots in a large field assay the degree of HR (in terms of yield) attained. Once developed, cultivars with HR could (1) be grown as they are, (2) offer background resistance for use with VR genes in pure-line cultivars (probably the situation with Portal and other HR cultivars tested by Heagle and Moore, 1970), or (3) be used as the recurrent parents in a multiline development program.

*Tolerance.*    Conceptually, tolerance to rust is like HR in being race nonspecific in its effect on the epidemic and, therefore, in the methods required to test its value; but tolerance is unlike HR in that a tolerant line must look susceptible.

In Indiana, in comparison with their respective resistant isolines, Benton cultivar was tolerant and Clinton susceptible to a massive crownrust attack, even though both appeared susceptible (Caldwell et al., 1958). Simons (1969) has run extensive tests for tolerance to crown rust by using paired sprayed and rusted hill plots, each one $ft^2$. His experiments incorporated rust spreaders identical to those used to test for VR. Such experiments are economical and convenient, but they consistently underrate a line's true tolerance or HR. Nevertheless, Simons (1969) has made marked progress in tolerance-testing methodology and has found tolerance to be inherited as a quantitative character, with heritability (calculated from components of variance) ranging up to 48, 75, and 52%, respectively, when measured through yield, kernelweight, and kernel-density response. Under polygenic inheritance, tolerant x tolerant crosses might yield segregates with tolerance superior to that of the parents. As with cultivars with superior HR, a tolerant cultivar could be grown on its own, offer background resistance for use with VR genes, or be the recurrent parent in a multiline development program.

## Stem Rust

*Combining Seedling and Adult-Plant Resistance.*   Since race 94 (old race 6AFH) has virulence for known hexaploid sources of seedling resistance, it is pointless to discuss a stem-rust resistance program incorporating several breeding alternatives. Until new genes are discovered and characterized, few choices exist.

Adult-plant (but not seedling) resistance to race 94 is conditioned by resistance gene *pg*-11 (from C.I. 3034). Since stem rust affects spring-sown oats only after the heading stage in the upper Puccinia Path, it seems that adult-plant resistance should provide adequate protection. However, Rothman (1970) recently combined the adult-plant resistance of C.I. 3034 (gene *pg*-11) with the seedling (but not adult-plant) resistance to all prevalent races of Kyto, C.I. 8250 (gene *pg*-12). How long genes *pg*-11 and *pg*-12 will "last" may depend on how they are managed.

*pg*-11) with the seedling (but not adult-plant) resistance to all prevalent

*Partial Resistance.*   Vertical resistance to stem rust in *A. sterilis* plants in Israel is rare; instead, wild oat populations are protected by slow rusting and tolerance (Wahl, 1967), manifested as "partial resistance." In Iowa under conditions where *A. sativa* cultivars with common major stem-rust resistance genes were dead from race 31 (old 6AF), *A. sterilis* remained green until maturity (Wahl and Browning, *unpublished data*). Rust did not develop on such *A. sterilis* plants until after heading, and then only part of the leaf sheath appeared susceptible. The remaining plant parts, normally rusted in *A. sativa,* remained rust free; hence the term "partial resistance." This highly effective resistance probably is race-nonspecific and polygenic in inheritance. The extent to which it can be transferred to cultivated oats is unknown, but investigation seems worthwhile. It should make excellent background resistance for *pg*-11 or some other VR gene.

## Yellow Dwarf

With systemically BYDV-infected plants, distinctions between moderate resistance and tolerance are fuzzy and probably meaningless. Contributing is the fact that cultivars rated tolerant in one test may not seem so in another. This can vary with cultivar-virus isolate-vector interactions, growth stage of the plant at the time of inoculation, and environment (Bruehl, 1961; Bruehl et al., 1962; Endo and Brown, 1963).

Sources of resistance or, probably better, tolerance to the BYDV, have been reported from all ploidy levels. Oswald and Houston (1953) observed that Kanota and Bond were YD tolerant and that seven other oat lines were in an "intermediate" class. Differences in cultivar response were reported by several authors from the 1959 pandemic experience. Bruehl et al. (1962) found tolerance to natural YD epidemics in *A. strigosa, A. sativa,* and *A. byzantina.* From epidemics induced artificially with four BYDV strains, Endo and Brown (1964) found tolerance in 13 selections of *A. strigosa* and either tolerance or only moderate susceptibility in 13 hexaploid lines. *A. barbata* and *A. sterilis* were resistant, based on leaf symptoms, to YD in California (Zillinsky and Murphy, 1967). Excellent tolerance to two strains of the BYDV was found among 100 *A. sterilis* selections tested in Illinois (Jedlinski and Brown, *personal communication*). I also have observed much variability in *A. sterilis* for response to YD and much apparent resistance, in nurseries at Ames, Iowa, and in natural populations in Israel.

Brown and Poehlman (1962), in the 1959 pandemic in Missouri, found that the tolerance from C.I. 7448 was inherited quantitatively, with a heritability of 51%. Endo and Brown (1964) found the tolerance of Albion and Fulghum heritable and obtained selections homozygous for tolerance. Continuing the Illinois work, Brown and Jedlinski (*personal communication*) made diallel crosses among Albion and other YD-tolerant lines and found that heritability is high, that there is transgressive segregation for YD reaction, and that diverse genes for tolerance are involved. Using transgressive segregation, Brown and Jedlinski built up YD tolerance in pure lines that has stood well against YD for several years in several states. They plan to release seed of this valuable YD-tolerant material as registered germ plasm for breeding purposes.

## Use of Heterogeneity Against Crown and Stem Rust

Most homogeneity contributed to the problem of ephemeral rust resistance, and heterogeneity promises to contribute to the answer. Heterogeneity in oats can be achieved via inter- and intra-regional diversity. Gene deployment is the practical approach to the former. The latter can be achieved by intra- and inter-cultivar diversification, but intra-cultivar diversification will have the greater reducing effect on epidemics. Intracultivar diversity can be achieved through composite crosses, Jensen's (1970) new diallel, selective mating system, or by use of multiline cultivars. Assuming equal diversity, these should have equal effect on disease development, though the first two offer the advantage of diversity for traits other than disease reaction. Since breeding multilines only involves a simple, straightforward backcrossing program, and since a uniform agronomic background will aid in assessing the true contribution of heterogeneity for disease reaction (which then should be applicable to other breeding systems), only multilines will be considered herein.

## Multiline Cultivars for Crown-Rust Control

Multiline cultivars are mechanical mixtures of isolines, each of which contains a different VR gene. The mixture of vertically resistant isolines gives a cultivated unit, a multiline cultivar, which behaves as if it were a pure-line cultivar with HR. Use of multiline cultivars as a means of disease control was reviewed recently by Browning and Frey (1969).

Using Van der Plank's (1968) concepts, they analyzed the action of multilines as follows: Let the pathogen increase from initial inoculum $x_o$ at rate $r$ in time $t$ and result in $x$ amount of disease. A cultivar with VR is selectively resistant to the race population: it reduced $x_o$. But with $r$ the same (and usually so high as not to be limiting), $x$ may be large, and destruction great, unless $x_o$ is very low for all races.

A cultivar with HR, by definition resistant to all races, reduces $r$, not $x_o$. With $r$ small, the epidemic is reduced to the point where the crop matures with small $x$ and little measurable damage.

Assume that a field is planted to a 10-component multiline and that a viable spore of a race virulent on only one of the 10 components lands on a plant. The probability is only 0.1 that the spore will land on a susceptible plant; thus, $x_o$ is reduced, as is characteristic of VR. If the spore lands on a susceptible plant and the fungus invades the tissue and sporulates, the progeny have a probability of 0.1 that an adjacent plant will be compatible; thus, $r$ is reduced also, as is characteristic of HR.

Multiline cultivars therefore share characteristics in common with both vertically and horizontally resistant cultivars and, also, with tolerant cultivars, and they result in population buffering called "population resistance."

Browning and Frey (1969) listed the following advantages of multiline cultivars: "(a) They provide a mechanism to synthesize instant, well-buffered, horizontally resistant cultivars which, unlike pure-line cultivars, can utilize without difficulty several resistance genes at the same locus or resistance genes which happen to be linked in the repulsion phase. (b) They should extend indefinitely the useful life of a given resistance gene and enable a resistance breeding program eventually to be reduced in size while the breeder carries on parallel improvement in the recurrent parent. (c) Removing the rust hazard should stabilize the cultivars used and enable farmers to optimize production for a given multiline cultivar on a given farm. (d) They offer a means whereby a center of variety development can distribute host cultivars far and wide without risk of homogenizing the pathogen population on a global scale."

Oat multilines have been developed in Iowa for production in the high crown-rust-hazard area of Iowa and contiguous states, and six multiline cultivars, containing from 7 to 11 components, have been released in two maturity classes, early and midseason. Implementation of this multiline-development program was described by Frey, Browning, and Grindeland (1971).

Extensive tests of isoline blends in 50 x 50 ft plots (large plots were necessary as for pure-line HR), in which the spore yield of the pathogen (estimated by daily trapping of spores outside the plots) and grain yield of the host were measures of the effect of multilineness under epidemic conditions, supported "the multiline theory that mixtures of near-isogenic lines effectively buffer the host population against the rust population" (Cournoyer, 1970). Even host populations with 10% resistance had a marked effect on yield of the pathogen population, and 40% resistance has tended to give population stability.

In 1970 at Ames, Iowa, commercially available cultivars Multiline E-68, Multiline M-68, and Jaycee were each sown in 10-acre fields (Browning, *unpublished data*). The three fields were exposed to inoculum of four virulent crown-rust races from an early-sown spreader range. Paired fungicide-sprayed and nonsprayed plots in eight replications were staked out in the fields for harvest at three distances from the spreader. Jaycee, a susceptible cultivar, served as a check of rust development. Crown rust reduced the yield of Jaycee by 30%. In contrast, the multilines were not affected measurably by rust, even though this was a test more severe than a multiline is likely to experience in a farmer's field.

Further indicating the possible merits of multilines in disease control, Ayanru (1970) found that heterogeneous oat populations helped limit epidemic development of *Helminthosporium victoriae*, a soil-borne pathogen.

## Gene Deployment for Crown- and Stem-Rust Control

Intracultivar diversity via multiline cultivars should control rust and conserve genes within a given state. Unfortunately, individual states within the Puccinia Path are not units for control of continental pathogens; the entire Puccinia Path constitutes unity. "Recognizing (a) that the Puccinia Path is now a single, functional epidemiological unit for cereal rusts, (b) that it must be maintained if the rust fungi are to remain serious threats, (c) that unity of the Puccinia Path will continue inviolate as long as the rust fungi can recycle between compatible hosts growing in the summer and winter areas, and (d) that forcing the fungi to recycle between incompatible hosts will break the cycle, suggests the logical sequel to barberry and buckthorn eradication; namely, inter-regional diversification by deployment of different resistance genes north and south to break the unity of the Puccinia Path" (Browning et al., 1969).

Obviously, only effective VR genes not currently being utilized are subject to deployment. Deployment north and south is indicated against those continental pathogens, like *P. coronata,* for which an adequate number of resistance genes is available. (Some 50 crown-rust resistance genes have been assigned *Pc* numbers — Simons, Zillinsky, and Jensen, 1966; Simons, *personal communication*). Browning et al. (1969) proposed that the Puccinia Path be divided into two or three zones for deployment of genes with VR for crown rust. Zone 1 would include the rust-overwintering area near the Gulf Coast. Zone 3 would coincide with the area where late spring oats are adapted—the Prairie Provinces of Canada and contiguous areas of extreme north-central U.S.A. Zone 2 would lie in between. Sets of VR genes tested and determined to be of approximately equal value should be deployed by agreement for use in each zone. How they were used would be the prerogative of workers at each experiment station. However they were used, spores that increased successfully in one zone should be avirulent on oats in another zone.

For stem rust, with an insufficiency of VR genes, it becomes even more imperative to use the genes available while conserving them via deployment. If the results of combining genes *pg*-11 and *pg*-12 continue promising, they should be deployed in the North. These and another strong VR gene, if it becomes available, should be deployed for use in Zone 3, or at most in Zone 3 and the northern portion of Zone 2. Hopefully, one or more additional VR genes will be found also for Zone 1. This would place the genes where they are needed most and manage them so they should last longest.

# The Coordinated Continental Control Program

The oat cultivar of tomorrow must be agronomically well adapted to the area—state and even field—of production. It must have high yield, high protein percentage, good kernel quality, desired maturity (usually early), and standing ability. It should be protected with fungicides during development so diseases do not mask agronomic improvement, but adequate parallel tests must be run to insure resistance to diseases (root rot, septoria black stem, and so forth) caused by residual pathogens. This agronomically improved cultivar, then, becomes the focus for disease-resistance improvement, some of which could be made concurrently with agronomic improvement.

Tolerance and HR cover all races, gain in effectiveness from extensive use, and are the indicated control measures for the cereal rusts. But these may be only useful working concepts; at any rate, combining desirable agronomic characteristics with polygenic HR and tolerance to three major diseases would be difficult to attain. Considering breeding realities, therefore, my proposal calls for using HR and tolerance for YD and stem rust, where adequate VR is not available, and then for crown rust, to manage VR genes so as to achieve the effect expected from pure-line HR.

## Yellow Dwarf

Using registered germ plasm, the Illinois tolerance to YD can be incorporated into agronomically well-adapted cultivars for commercial production in low rust-hazard areas (such as eastern Washington) or serve as parents for additional crosses to incorporate rust resistance. Being polygenic, the Illinois tolerance offers promise of stability against the BYDV. Contributing to this is the fact that BYDV is obligately part of a host-host-vector-pathogen system, and the non-oat host should tend to stabilize the BYDV on oats. Therefore, control of YD with extensive use of the Illinois tolerance in the Puccinia Path should be safe and is indicated.

## Stem Rust

If the partial resistance of *A. sterilis* proves polygenic, it should be safe to use extensively. It, then, should be added to the locally adapted YD-tolerant cultivars for use in the Puccinia Path. If the resistance obtained from combining genes *pg*-11 and *pg*-12 continues promising, these genes should be deployed for use only in Zone 3, or in Zone 3 plus the northern part of Zone 2, where the stem-rust hazard is greater. Additional VR genes, when available, can be deployed in the South. Prospects are good for permanence of the polygenic partial resistance and for the VR genes if they are deployed.

## Crown Rust

Of the three major oat diseases, there is a sufficiency of VR genes only for crown rust. These should be tested and groups of equal value deployed for use in the different zones. Locally adapted cultivars with the Illinois tolerance to YD and partial resistance to stem rust would constitute recurrent parents to which VR genes for crown-rust resistance would be added as local workers saw fit, probably via backcrossing. Undergirded with gene deployment, pyramiding two or three VR genes into a single pure line should offer protection with promise for conservation and stability. Incorporating the deployed genes singly into the recurrent parent and blending the resultant isolines into a multiline, however, has many advantages and is recommended.

## Summary and Concluding Remarks

Previous "coordinated" attempts at controlling oat rusts in the Puccinia Path consisted of widespread use of the same superior VR genes in an attempt to "shut out" the fungi entirely. These genes did their job; there was no failure of resistance per se. But there was a failure to manage the genes so their effect could endure. Probably today's workers would have done the same. But in light of this experience, we must apply current resistance theory and use available genes in a control program that can be continental in scope and still give lasting resistance.

The suggested program consists first of the development of pure-line cultivars with polygenic HR or tolerance to YD to which HR or partial resistance to stem rust would be added. Such HR and tolerance should be more stable than high VR because they should provide no mechanism for new races to prevail over established ones. Of course, only time and research can tell if HR and tolerance will provide the requisite panracial protection from YD and stem rust. But they are promising, and to utilize them widely over time will be to conduct the research.

Meanwhile, multiline cultivars offer "instant" and predictable HR to crown rust. They offer a means of utilizing all types of resistance in a coordinated, integrated control program. Indeed, "HR and VR supplement one another and are miscible in all proportions" (Van der Plank, 1968). Therefore, multiline cultivars with heterogenic HR to crown rust are proposed. Since a multiline cultivar contains individual lines, each with a different crown rust VR gene, the cultivated unit (that is, the heterogeneous population of plants) has quantitative population resistance to the rust population similar to that possessed by a pure-line cultivar with tolerance or HR. If the VR genes in the multiline cultivar are incorporated on a background of HR to crown rust, the combined effect of the polygenic HR (in the recurrent parent) and of the heterogenic HR (from the blended isolines) should reduce still more the rate of increase and final amount of crown rust. This should constitute population resistance at its best—a situation in which the fungus population cannot increase and damage the well-buffered host population that is protected by VR, HR, tolerance, and combinations of them.

Workers can decide independently whether to use HR or tolerance in their programs. But in the Puccinia Path, workers' agreement should delimit the area of VR genes. Thus, this program must be "coordinated." If only a few VR genes are known, as is the case for oat stem rust, they should be used where disease hazard is greatest. With stem rust, this is in the North. Failure to delimit their use is to repeat history, with loss of benefit to all. If many VR genes are available, as for crown rust, then groups of equal size and value should be deployed for use in each zone. In the philosophy behind gene deployment, the decision *not to use* a given gene in one area is as important as the decision *to use* it in another. Full cordination between workers North and South is of fundamental importance to the program.

If the work load of protecting oats in the Puccinia Path from major diseases is shared among pure-line polygenic horizontal resistance, multiline heterogenic horizontal resistance, and tolerance, and if these are undergirded with deployment of VR genes, the resultant level of population buggering intra- and inter-regionally should keep the major pathogens of oats at a level of little or no economic importance.

## Acknowledgments

I acknowledge, with gratitude, many constructive suggestions on the manuscript by Drs. Henryk Jedlinski, Marr D. Simons, and K. J. Frey.

# References

Asai, G. N. 1960. Intra- and inter-regional movement of uredospores of black stem rust in the upper Mississippi River Valley. Phytopathology 50:535-541.

Ayanru, D. K. G. 1970. Development of *Helminthosporium victoriae* in blends of resistant and susceptible isogenic oat lines. Phytopathology 60:1282. (Abstr.)

Brown, G. E., and J. M. Poehlman. 1962. Heritability of resistance to barley yellow dwarf virus in oats. Crop Sci. 2:259-262.

Browning, J. A., and K. J. Frey. 1969. Multiline cultivars as a means of disease control. Ann. Rev. Phytopathol. 7:355-382.

————, M. D. Simons, K. J. Frey, and H. C. Murphy. 1969. Regional deployment for conservation of oat crown-rust resistance genes, p. 49-56. *In* J. A. Browning [ ed.] Disease consequences of intensive and extensive culture of field crops. Iowa Agr. and Home Econ. Exp. Sta. Spl. Rep. 64. 56 p.

Bruehl, G. W. 1961. Barley yellow dwarf. Am. Phytopathol. Soc. Monograph 1. St. Paul, Minn. 52 p.

————, V. D. Damsteegt, H. M. Austenson, and P. C. Crandall. 1962. Resistance to yellow dwarf of oats in Washington. Plant Dis. Reporter 46:579-582.

Caldwell, R. M., J. F. Schafer, L. E. Compton, and F. L. Patterson. 1958. Tolerance to cereal leaf rusts. Science 128:714-715.

Coffman, F. A. [ ed.] 1961. Oats and oat improvement. Amer. Soc. Agron. Monograph 8. Madison, Wisc. 650 p.

Cournoyer, B. M. 1970. Crown rust epiphytology with emphasis on the quantity and periodicity of spore dispersal from heterogeneous oat cultivar-rust race populations. Ph.D. Thesis. Iowa State Univ. (Diss. Abstr. 31:3104-B).

Crowder, L. V., J. Lotero, J. Fransen, and C. F. Krull. 1967. Oat forage production in the cool tropics as represented by Colombia. Agron. J. 59:80-82.

Endo, R. M., and C. M. Brown. 1963. Effects of barley yellow dwarf virus on yield of oats as influenced by variety, virus strain, and developmental stage of plants at inoculation. Phytopathology 53:965-968.

———— ————. 1964. Barley yellow dwarf virus resistance in oats. Crop Sci. 4:279-283.

Frey, K. J., J. A. Browning, and R. L. Grindeland. 1971. Implementation of oat multiline cultivar breeding, IAEA PL412/17, p. 159-169. *In* Panel on mutation breeding for disease resistance. Int. Atomic Energy Agency. IAEA STI/PUB/271. Vienna. 249 p.

Gill, C. C. 1969. Annual variation in strains of barley yellow dwarf virus in Manitoba, and the occurrence of greenbug-specific isolates. Can. J. Bot. 47:1277-1283.

Hamilton, L. M., and E. C. Stakman. 1967. Time of stem rust appearance on wheat in the western Mississippi basin in relation to the development of epidemics from 1921 to 1962. Phytopathology 57:609-614.

Heagle, A. S., and M. B. Moore. 1970. Some effects of moderate adult resistance to crown rust of oats. Phytopathology 60:461-466.

Jedlinski, H., and C. M. Brown. 1965. Cross protection and mutual exclusion by three strains of barley yellow dwarf virus in *Avena sativa* L. Virology 26:613-621.

Jensen, N. F. 1970. A diallel selective mating system for cereal breeding. Crop Sci. 10:629-635.

Johnson, T. 1961. Man-guided evolution in plant rusts. Science 133:357-362.

Leonard, K. J. 1969. Selection in heterogeneous populations of *Puccinia graminis* f. sp. *avenae*. Phytopathology 59:1851-1857.

Martens, J. W., R. I. H. McKenzie, and G. J. Green. 1970. Gene-for-gene relationships in the *Avena:Puccinia graminis* host-parasite system in Canada. Can. J. Bot. 48:969-975.

Michel, L. J., and M. D. Simons. 1966. Pathogenicity of isolates of oat crown rust collected in the USA, 1961-1965. Plant Dis. Reporter 50:935-938.

Murphy, H. C. 1965. Protection of oats and other cereal crops during production, p. 99-113. *In* G. W. Irving, Jr. and S. R. Hoover [ed.] Food quality. Effects of production practices and processing. Amer. Assoc. Adv. Sci. Publ. 77. 298 p.

———, and F. A. Coffman. 1961. Genetics of disease resistance, p. 207-226. *In* F. A. Coffman [ed.] Oats and oat improvement. Amer. Soc. Agron. Monograph 8. Madison, Wisc. 650 p.

———, K. Sadanaga, F. J. Zillinsky, E. E. Terrell, and R. T. Smith. 1968. *Avena magna*: an important new tetraploid species of oats. Science 159:103-104.

Oswald, J. W., and B. R. Houston. 1953. The yellow-dwarf virus disease of cereal crops. Phytopathology 43:128-136.

Rochow, W. F. 1961. The barley yellow dwarf virus disease of small grains. Adv. Agron. 13:217-248.

———. 1969. Biological properties of four isolates of barley yellow dwarf virus. Phytopathology 59:1580-1589.

———, and H. Jedlinski. 1970. Variants of barley yellow dwarf virus collected in New York and Illinois. Phytopathology 60:1030-1035.

Rothman, P. G. 1970. Combining seedling and adult resistance to *Puccinia graminis* f. sp. *avenae*. Phytopathology 60:1311. (Abstr.)

Rowell, J. B., and R. W. Romig. 1966. Detection of urediospores of wheat rusts in spring rains. Phytopathology 56:807-811.

Simons, M. D. 1969. Heritability of crown rust tolerance in oats. Phytopathology 59:1329-1333.

———. 1970. Crown rust of oats and grasses. Amer. Phytopathol. Soc. Monograph 5. St. Paul, Minn. 47 p.

———, and H. C. Murphy. 1961. Oat diseases, p. 330-390. *In* F. A. Coffman [ed.] Oats and oat improvement. Amer. Soc. Agron. Monograph 8. Madison, Wisc. 650 p.

———,———. 1968. Oat diseases and their control. U.S. Dept. Agr., Agr. Handbook 343. 15 p.

———, F. J. Zillinsky, and N. F. Jensen. 1966. A standardized system of nomenclature for genes governing characters of oats. U.S. Dept. Agr., Agr. Res. Serv. Publ. ARS 34-85. 22 p.

Stewart, D. M., and B. J. Roberts. 1970. Identifying races of *Puccinia graminis* f. sp. *avenae*—a modified international system. U.S. Dept. Agr. Tech. Bul. 1416. 23 p.

———, and P. G. Rothman. 1971. Distribution, prevalence, and new physiologic races of *Puccinia graminis* in the USA in 1969. Plant Dis. Reporter 55:187-191.

Van der Plank, J. E. 1968. Disease resistance in plants. Academic Press, New York. 206 p.

Wahl, I. 1967. The screening of collections of wild oats for resistance and tolerance to oat crown rust and stem rust fungi. Final Rep., U.S. Dept. Agr. PL 480 Project A10-CR-20 (FG-Is-138).

Wallin, J. R., D. Peters, and L. C. Johnson. 1967. Low-level jet winds, early cereal aphid and barley yellow dwarf detection in Iowa. Plant Dis. Reporter 51:527-530.

Zillinsky, F. J., and H. C. Murphy. 1967. Wild oat species as sources of disease resistance for the improvement of cultivated oats. Plant Dis. Reporter 51:391-395.

# 12 Cotton
## L. S. Bird[1]

## Introduction

Cotton, *Gossypium* spp., is one of the marvels of the plant kingdom in providing the needs of mankind. It produces the basic raw materials cellulose, protein, and oil in quantity and quality surpassed by few plant species. The cellulose is pure and in the form of a natural fiber.

Pigment glands in cotton tissues accumulate gossypol, which is in cottonseed meal. Gossypol is toxic to nonruminant animals, and consequently meal from glanded seed has limitations as a source of high quality protein. Two genes, $gl_2$ and $gl_3$, make it possible to breed glandless cottons (McMichael, 1960). Meal from glandless seed is naturally free of gossypol, which makes it a new prime source of protein for humans. Glandless cottons, native to tropical and subtropical climates where population centers abound, may become the key to successfully clothing and feeding large masses of people.

1. Professor of Plant Pathology and member of the Genetics Faculty, Department of Plant Sciences, The Texas Agricultural Experiment Station, Texas A & M University; and Collaborator, Crops Research Division, ARS, U.S. Department of Agriculture, College Station, Texas.

As cotton becomes more important to man, consistent means of controlling its diseases must be found. The critical measure of control must be the degree to which economic loss is prevented in field grown cotton. Cotton is subject to economic loss from three to seven major and minor diseases, depending on where it is grown. Consequently, varieties with resistance to one or two diseases are subject to economic loss wherever they are grown. For this reason, genetic improvement should be for varieties with resistance and escape from all potential major and minor diseases.

## The Cotton Plant

Cotton is naturally a perennial. The cultivated *G. arboreum* L. and *G. herbaceum* L. (Old World group, $n=13$), and *G. hirsutum* and *G. barbadense* L. (New World group, $n=26$) species are largely managed as annual crops. Determinant and indeterminant growth types are recognized (Brown and Ware, 1958). The main stem and fruiting branches of determinant types remain relatively short. Determinant early types are becoming more important in new management programs with high plant population per acre for reducing production cost.

Cotton bolls are referred to as open, storm-resistant, or storm-proof. This classification is based on the degree to which locks of seed cotton remain compact and stay in the boll (Brown and Ware, 1958).

Cotton is naturally self-pollinated, and outcrossing is in proportion to the level of insect activity. Self-pollination may be assured by preventing the corolla from opening by sealing it with paper clips, small-gauge wire, cellulose acetate, or other means the day before opening (Brown and Ware, 1958).

Emasculation for controlled crossing is accomplished by removal of the corolla and picking off the stamens from the staminal column or by simultaneous removal of the corolla and staminal column (Brown and Ware, 1958). Pollinations are made during the morning on flowers emasculated the previous afternoon.

## Concepts and Breeding Procedures

Most genetic studies identifying genes with major effect offer evidence that the major gene is dependent on minor or modifying genes conditioning phenotypic expression. Minor genes, in these discussions, are those with small effect; but the presence of several can be readily measured. Their effect is additive with major genes. The presence of modifier genes is difficult to measure in the absence of the major genes being modified. Modification of a major gene may be favorable or unfavorable. Genes of small effect may be the minor ones for some traits and modifiers for others, and vice versa. Thus, favorable backgrounds for major disease resistance genes are those with minor-modifier genes which are additive and provide favorable modifications.

Economic traits in cotton are governed by major-minor-modifier gene complexes which can only be dealt with by using quantitative breeding procedures. Combining ability studies must be used to find major, minor, and modifier gene combinations that are complementary. The total improvement of the plant must be considered in these evaluations.

Isolates of pathogens varying in levels of pathogenicity must be sought and welcomed to an advanced genetic improvement program. Several races of a pathogen or, if none are known, isolates from different environments should be obtained and used in artificial inoculations. Progenies should be planted in duplicate disease nurseries located in different environments. The specific intent is to deal only with material performing average to above average in all environments. Utilizing representative pathogen and environmental variability is the key to establishing selection pressure for identifying the best gene combinations. It is in this manner that host-pathogen and host-pathogen-environmental interactions are reduced to nil and hereditary disease control is achieved.

The term "resistance" is used here to include mechanisms which, once the pathogen enters the host, prevent colonization, effective growth, or reproduction. The term "immunity" is used when no phenotypic symptoms occur, even after inoculation with virulent isolates of pathogens. The term "escape" is used to include all mechanisms, whether they result in reduced penetration, protection, klenducity, or alterations of the environment creating an unfavorable host-pathogen interaction. Three ideas govern these definitions: first, the actual mechanism of disease control has not been identified for most diseases; second, limited restrictions may cause one to overlook an effective mechanism of reducing economic loss; and third, it is highly possible that more than one gene-conditioned mechanism is available for controlling a disease. Levels of selection pressure, evaluation techniques, and breeding procedures must be such that all hereditary mechanisms for disease control will be combined into single strains.

One must recognize that favorable varieties of today will be poor ones tomorrow. Thus, backcrossing to current favorable varieties can be an obstacle to developing new highly improved disease-resistant cottons. Backcrossing is placed in proper perspective when it is used to transfer genes from strains and wild types into balanced breeding stocks.

Quantitative breeding procedures with multiple parents should be used. The parents chosen should represent at least two sources of resistance for each disease being considered. Several years spent shaping parental material can pay big dividends in the end. Once the parental pool is established, breeding procedures along the lines of "delayed" convergent improvement and cumulative selection should be followed. Handling of major genes is not difficult, but the challenge lies in manipulation of minor-modifier gene complexes.

# Diseases of Cotton

## Bacterial Blight

Resistance to *Xanthomonas malvacearum* (E. F. Sm.) Dowson is one of the best understood traits in cotton. Thirteen independent genes of measurable effect have been identified (Knight, 1956; Innes, 1965). Other genes and sources of resistance have been identified (Brinkerhoff, 1970, a review; Blank and Hunter, 1955), but homology tests with Knight's "B" genes have not been completed.

All investigators have emphasized the importance of minor and modifying genes combined with major genes for obtaining highly resistant or immune strains. In *G. barbadense,* Sakel types, the gene combinations $B_2B_6$, $B_4B_6$, $B_1B_{9K}$, and $B_6B_{11}$ gave the best levels of resistance (Innes, 1965). In *G. hirsutum,* the $B_4B_{Sm}$ combination gave near immunity in the presence of race mixtures of the pathogen (El-Zik and Bird, 1970). Genes $B_2$, $B_3$, and $B_7$ are ineffective with $B_{Sm}$, but the combinations of $B_{Sm}$ with $B_2B_3$, $B_2B_3B_7$, and $B_2B_6$ give near immunity. The $B_{9L}B_{10L}$ combination in Reba varieties gives good resistance (Carvalho, 1969).

*X. malvacearum* is quite variable, and numerous races can be identified as long as adequate host differentials are available (Hunter, Brinkerhoff, and Bird, 1968). The race problem became serious in United States programs when strains with the $B_7$ gene were attacked by a new race. Weak or vertical resistance and a narrow representation of pathogen variability were the two factors causing failure of the early work. There has been no problem with new races since we began using several known races and more of the "B" genes, especially $B_2$, $B_3$, $B_4$, and $B_{6_m}$ in our programs. Several races of the pathogen provide selection pressure for identifying gene combinations that give immunity. Programs the world over are reporting success in developing desirable types highly resistant to or immune from *X. malvacearum*. Concepts and procedures for developing bacterial blight-resistant cottons may become classic as examples to follow in obtaining hereditary disease control.

## Seedling Disease

Seedling disease may be caused by *Rhizoctonia solani* Kuehn, *Pythium* spp., *Thielaviopsis basicola* (Berk. and Br.) Ferr., *Colletotrichum gossypii* South., *Fusarium* spp., *Ascochyta gossypii* Syd., and *Meloidogyne incognita* (Kofoid and White) Chitwood. Any one of any possible combinations of primary pathogens cause seed rot, pre- and postemergence damping-off, stunting, and root damage to seedlings. A broad spectrum approach considering seed and seedling characteristics must be used to achieve effective resistance and escape from seedling disease.

When cotton matures in dry areas with low humidity, seed are unconditioned. These are characterized by a reduced rate and percentage of germination at temperatures below 20 C. Exposure to moisture and heat gives conditioned seed, which have increased rates and percentages of germination. Continued exposure to moisture and heat gives peak conditioning (as occurs in humid rainfall areas), and thereafter cottonseed begin to deteriorate. At lower temperatures deteriorated seed have reduced rates and percentage germination (Bird and Reyes, 1967). With conditioning and deterioration, seed lose a natural resistance to mold growth and require progressively higher temperatures for germination. Inherent cold tolerance is lost when seed deteriorate. Progressive deterioration is associated with increased levels of pre- and postemergence damping-off, and seedlings are more apt to be attacked by moderate pathogens, although undeteriorated seed and their seedlings are still susceptible to primary pathogens (Bollenbacher and Fulton, 1959).

Cold tolerance has been suggested as an aid in getting stands under early season conditions. Unconditioned seed of cold tolerant strains can be used for controlling seedling disease in plantings made at temperatures from 15 to 20 C (Bird and Reyes, 1967).

Tolerance to several primary pathogens was found in strains of

Arkot, *G. thurberi, G. arboreum,* and Yugoslav lines. No strain showed high tolerance to all pathogens, and tolerance was independent of low temperature performance (Fulton, Waddle, and Bollenbacher, 1962). Tolerance to *R. solani* was found in Acala 1517 types (Hefner, 1968). Tolerance to *T. basicola* was found in some Acala 4-42 types (Garber, 1966). The ability to give a stand of cotton is inherited, with dominance and three to four effective factors involved.

Strains and varieties differ in rate of seed deterioration. Absence of mold growth on seed under wet-cool conditions indicates resistance to deterioration (Bird and Presley, 1965). Resistance to mold growth is inherited and is conditioned by three effective factors with dominance being incomplete. Sources of cold tolerance are available in a number of strains (Fulton et al., 1962; Bird and Presley, 1965). The use of hard seed coat to prevent seed deterioration appears to be impractical. Seed must have the ability to resist deterioration after treatment with hot water to nullify the hard coat. Strains showing resistance to seedling pathogens became susceptible after hot water treatment to eliminate seed dormancy (Fulton et al., 1962).

The danger of having vigor at the expense of tissue hardening should be emphasized. In our own research, we repeatedly find that strains giving rapid germination at reduced temperatures and rapid emergence in field tests have greater losses from postemergence damping-off. Thus, one should not breed for seedling vigor in the sense of achieving quick emergence and rapid hypocotyl elongation.

Genetic variability for resistance to seed deterioration, cold tolerance, and tolerance to primary pathogens provide the characters needed for developing cottons which resist and escape seedling disease. Laboratory-greenhouse procedures for simultaneously selecting resistance to seed deterioration and cold tolerance have been developed (Bird and Presley, 1965). Screening and identification of strains having tolerance to specific pathogens were done in artificially infested soils (Fulton et al., 1962; Hefner, 1968). It would not be difficult to combine the various methods. One would need merely to choose the proper parents, create hybrid populations, and submit them to the selection pressure of the combined method.

## Fusarium Wilt and/Wilt-Nematode Complexes

Resistance to Fusarium wilt, wilt-nematode complex and nematode resistance — *Fusarium oxysporum* f. sp. *vasinfectum* (Atk.) Snyd. & Hans., *Meloidogyne incognita* (Kofoid & White) Chitwood, *Rotylenchulus reniformis* Linford and Oliveira, and *Belonolaimus gracilis* Steiner — involve individual disease and disease complexes. Practical resistance for control of Fusarium wilt must also include resistance to nematodes. Resistance reduces losses from nematodes as well as losses from the wilt-nematode complex.

Genes conditioning resistance to wilt and nematodes are identified by strain, variety, or species in which they are found. In Egyptian cotton (*G. barbadense*) resistance is conditioned by one dominant and one or more minor genes. A dominant gene was found in *G. herbaceum* and two complementary dominant genes and a third one having an inhibitory effect in *G. arboreum* (Kelkar et al., 1947). Cook 307 carries a major dominant with modifying genes for wilt resistance and additional genes for root-knot nematode resistance. Seabrook, a variety of Sea Island, has two dominant genes which are additive in conferring a high level of wilt resistance (Smith and Dick, 1960). In using Delfos 425 and Coker 100 Ga with Half and Half in the presence of wilt and the reniform nematode it was concluded that resistance in Upland cotton is conditioned by two to three genes. Root-knot resistance in Clevewilt-6 is a quantitative character with relatively few genes involved (Jones and Birchfield, 1967). A high level of root-knot resistance is found in *G. barbadense* var. *darwinii*. At least two recessive genes are present in this source of resistance (Wiles, 1957; Turcotte et al., 1963).

The reniform nematode causes injury to resistant and susceptible varieties; however, increases in incidence of wilt are caused only in wilt-susceptible varieties (Jones et al., 1959). The wilt fungus is known to be variable in pathogenicity. The root-knot nematode is variable, with *M. incognita* var. *acrita* generally causing more injury than *M. incognita* (Wiles, 1957).

The first superior wilt-nematode resistance variety was Auburn 56, developed from the cross of Cook 307 x Coker 100 wilt. Bayou strains, developed from the cross Deltapine 15 x Clevewilt-6, are superior to Auburn 56, especially in preventing build-up of nematode populations during the season (Jones and Birchfield, 1967). Auburn 623 RNR is the newest wilt-root-knot resistant type and was developed from the cross of Clevewilt-6 x La Mexico wild. Evidence suggests that Auburn 623 RNR will prevent a build-up or even reduce nematode populations during the season (Shepherd, release announcement Auburn University and U.S.D.A. Crops Research Service, 1970).

The response of 50 strains to the Fusarium root-knot and the Fusarium reniform complexes in Louisiana revealed that Mexico wild, Sikes 10-38, Auburn 56, and Clevewilt-6 were superior to other types, including Coker 100 wilt and Cook 307, for Fusarium root-knot resistance. Two of these went into the development of Auburn 623 RNR. Coastland, Delfos 425, and La. Hybrid 33 x 14 gave the best resistance to wilt when grown in soil infested with the wilt-reniform complex. Whether or not hybrids among these can give superior resistance to reniform nematodes remains to be seen. There is need for either discovering or developing new sources of resistance to this nematode.

One can anticipate what will come from crosses involving Auburn 56 (Cook 307 x Coker 100 wilt), Bayou (Deltapine 15 x Clevewilt-6), Auburn 623 RNR (Clevewilt-6 x La. Mexico wild) and *G. barbadense* var *darwinii*. Immunity from the Fusarium wilt root-knot nematode complexes should be possible with such genetic material and modern breeding methodology.

## Verticillium Wilt

Verticillium wilt (*Verticillium albo-atrum* Reinke and Berth., MS) is controlled by inherent means by two approaches. One is the development of strains capable of giving high yields in the presence of the pathogen. The other is the development of types which prevent invasion and symptom development, and which yield well in the presence or absence of the pathogen.

Little information is available on the number of genes and the action of genes conditioning *Verticillium* resistance in cotton. In a study with Upland varieties and the OK141-5 resistant strain (a selection from New Mexico 8060-3) the statistics suggest that resistance is recessive (Brinkerhoff et al., 1970). In crosses involving resistance of 250 x Coquette and an Acala selection with Hopicala, Acala 1517C and Lankart 57 resistance was recessive and two genes were detected (Roberts, 1969). Resistance was dominant and conditioned by one effective factor in studies involving A9519, a strain with the Hartsville source of resistance (Barrow, 1970). Performance of the $F_1$ of resistant KP28 x B181 with susceptible material indicates the presence of a dominant gene in this material. Crosses involving Seabrook Sea Island (*G. barbadense*) suggested the presence of one dominant gene in Seabrook (Wilhelm et al., 1970). Thus, it appears that dominant and recessive genes for resistance are available. Homology tests must be conducted before the actual number can be estimated.

Host response to *V. albo-atrum* can be altered drastically by a number of variables. Levels of nitrogen and potassium and sources of nitrogen (Ranney, 1962) influence resistance. In fact, a balanced N-K nutrition is extremely important. The temperature, especially that of the soil, has a marked influence on resistance. Adaptation or lack of adaptation of a strain to local environments influences resistance

(Fisher, 1968). Thick stands enhance resistance in comparison with thin stands in the row. The pathogen is also quite variable for both cultural characteristics and pathogenicity. Loss of resistance resulting from an unfavorable host-variable interaction identifies, in many cases, a weak source of genetic resistance.

Greenhouse root-dip, stem-puncture, and root-wounding methods of artificial inoculation have been developed. A rapid stem-puncture technique is effective for both greenhouse and field inoculations (Bugbee and Presley, 1967). Evaluation in naturally infested field nurseries has been the primary means of breeding resistant types. This may be the best approach, in view of the many variables influencing resistance, especially if nurseries at different locations are used.

It must be recognized that weak genetic resistance only complicates matters when host response is influenced by so many variables. For this reason sources of strong genetic resistance which are stable across many natural environments must be identified and used in developing superior varieties.

Many of the early Acala types were resistant in Arizona and New Mexico but susceptible to the extent of being completely defoliated in Mississippi. Cook 307 types are resistant in Mississippi (Wiles, 1953) but apparently susceptible in Arizona (Presley, 1950). Acala 4-42 was irregular in its behavior in the Western uniform Verticillium wilt test (Chilton and Blank, 1955). These are examples of unstable types which should not be relied on too heavily in breeding programs.

Many sources of resistance have been listed (Cotton, 1965; Fisher, 1968; Fulton and Waddle, 1957; Presley, 1950; Wiles, 1953). It seems clear that Hartsville is the best source of genetic resistance in G. hirsutum that is stable across environments (Cotton, 1965; Fisher, 1968; Presley, 1950; Wiles, 1953). KP28 x B181 Uganda material is preferred as a source of resistance over that of Tanguis. This same KP material performed well in New Mexico and in Arizona. The Arkansas D x K3131 strain performs well in Arkansas, New Mexico, and Arizona. The consistent high resistance of Seabrook Sea Island and other G. barbadense types across environments is well established.

Early varieties for Verticillium wilt resistance were released by the U.S. Cotton Field Station in New Mexico (Sherbakoff, 1949, a review). More recent releases came from crosses made to obtain gene recombinations for higher tolerance. These include Acala 1517V from the cross Acala x Coquette (Cotton, 1965); Hopicala from the cross Acala x Hopi x Acala (Fisher, 1968), developed in California, evaluated in New Mexico as Hopi Acala 4447, and released by Arizona; and Paymaster 909 from the cross Paymaster 101 with a

California Acala (probably Hopicala) type (personal discussion with Quentin Adams, ACCO Seed Farm, Aiken, Texas).

Key indications of progress come from the fact that very susceptible materials have been eliminated from genetic improvement programs (Fisher, 1968), and that selections from complex crosses are giving higher levels of resistance. Crosses involving *G. barbadense*, Acala 49, Acala C108, and KP Uganda gave some of the most resistant material obtained in New Mexico (Cotton, 1965). Paymaster 266, another new highly resistant source, came from crossing Deltapine 554 with Arizona Acala 6024, and then crossing this material with Paymaster 101 x 105 hybrid material carrying the $B_4$ gene for blight resistance (personal discussion with Delbert Hess and Quentin Adams, ACCO Seed Farm, Aiken, Texas). Arizona breeders have developed a number of strains, such as 5225 (Fisher, 1968), with high levels of resistance. Other strains, such as 6016, 6017, 6019, and 6020, involved complex crosses among resistant Acala and Hartsville-Acala material. Strain 6022 came from Acala, Hopicala, and Coquette crosses, while 6024 came from Acala, Hopicala, Roxe, and KP Uganda crosses (L. M. Blank, *personal communication*). In a yield test, where Verticillium wilt is severe, 6024 and 6020 strains performed well for both yield and wilt grades (Ray et al., 1966). The Arizona strains perform well in at least two different environments.

In the relatively short time of 25 years, progress has been made from having susceptible types that yield fairly well in the presence of the pathogen to having types which yield well and express resistance under a severe disease situation. Further progress will certainly come as crosses are made to combine more of the known and yet unidentified genes into single strains.

## Boll Rot

Boll rot has been attributed to a wide range of organisms, and listings may be found elsewhere (Cauquil and Ranney, 1967). Resistance is known only for *X. malvacearum,* which is discussed above. Inherent means of boll-rot control involve morphological characteristics having to do with entrance into the boll and those influencing the environment around the boll.

The soundness of sutural areas, especially at the boll apex, affects waterproofing. Boll shape is associated with the degree of waterproofing, and varieties with boll shapes approaching the desired type had less boll rot (Cauquil and Ranney, 1967).

Entrance of fungi into the boll occurs via the inner and outer involucral nectaries. Removal of nectaries would eliminate this avenue of penetration (Cauquil and Ranney, 1967). Absence of nectaries in cotton is conditioned by two recessive genes.

Removal of normal bracts from bolls reduced boll rot (Luke and Pinchard, 1970). Comparisons of frego (a strap-shape leaving much of the boll base uncovered) with normal (a triangular shape covering all of the boll base) revealed that reduced boll rot was associated with frego bract (Jones et al., 1968-69). Frego bract is inherited as a simple recessive character.

Boll rot is less in strains having okra-shaped leaves. Okra leaf types are earlier and tend to give higher yields with higher plant populations in comparison with normal leaf types (Andries et al., 1970).

Certain boll shapes and absence of nectaries reduce easy avenues of penetration into the boll. Frego bract and okra leaf plants are more open, permitting greater movement of air and penetration of sunlight. Consequently, wet environments favorable for boll-rotting organisms are present for shorter periods of time.

It should be pointed out that frego bract and nectariless also aid in reducing losses from insects. Thus, a dual role in reducing losses from adversities may be expected.

## Phymatotrichum Root Rot

Phymatotrichum root rot [ *Phymatotrichum omnivorum* (Shear) Dug.] control by inherent means is approached by using the traits of earliness and delay or slow rate of plant kill by the pathogen. The two traits together often permit maturing a full crop before plants are killed.

Scientists involved in the earlier phase of root-rot control recognized that resistance or immunity might never be found, but that varieties capable of delaying plant kill until late in the season would be worthy of use for control (Taubenhaus and Killough, 1923). Progress in selecting for delay in kill was reported (Goldsmith and Moore, 1941). Cotton seedlings tend to be resistant but become susceptible with age (Blank, 1940). Hubbard (1960, *unpublished data*) found one progeny with marked ability to delay kill. Further evaluation by the author revealed the delay trait was less effective in some environments than in others.

It seemed that the delay-in-kill concept might be valid. If so, and if it were combined with rapid maturity traits, it was concluded that inherent means could be used to prevent economic loss. Work by the author indicated that the concept was valid. The procedure was to develop strains and evaluate them in replicated lattice experiments in field areas having known histories of continuous *P. omnivorum* infestation. Each year the better strains were carried forward for re-evaluation along with new strains the next year. At the same time crosses were made between the better strains. Continued progress was made with this procedure. One problem was that defruited cotton plants are resistant to root rot, and it was necessary to be sure that delayed kill was occurring in strains giving above-average yields. By 1970 it was apparent that earliness, delay-in-kill by *P. omnivorum,* and yielding ability were being combined in the same strains.

## Southwest Cotton Rust

Southwest cotton rust (*Puccinia cacabata* Arth. and Holw.) occurs in the southwestern United States and Mexico where desert grama grass grows. Cotton is the alternate host, and inherent control must be based on resistance to germinating sporidia.

Naturally occurring rust in the field and inoculation procedures conducted in the greenhouse are used to search and breed for resistance. No resistance has been found in *G. hirsutum* and *G. barbadense.* A high degree of resistance was found in *G. arboreum, G. herbaceum, G. anomalum, G. bickii,* and *G. aridum,* and near immunity was found in *G. barbosanum.* Strains developed from interspecific crosses involving the above species with *G. hirsutum* were screened for resistance. Resistant plants from two interspecific combinations were obtained. This resistance is being transferred to Arizona Upland varieties (L. M. Blank in cooperative work with W. D. Fisher and L. S. Stith, Plant Breeding Department, University of Arizona) and progress is encouraging.

Information obtained from $F_1$ and $F_2$ populations suggest that resistance is incompletely dominant or possibly additive and conditioned by one major gene. Thus, it is now possible to control southwestern cotton rust by inherent means. (This section is based on personal observations, discussions with and a personal communication from Lester M. Blank, USDA Plant Pathologist, Phoenix, Arizona.)

## Multiple Disease Resistance and Escape

Multiple disease resistance, escape and environmental neutrality (seedling disease, bacterial blight, wilts, wilt-nematode complex, Phymatotrichum root rot, and boll rots) involve developing strains which resist and/or escape diseases caused by eight or more pathogenic organisms. For practical purposes strains should have the genetic potential of being neutral to environments which may affect disease control as well as production mechanisms.

As early as 1957, several scientists noted that strains resistant to Verticillium wilt were also resistant to the Fusarium wilt root-knot nematode complex. It was postulated that the same genes condition resistance to both diseases. Within a set of breeding strains and varieties evaluated for both wilts, positive correlation coefficients were obtained between the percentage wilt values for each wilt (Bird, 1966).

A higher frequency of plants resistant to Fusarium wilt was obtained from bacterial blight-resistant than from blight-susceptible strains. Correlations between resistance to bacterial blight and both wilts were obtained. The size and sign of the coefficients varied with the race of $X$. malvacearum used in the study (Bird, 1966). Using individual plants of parents, $F_1$'s, backcrosses, and $F_2$ generations, high correlation coefficients between wilt and blight grades which varied in size with the race of $X$. malvacearum used, were obtained. The data suggested the same genes conditioned resistance to both diseases.

Using a number of strains, positive correlation coefficients were obtained among grades representing resistance to bacterial blight, Fusarium and Verticillium wilts, and escape from seedling disease and Phymatotrichum root rot. Strains with resistance to bacterial blight and Fusarium wilt had higher levels of escape from Phymatotrichum root rot. Positive correlations between Fusarium wilt resistance and root-rot escape were obtained (Bird, 1966).

In connection with seedling disease escape research, correlation studies revealed favorable associations between seed and seedling measurements and resistance and escape from bacterial blight, seedling disease, wilts, and Phymatotrichum root rot. A more formal study revealed associations between mold growth on seed and resistance to bacterial blight and Fusarium wilt. Further studies revealed favorable correlations between the seed and seedling traits, earliness, and yielding ability.

In applying the interrelation information in a genetic improvement program, it became apparent that it was easier to make progress by selecting simultaneously for all adversity traits than it was to deal separately with each (Bird et al., 1968). Interrelation information indicated that resistance to bacterial blight and mold growth on seed were key characters to use in selecting for multi-resistance (Bird, 1966). The seed and seedling measurements are key traits for maintaining favorable relations among multi-resistance, environmental neutrality, yield, and earliness.

Evidence indicating the existence of genes conditioning a broad spectrum biological resistance continued to accumulate. This, along with results obtained in a genetic improvement program, led us to develop the following concept. Adversity-neutrality involves the use of genes that condition a mechanism giving resistance to several different pathogens and their variants (multi-resistance genes); genes that are more specific for individual pathogens and their variants (mono-resistance genes—mono-resistance includes genes for both vertical and horizontal resistance as defined by Van der Plank, 1968); genes that govern morphological and growth characteristics that aid in escape from two or more diseases (escape genes); and genes conditioning yield, fiber, and seed quality attributes (economic genes). In order to be effective, resistance, escape, and economic genes must function in different environments. This is accomplished by combining adversity and economic genes into genotypes that are neutral to fluctuations within the environmental limitations known for the cultivated species (neutrality genes). Genes conditioning all of the above traits are referred to as adversity-neutrality genes. Evidence suggests that minor and modifier genes are of primary importance in conditioning the adversity-neutrality traits.

Varieties with a balanced complex of adversity-neutrality genes will behave as plants native to most areas where they are cultivated. Few if any production problems will be encountered when varieties having measurable levels of the adversity-neutrality traits are grown in systems managed for minimizing adversities and optimizing yield. Cost of production will approach an absolute minimum under such circumstances, economic loss from disease will be nil, and use of pesticides will be minimized.

Strains developed under the adversity-neutrality concept should perform well wherever grown. They should perform as well as the best variety in the absence of adversities and better than the best varieties when adversities are present. Formal yield tests and demonstration plantings have indicated this is the case, even for prototype strains. In 21 trials where no adversities were present, the MDR strains performed as well as the best variety in the planting. In 15 trials where adversities (both diseases and insects) were present, the MDR strains performed better than the highest yielding variety. Apparently, earliness of the MDR strains aids in reducing losses from some insects. The strains have desirable fiber and seed attributes.

The performance of strains developed in the adversity-neutrality program gives strong support to the concepts and procedures used in their development. Thus, a new way has been found simultaneously to improve cotton for resistance to diseases and for general agronomic performance.

# Conclusion

The challenge of developing highly improved cottons which resist and escape all diseases is a real one. Surprisingly, the task does not appear to be as difficult as it did only recently.

That sources of resistance to many diseases are present in a few strains of *G. arboreum* and *G. herbaceum* (Old World, $n = 13$ cottons) indicates the same situation can be established in *G. hirsutum* and *G. barbadense*.

It is encouraging that progress for each disease has already been made. The progress made with bacterial blight in a relatively short period of 30 years points to the importance of identifying, determining the mode of inheritance, and assigning symbols to genes (as done by R. L. Knight). This simplifies identifying and combining key genes into the same material. Considerable progress has been made in breeding Fusarium wilt-nematode and Verticillium wilt resistant varieties. One speculates on whether or not progress might have been more rapid if genetic studies similar to those for blight had been made.

Resistance to and escape from seedling pathogens can be accomplished. However, the effort is in its infancy. Boll rot escape is a definite possibility, and actual resistance should not be overlooked. It should be emphasized that the boll rot escape traits also aid in reducing losses from insects. Phymatotrichum root rot escape can be accomplished, but recent progress suggests possibilities of resistance. The recent discovery of genes for resistance to southwestern cotton rust encourages the belief that resistance to any pathogen can be found.

The challenge is to combine resistance and escape from all pathogens into the same varieties. The adversity-neutrality concept of breeding offers more efficient means of accomplishing this. However, one should not begin with the objective of duplicating old or existing varieties. Instead, the objective should be to develop new highly improved varieties with plant and boll types, fiber attributes, seed quality, and production systems of the future in mind.

Successful cotton varieties of the future will comply with the adversity-neutrality concept for disease and insect resistance, and they will be glandless. They will have high yielding ability with fiber and seed properties desirable for marketing and end product use. Such varieties will benefit producers, and they will provide important raw materials for man to use for clothing and nutrition as he works and enjoys the world in which he lives.

## Acknowledgments

Publication policy limits the number of references, and for this reason, I have freely used work of others without a citation. A complete list of references may be obtained by contacting the author.

Many listings of sources of resistance may be found in committee reports for the respective diseases, which are published by years from 1936 to the present, in the Proceedings of the Cotton Disease Council.

My sincere thanks go to the following people for reviewing the original manuscript and offering suggestions for improvement: William E. Batson, Jr., Robert G. Davis, G. M. Watkins, and R. A. Frederiksen of Texas A & M University; Lester M. Blank, Research Plant Pathologist, USDA, Cotton Research Center, Phoenix, Arizona; L. A. Brinkerhoff and R. E. Hunter, Research Plant Pathologists, USDA, Oklahoma State University; and Jack E. Jones, Louisiana State University.

This chapter is dedicated to Dr. Lester M. Blank for his encouragement and concern for fledgling cotton scientists and for his valuable advice on breeding disease resistant cottons.

## References

Andries, J. A., J. E. Jones, L. W. Sloane, and J. G. Marshall. 1970. Effects of okra leaf shape on boll rot, yield, and other characters of Upland cotton *Gossypium hirsutum* L. Crop Sci. 10:403-407.

Barrow, J. R. 1970. Heterozygosity in inheritance of Verticillium wilt tolerance in cotton. Phytopathology 60:301-303.

Bird, L. S. 1966. Interrelation of resistance and escape of cotton from five major diseases. Proc. Cott. Dis. Council 26:92-107.

———, K. M. El-Zik, E. Free, and R. Arnold. 1968. Concepts and procedures for developing cottons with multiple disease resistance. Beltwide Cott. Prod. Res. Conf., Proc. Cott. Dis. Council 28:158-162.

———, and J. T. Presley. 1965. Cotton seedling disease escape-resistance to seed deterioration and low temperature germination and growth. Proc. Cott. Dis. Council 25:88-98.

———, and A. A. Reyes. 1967. Effects of cottonseed quality on seed and seedling characteristics. Proc. Cott. Dis. Council 27:199-206.

Blank, L. M. 1940. The susceptibility of cotton seedlings to *Phymatotrichum omnivorum*. Phytopathology 30:1033-1041.

———, and R. E. Hunter. 1955. *In* Report of the bacterial blight committee. Proc. Cott. Dis. Council 16:18-19.

Bollenbacher, K., and N. D. Fulton. 1959. Disease susceptibility of cotton seedlings from artificially deteriorated seeds. Plant Dis. Reporter Suppl. 259.

Brinkerhoff, L. A. 1970. Variations in *Xanthomonas malvacearum* and its relation to control. Ann. Rev. Phytopathol. 8:85-109.

————, I. M. Verhalem, and K. C. Fun. 1970. Inheritance of resistance to Verticillium wilt in ten Upland lines of cotton. Beltwide Cott. Prod. Res. Conf., Proc. Cott. Impr. Conf. 22:65-68.

Brown, H. B., and J. O. Ware. 1958. Cotton, 3rd ed. McGraw-Hill Book Co., New York. 566 p.

Bugbee, W. M., and J. T. Presley. 1967. A rapid inoculation technique to evaluate the resistance of cotton to *Verticillium albo-atrum*. Phytopathology 57:1264.

Carvalho, P. P. 1969. Susceptibilidade de cultivares de algodão à bacteriose. Técnicas para pesquisa da doenca. Agron. Mocamb. Lourenco Marques 3:27-48.

Cauquil, J., and C. D. Ranney. 1967. Internal infection of green cotton bolls and the possibility of genetic selection. Miss. Agr. Exp. Sta. Tech. Bull. 53. 24 p.

Chilton, J. E., and L. M. Blank. 1955. *In* Report of the Verticillium wilt committee. Proc. Cott. Dis. Council 16:7-14.

Cotton, J. R. 1965. Breeding cotton for tolerance to Verticillium wilt. ARS 34-80, Crops Research Division, ARS, USDA. 18 p.

El-Zik, K. M., and L. S. Bird. 1970. Effectiveness of specific genes and gene-combinations in conferring resistance to races of *Xanthomonas malvacearum* in Upland Cotton. Phytopathology 60:441-447.

Fisher, W. D. 1968. Breeding cotton for tolerance to Verticillium wilt. Beltwide Cott. Prod. Res. Conf., Cott. Impr. Conf. 20:230 (Abstr.).

Fulton, N. D., and B. A. Waddle. 1957. *In* Report of the Verticillium wilt committee. Proc. Cott. Dis. Council 18.6-18.

———— ————, and K. Bollenbacher. 1962. Varietal resistance to seedling disease in cotton. Phytopathology 52:10 (Abstr.).

Garber, R. H. 1966. *In* Report of the seedling disease committee. Proc. Cott. Dis. Council 26:71-74.

Goldsmith, G. W., and E. J. Moore. 1941. Field tests of the resistance of cotton to *Phymatotrichum omnivorum*. Phytopathology 31:452-463.

Hefner, J. J. 1968. Screening cotton for resistance to damping-off by *Rhizoctonia solani*. Beltwide Cott. Prod. Conf., Joint Meeting Cott. Impr. Conf. and Cott. Dis. Council 20-28:164-165.

Hunter, R. E., L. A. Brinkerhoff, and L. S. Bird. 1968. The development of a set of Upland Cotton lines for differentiating races of *Xanthomonas malvacearum*. Phytopathology 58:830-832.

Innes, N. L. 1965. Resistance to bacterial blight of cotton: The genes $B_9$ and $B_{10}$. Exp. Agr. 1:189-191.

Jones, J. E., J. A. Andries, L. W. Sloane, and S. A. Phillips. 1968-69. Frego bract reduces cotton boll rot. Louisiana Agr. Exp. Sta., Louisiana Agriculture 12:8-11.

————, and W. Birchfield. 1967. Resistance of the experimental cotton variety, Bayou and related strains to root knot nematode and Fusarium wilt. Phytopathology 57:1327-1331.

——, L. D. Newsom, and E. L. Finley. 1959. Effect of the reniform nematode on yield, plant characters, and fiber properties of Upland Cotton. Agron. J. 51:353-356.

Kelkar, S. G., G. S. Chawdhari, and N. B. Hiremath. 1947. Inheritance of Fusarium resistance in Indian cotton. Proc. Third Conf. Cott. Growing Publ. India I.C.C.C., Bombay. p. 125-162.

Knight, R. L. 1956. Plant protection conference, p. 53-59. Academic Press, New York.

Luke, W. J., and J. A. Pinchard. 1970. The role of the bract in boll rots of cotton. Cott. Growing Rev. 47:20-28.

McMichael, S. C. 1960. Combined effect of glandless genes $gl_2$ and $gl_3$ on pigment glands in the cotton plant. Agron. J. 52:385-386.

Presley, J. T. 1950. Verticillium wilt of cotton with particular emphasis on variation of the causal organism. Phytopathology 40:497-511.

Ranney, C. D. 1962. Effects of nitrogen source and rate on the development of Verticillium wilt in cotton. Phytopathology 52:38-41.

Ray, L. L., E. B. Minton, and R. W. Berry. 1966. Evaluation of cotton strains and varieties for resistance to Verticillium wilt and bacterial blight. Texas Agr. Exp. Sta. Prog. Report 2410. 7 p.

Roberts, C. L. 1969. Heritability of tolerance to *Verticillium albo-atrum* in American Upland Cotton. Ph.D. Thesis. New Mexico State Univ. 52 p.

Sherbakoff, C. D. 1949. Breeding for resistance to Fusarium and Verticillium wilts. Bot. Rev. 15:377-422.

Smith, A. L., and J. B. Dick. 1960. Inheritance of resistance to Fusarium wilt in Upland and Sea Island cottons as complicated by nematodes under field conditions. Phytopathology 50:44-48.

Taubenhaus, J. J., and D. T. Killough. 1923. Texas root rot of cotton and methods of its control. Texas Agr. Exp. Sta. Bull. 307. 98 p.

Turcotte, E. L., H. W. Reynolds, J. H. O'Bannon, and C. W. Feaster. 1963. Evaluation of cotton root-knot nematode resistance of a strain of *G. barbadense* var. *darwinii*. Cott. Impr. Conf. Proc. 15:36-44.

Van der Plank, J. E. 1968. Disease resistance in plants. Academic Press, New York. 206 p.

Wiles, A. B. 1953. *In* Report of the Verticillium wilt committee. Proc. Cott. Dis. Council 13:19.

——: 1957. Resistance to root-knot nematode in cotton. Phytopathology 47:37 (Abstr.).

Wilhelm, S., J. E. Sagen, and H. Tietz. 1970. Seabrook (*Gossypium barbadense)* x Rex (*Gossypium hirsutum*) crosses give Verticillium wilt resistant, Upland-type, all fertile offspring. Beltwide Cott. Prod. Res. Conf., Proc. Cott. Impr. Conf. 22:70-76.

# 13 Tobacco

## T. W. Graham[1]
## L. G. Burk[2]

## Introduction

International trade in tobacco and its products among the 79 tobacco producing countries on all continents represents an important segment of the world's commerce in agricultural products. In 1968 the United States produced about one-sixth of the world's supply, or about 1.7 billion pounds. Gradual intensification of tobacco culture in the U.S. since colonial times has resulted in distinct morphological types grown in localized and specialized production areas principally in the southeastern and mid-Atlantic states. Tobacco types in the U.S. are referred to as flue-cured, Burley, Maryland, cigar wrapper (shade grown), cigar filler, cigar binder, dark fired, and dark air-cured. Similar types and production areas are found in other countries producing

1. Research Pathologist, Agricultural Research Service, Plant Science Division, U.S. Department of Agriculture and South Carolina Agricultural Experiment Station, Florence, South Carolina. Tech. paper No. 856.

2. Research Geneticist, Agricultural Research Service, Plant Science Division, U.S. Department of Agriculture and Associate Professor, Department of Genetics, North Carolina State University, Oxford Tobacco Research Station, Oxford, North Carolina.

tobacco. Aromatic types are produced in Turkey and other mid-Eastern European countries bordering the Black Sea.

Losses from tobacco diseases have increased along with intensified culture of the crop. Major producing areas repeatedly are threatened by disease outbreaks. The total losses from 1951-1960 in the U.S.A. are estimated at more than 132 million dollars, or about 11% of total production. In 1960 an epiphytotic of blue mold in Europe cost the growers an estimated 25 million dollars where three years earlier the disease was unknown (Lucas, 1965). Although most other diseases are less spectacular, they often represent long standing problems and generally cause heavy overall losses. Some 20 major diseases caused by fungi, bacteria, viruses, and nematodes occur on the growing crop. Resistant varieties now offer an important means of disease control in the U.S.A. and other countries. Plant breeders and pathologists deserve major credit for contributing to the solution of many tobacco disease problems.

## The Tobacco Plant

Common tobacco, *Nicotiana tabacum L.* is a native of South America and is believed to be a natural hybrid between two species resembling present day *N. sylvestris* Speg. and Comes and *N. tomentosiformis* Goodsp. (Goodspeed, 1954). Tobacco is an allotetraploid but has undergone considerable diploidisation. For purposes of convenience it is usually referred to as a diploid ($2n = 48$). Sixty-five species in the genus *Nicotiana* (*Solanaceae*) are now recognized. Most are found in western South America, Central America, and western North America. A subgeneric Section, *Suaveolens,* includes 20 species native to Australia and one native to the Marquesas Islands.

Cultivated American types of *N. tabacum* produce from 18 to 30 large leaves (30-37 x 35-42 cm) borne on plants 120-345 cm tall. The inflorescence is made up of typically self-pollinated flowers borne on a terminal panicle which may produce 100 or more seed capsules, each containing as many as 3,000 seeds. These characteristics provide a plant in which breeding and genetic investigations are greatly facilitated.

## Sources of Disease Resistance

All tobacco types with *N. tabacum* normally yield fertile $F_1$ progeny when intercrossed. Potential sources of resistance include a collection of more than 1,000 T.I. (Tobacco Introduction) lines obtained mostly from Central and South America. These were started originally under the direction of Dr. E. E. Clayton and are maintained by the U.S. Department of Agriculture. Current benefits to the tobacco industry derived from disease resistance are a tribute to Clayton's far-sighted concept of a germ plasm bank. Clayton and Smith (1942) found, among the T.I. lines, resistance to two major diseases which proved to be of great value to flue-cured tobacco.

The wild relatives of *N. tabacum* represent proven sources of resistance. Burk and Heggestad (1966) recorded responses among 65 *Nicotiana* species, of which 54 had potential value for controlling one or more of 13 major diseases. Breeding stocks of these *Nicotiana* species are maintained by state experiment stations and universities in the U.S.A. and by the U.S. Department of Agriculture. In European and Asiatic countries more than 20 tobacco institutes engage in genetic and breeding programs and maintain seed supplies.

## Use of N. tabacum

Disease resistance derived from *N. tabacum* is often inherited polygenically. The involvement of several genes tends to complicate the recovery of high levels of resistance. Occasionally, problems with "block inheritance" or "linkage" are encountered, where unwanted growth characters are difficult to separate from those of disease resistance. Despite these problems, hybridization and selection within introduced and commercial tobaccos have been vitally important in establishing and maintaining resistant cultivars.

## Use of Noncultivated Nicotiana species

A major barrier to interspecific hybridization is sterility in the $F_1$ progeny. The chromosomes of *N. tabacum* consist of 24 pairs, and with the exception of the progenitors, are only remotely related to chromosomes of the remaining *Nicotiana* species, which range in number from 9 to 32 pairs. Thus, normal chromosome pairing between most species and *N. tabacum* is rare. Successful interspecific transfer of disease resistance also depends on traits that display a high degree of dominance and a monogenic pattern of inheritance. This simple pattern of inheritance, once established in the tobacco genome, provides a convenient and rapid means of transferring resistance to any commercial tobacco type or variety. To date, four wild species of *Nicotiana* have been used to incorporate germ plasm into *N. tabacum*, resulting in acceptable resistance to six major diseases. After genes for resistance are transferred from the wild species to chromosomes of tobacco prototypes, conventional intraspecific breeding procedures are followed. The back-cross method is preferred. Clayton (1954) suggests the following steps to isolate and utilize resistance from the *Nicotiana* species: "1. By repeated evaluation of the disease resistance of available species, narrow the choice to those that are uniformly highly resistant. 2. Measure the disease reaction of these species in direct comparison with the allopolyploids or $F_1$s that they form with the crop plant; in this case tobacco. 3. If possible, use for interspecific work a species that produces an allopolyploid or $F_1$ which shows the full resistance of the resistant parent."

## Interspecific Hybridization

In attempts to hybridize commercial *N. tabacum,* represented by the genetic symbol $(T/T)$, and the species *N. glutinosa* $(S/S)$, the resulting $F_1$ hybrid is a sterile hybrid amphihaploid $(T/G)$. To overcome sterility, seeds of the hybrid are treated with colchicine (soak germinating seeds in 0.2% aqueous solution 4-6 hr). As a result, the chromosomes of some plants are often doubled $(T/T\text{-}G/G)$ and fertility is restored. The next step is to cross the amphidiploid hybrid $(T/T\text{-}G/G)$ with the original *N. tabacum (T/T)*. The result is a "secondary," relatively uniform, hybrid $(T/T\text{-}G)$, the sesquidiploid. The progeny of the back-cross $(T/T\text{-}G)$ by $(T/T)$ is morphologically variable and is sometimes called the "break-down" generation. This is the critical generation in interspecific gene transfer. The usual breeding procedure is then followed by making repeated backcrosses of resistant selections to the recurrent parent. Chromosomes are now unequally distributed in the hybrid progeny because of the unpaired genome $(G)$ of the sesquidiploid. Tobacco-like segregants are selected that are disease-resistant in order to isolate plants possessing the wild species chromosomes carrying genes for resistance. Other special methods, such as the bridge cross, have been used to overcome incompatibilities between *Nicotiana* species (Burk, 1967; Chaplin and Mann, 1961). Further discussions on inter-specific hybridization are given by Clayton (1954), Burk and Hegge-stad (1966), and Schweppenhauser (1968).

## Male Sterility and $F_1$ Hybrids

Interspecific hybridization may result in cytoplasmic male sterility, as first reported by Clayton and later by other workers. Male-sterile plants were found in the sesquidiploid generation of hybrids of *N. debneyi, N. megalosiphon,* and other species (as maternal parents) crossed with *N. tabacum.* The reciprocal cross (*N. tabacum* as the maternal parent) was found by Burk to produce no male-steriles. Male sterility is believed to result from residual cytoplasm of the alien species retained in the hybrid after introgressive introduction of chromosomes of *N. tabacum* plus the concomitant loss of certain chromosomes of the alien species. Such plants have distorted anthers that produce no pollen. Distorted anthers, which are characteristic of the cytoplasmic parent involved, may be stigmatoid, petaloic, feather-like, or with other abnormal formations as reported by Chaplin. Male-steriles are perpetuated by hand-pollinating with a related male-fertile plant.

Male-sterile lines may be used to produce first generation $F_1$ hybrids; however, in flue-cured tobacco such $F_1$ hybrids do not show significant yield increases above parental lines. Evidence now available generally supports the superiority of homozygous genotypes derived by conventional breeding methods. On the other hand, $F_1$ hybrids in some tobacco types have been reported with improved quality, more uniformity, and earlier maturity than their respective parents. Evidence by Dean et al. (1968) suggests that general combining ability of properly selected parental lines may offer other advantages from $F_1$ hybrids. In the event of sudden disease epiphytotics, $F_1$ hybrids might be used rapidly to meet emergency needs. Dean et al. (1968) described studies on hand-pollinated $F_1$ cigar tobacco hybrids for control of blue mold under shade. $F_1$ hybrids have been advocated and used to some extent in Europe and Australia against blue mold. A disadvantage of $F_1$ hybrids for commercial use is the cost of seed produced by hand pollination. Despite these limitations, commercial production of $F_1$ hybrid seed has been under way since 1967 in the Florida cigar wrapper area and in the Burley area since 1963. Potential merits of $F_1$ hybrids in tobacco have not been fully explored.

## Disease Resistance

Resistance is sufficiently stable when it is expressed through a series of five or six selfed generations. Disease development depends on many variables affecting both pathogen and host. The breeder should be well informed about these factors and be able to devise methods to assure critical screening. Circumstances and the disease involved will determine the specific approach to each problem. Host-pathogen relations are discussed later with respect to individual diseases.

## Agronomic Performance

New varieties are generally acceptable to growers if they yield well. Characteristics of concern to breeders include leaf number, plant height, leaf spacing, and handling and curing properties. These properties are quantitatively inherited and poorly defined genetically; they are, however, responsive to modification by selection. Replicated plantings of breeding lines and standard varieties in disease-free plots are standard procedures that help to determine yields and quality. Uniform cooperative regional programs to measure disease levels and agronomic performance of candidate varieties over widely separated locations are in operation in the flue-cured and Burley areas of the United States. Resulting data help to identify lines with superior adaptability and resistance to disease under varying conditions of climate and soil.

## Chemical Constituents and Quality Considerations

Quality in tobacco is not entirely measurable by the usual methods of recording objective data. Characteristics that determine quality are quantitative and not well defined genetically. Various factors affect them, including cultural practices and methods of harvesting and curing; all are essential steps for a quality product. Quality is based in part on chemical make-up of the cured leaf, for which desired levels of nicotine, sugar, nitrogen, and other constituents have been established. Finally, the manufactured end product is assessed by experienced smoke panelists who appraise smoking taste and aroma. Some measure of quality is determined by physical data based on body, color, and texture of the cured leaf and by its position on the stalk. Thus, quality in the growing plants is not measurable only by the plant breeder, but depends greatly on cooperation from various other specialists.

# Tobacco Diseases

### Black Shank

*Nature and Importance.* Black shank, incited by *Phytophthora parasitica* var. *nicotianae* (Breda de Hann) Tucker, is a very destructive tobacco disorder. Lucas (1965) reported its occurrence in 23 countries representing North and South America, Europe, Asia, and Africa. In the United States it appeared first in 1915 in the Florida shade area and became widespread by 1922. It appeared in North Carolina in 1931 and spread by 1951 throughout the flue-cured and Burley areas of all the mid-Atlantic states, including Maryland and Pennsylvania. Black shank has not been found in the northern states of Massachusetts, Connecticut, and Wisconsin. The disease is more prevalent in warmer latitudes: temperatures in excess of 20 C are most favorable for its development. The pathogen is soil-borne and remains viable for 5 years or more in the absence of the host. It spreads quickly from the initial infestation site by means of surface drainage water, transported farm equipment, or the feet of workers. New areas may be infested by the pathogen carried on transplants from an infested seed bed. Wind-borne sporangia may cause aerial leaf infection. Principal symptoms on mature plants are sudden wilting, followed by yellowed and drying leaves and finally death of the plant. A dark lesion, sometimes 15-30 cm long, girdles the stalk near the ground level. A comprehensive treatment of the disease and the pathogen is given by Lucas (1965). Resistant varieties now generally planted in the southeastern United States provide a major means of control.

*Resistance From N. tabacum.*    Valleau (1952) reviewed the early breeding work. Most varieties in the United States derive their resistance from *N. tabacum* (the cigar shade variety Fla. 301). Burley and flue-cured varieties originating from Fla. 301 have sustained a remarkably useful level of resistance over many years. It is not uncommon, however, to find isolated field locations in some years where the disease causes unexplained heavy losses on resistant varieties.

Inheritance of resistance from Fla. 301 has been variously interpreted, but most investigators agree that more than a single dominant gene is involved, or that modifiers may be present. Smith and Clayton (1948) regarded the resistance as polygenic in behavior. Moore and Powell (1959) described resistance from Fla. 301 as a partially dominant single factor which can be accentuated by modifier genes from certain susceptible parents. Chaplin (1966) concluded that inheritance of black-shank resistance under field conditions is complex, and that segregating generations are influenced by susceptible parents, by age of plants, and by levels of soil infestation.

*Resistance From Nicotiana Species.*    Valleau et al. (1960) used *N. longiflora* (lgf) as a source of resistance to black shank for Burley tobacco. The related species *N. plumbagnifolia* (pbg) was hybridized with flue-cured varieties by Chaplin (1962) and by Apple (1962a). Smith and Earley (1964), using the backcross method and lgf resistance, developed two flue-cured varieties, McNair 20 and 30, which were released in 1963. Chaplin and Apple, in independent studies, were in agreement that resistance from pbg in flue-cured varieties was governed by a single dominant factor enhanced by modifiers of varying intensity, depending on the variety.

Valleau found that resistance arising from lfg in his L8 lines was of limited use because of an associated physiologic leaf spot and because of its failure to protect against a certain race of the black shank fungus (later designated by Apple as race 1). Valleau found evidence of simple dominant inheritance from lgf in Burley crosses, but modifying genes were involved. The association in flue-cured varieties of lowered quality and black shank resistance from pbg has been of concern to breeders.

*Variability of P. parasitica var. nicotianae.* Little is known about the genetics of tobacco pathogens, although races of the black shank organism are recognized. Apple (1962b) described two pathogenic races. Hybrids involving pbg and the flue-cured variety 402 were highly resistant to North Carolina isolates designated as common race 0, but were susceptible to three isolates from Kentucky designated as race 1. However, Apple recovered two genetic behavioral types from pbg. The first included the tobacco-like lines with resistance to common race 0, with simple dominant inheritance of susceptibility to race 1; a response the same as the pbg parent. The second type was described as a "Segmental chromosome substitution" which did not show simple dominant inheritance and was resistant to both races 1 and 0. This was the source material for the successful variety NC2326, released in 1965.

Race 1 was found in only one location in Kentucky in 1962 but was later reported from other locations in that state. Differentiation between races 1 and 0 was based on response of varieties with resistance derived from pbg or lgf.

*Other Factors Affecting Black Shank.*    Black shank often occurs in fields that are infested with root knot and other nematodes parasitizing tobacco. Severity of the disease increases in the presence of nematode root injury. A moderately resistant variety appears to lose some resistance when nematodes are present (Powell and Nusbaum, 1960). Black shank may be found in the same fields with bacterial· wilt, Fusarium wilt, or other soil-borne disease complexes. Therefore, multiple disease resistance is an important objective of many tobacco breeders (see section on multiple disease resistance).

Soil *p*H and other environmental influences affect black shank. Ranges of *p*H 5.0-6.0 are more favorable for the disease than lower *p*H levels. These and other relative factors were studied and discussed by Troutman and LaPrade (1962), and Dukes and Apple (1968).

*Testing for Resistance: Field Plots.*    Evaluation of resistance among breeding lines under field conditions offers such advantages as exposure throughout the season to the pathogen, opportunity to select desirable agronomic types from surviving plants, and sufficient land area to screen large plant populations. Within individual breeding plots, however, soil may not be uniformly infested, and as a result equal exposure to the pathogen may not be possible. Other limitations include mixed infestation with other pathogens or unfavorable weather for normal growth and disease development. A two-year cropping system with a susceptible variety in alternate years is a land management practice that is sometimes helpful. Frequent interplanting with a susceptible variety is an alternative means of raising inoculum potential. Macerating and scattering diseased stubble with a rotary cutter and turning the stalk residues into the soil has been employed. The normal practice of transplanting tobacco is adaptable to inoculation at transplanting time. Laboratory cultures of the pathogen can serve effectively as sources of inoculum where small plots or special tests are involved.

*Small-Plant Methods.*    Techniques involving the inoculation of small plants in the greenhouse or growth chambers offer a saving in time, precise control of inoculum levels, and more uniform results. However, certain limitations are inherent in such methods. Young succulent plants are more susceptible than mature plants and may not reflect field response. Those grown in small containers under reduced light intensities and with a short exposure to the pathogen offer a sharp contrast to plants grown under field conditions. Small-plant techniques contribute perhaps most significantly in tests of early generations and where selection for plant type is not critical. Generally, however, both greenhouse and field-testing methods should be considered as important aspects of breeding for disease resistance.

## Tobacco Nematode Diseases

*Nature and Importance.*    The nematode diseases of tobacco are major problems in the southeastern United States and in all tobacco-producing countries with warm or subtropical climates. The root knot nematode, *Meloidogyne,* Goeldi 1892, and root lesion nematodes, *Pratylenchus,* Filipjev 1934, parasitize all commercial tobaccos and often cause heavy losses if controls are not practiced. Yield reductions up to 30% occurred before the advent of effective control practices in the United States. Losses in North Carolina in 1969 were reported to be about 4.5 million dollars, or 1% of the crop value; yet North Carolina growers practice extensive control measures. Daulton (1962) estimated that the root knot nematode, *M. javanica,* destroyed 9-12% of the crop value in Rhodesia.

The root lesion nematode is present in the Southeast and in the Connecticut Valley of the United States. In the South it coexists with the root knot nematode and often contributes to root injury as part of a disease complex. The cyst nematode, *Heterodera tabacum,* Lownsby and Lownsb, was found on shade tobacco in Connecticut in 1951, and by 1968 it was also found in limited areas of Massachusetts. Osborne's cyst nematode (undescribed) was reported on tobacco in Virginia in 1961 but has not extended more than 20 miles from the original site. The-horsenettle cyst nematode, *H. virginiae,* Miller and Gray 1968, was found in 1958 near Suffolk, Va. on roots of the wild horsenettle, *Solanum carolinense.* Although not found on commercial plantings, it is a known tobacco parasite. Little attention has been given to breeding for resistance to nematode diseases except root knot.

*Root Knot Nematode.*    The root knot nematode *Meloidogyne incognita* (Kofoid and White 1919), Chitwood 1949, is recognized by characteristic galls on roots of host plants. Several life cycles of the pathogen are completed each year, depending on soil temperature and length of the growing season. The nematode female feeds in a stationary position within the root tissue and releases abundant eggs at the root surface. The new generation of larvae enter root tips and initiate another cycle of gall formation. Eggs may remain dormant in the soil over winter. The nematode has limited mobility in soil and is distributed principally by surface water movement and cultivating tools. Tobacco transplants pulled from infested seed beds may also spread the pathogen to the field. A complete review on root knot and other nematode diseases of tobacco is presented by Lucas (1965).

*Breeding for Resistance.*    The most sustained effort in breeding for resistance against the root knot nematode *M. incognita* was that of Clayton and his coworkers. This work is considered a classic contribution to plant breeding: it was a program extending over a 30-year period beginning in 1935. This work, Clayton et al. (1958), is reviewed here in some detail because of its significance to continued efforts in tobacco breeding. After screening 1,000 or more T.I. lines, Clayton and his colleagues selected T.I. 706 as the most suitable resistant parent. Crosses between susceptible flue-cured varieties and T.I. 706 indicated multiple-factor inheritance. After four backcrosses to the recurrent susceptible parent, the progeny had higher resistance than T.I. 706, and some of the undesirable characteristics of this parent had been eliminated. However, the small leaf size of T.I. 706 appeared to be tightly linked with nematode resistance. The solution was found when a root knot resistant breeding line was crossed with the allopolyploid hybrid 4n (*N. sylvestris* x *N. tomentosiformis*), composed of the progenitor species. After two backcrosses to a susceptible flue-cured tobacco, c.v. 402, several $F_3$ lines with larger leaves and improved resistance were obtained. Furthermore, resistance now behaved as a simple dominant factor with the progeny segregating into only two classes: resistant or susceptible. This was a remarkable achievement, since neither the allopolyploid nor *N. sylvestris* is resistant. Moderate resistance is present in *N. tomentosiformis*. Thus an increase in resistance could not be attributed to either of these species. The authors offer the following as a possible genetic explanation. Studies of Goodspeed and Clausen (1928) suggested that cultivated tobacco originated as a natural cross between *N. sylvestris* and a member of the Tomentosae group. They postulated that extensive chromosomal and structural changes took place in the evolution of present day tobacco. In spite of these changes, however, pairing still occurred between chromosomes of tobacco and progenitor species. The disrupting effect of such pairing and reassortment of chromosomes acted as an effective means to break linkages between small leaf type and resistance to root knot in T.I. 706. This explanation of a sudden change from polygenic to monogenic dominant resistance, however, was not confirmed by cytogenetic study. On the other hand, there is some evidence supporting the concept of residual (ancestral) chromosome homology between present day tobacco and its related or progenitor species (Chaplin and Mann, 1961).

The first variety with resistance to root knot, NC95, was released in 1961 by Moore et al. (1962). Other American varieties (released 1967-1969) with genetically similar resistance include SC66, Coker 254, Coker 258, NC2512, Speight G-28, and McNair 133. Resistance is effective against the common root knot species *M. incognita* but not against *M. javanica, M. arenaria,* or *M. hapla.* The latter species are not commonly found on tobacco in the southeastern United States. However, *M. javanica* is the prevalent species on tobacco in Rhodesia and in other African regions and also in Australia and New Zealand.

*Interspecific Bridge Cross.*   *N. repanda* has resistance to several species of root knot nematodes, including *M. javanica* (Burk and Dropkin, 1961). Although the interspecific cross *N. tabacum* x *N. repanda* has never been made directly, Burk (1967) accomplished this indirectly by means of a bridge cross using a fertile amphiploid 4n(*N. repanda* x *N. sylvestris*) backcrossed to *N. sylvestris*. Additional backcrosses were then made to successive progenies to obtain *sylvestris*-like plants. Those proving to be resistant were crossed with *N. tabacum*. Further studies of this source of resistance are underway.

Schweppenhauser (1968) reported a similar transfer of an alien univalent into *N. tabacum* from the interspecific hybrid 4n (*N. repanda* x *N. sylvestris*) x *N. longiflora* to obtain a source of resistance to *M. javanica*.

*Pathogenic Races of Root Knot.*   Graham (1969) reported a pathogenic race of *Meloidogyne incognita* parasitizing NC95 and other genetically similar varieties. The new race was morphologically indistinguishable from *M. incognita,* yet was parasitic on varieties with resistance derived from T.I. 706. Apparently the new race arose as a mutation in the breeding plots where it was first observed. Therefore, a new threat is posed to the new nematode-resistant varieties by this mutant, and also by a similar race found in North Carolina in 1969 (*unpublished report*).

*Methods of Testing.*   Field breeding plots, naturally infested with root knot nematodes, are frequently used for screening breeding lines as for other diseases caused by soil-borne pathogens. High infestation levels are maintained by two-year rotations with susceptible crops or with other management systems. Satisfactory greenhouse inoculation is effected by adding a small amount of chopped galled roots or infested soil to the root area of test plants. Root-knot nematodes may be maintained on susceptible hosts in greenhouse soil cultures to provide inoculum as needed. More precise control of inoculum levels is achieved by extracting root knot nematode eggs or larvae from root galls.

## Tobacco Mosaic Virus

*Nature and Importance.*   Mosaic disease of tobacco (TMV), *Marmor tabaci* Holmes, is the most important and common of the disorders incited by a virus and one of about 18 virus diseases on tobacco. TMV has a wide host range and is found worldwide wherever tobacco is grown. According to Lucas (1965), the estimated average annual loss from the disease in the U.S.A. is approximately 1% of the crop. Losses in other countries may be greater. Mosaic is rarely epiphytotic but is found every year, usually in a pattern of sporadic distribution. TMV appears to be causing increased damage to flue-cured tobacco, whereas it has been a continuing problem in the Burley and Maryland areas. Mosaic is easily spread from infected to healthy plants by the usual farm operations of transplanting, suckering, topping, and cultivating. The virus persists in perennial weed hosts, in infected tobacco roots, or dried tobacco trash. Greatest losses from depressed yields and poor quality (up to 35%) occur when young plants are infected. Infection of mature plants causes little or no leaf distortion and virtually no yield reduction, but some loss of quality. Extensive references on TMV relate to pathogenesis and to basic studies on the nature of the viral agent (Lucas, 1965).

Resistance to TMV has been incorporated into all major types including flue-cured. This is an important contribution toward control of the disease. A review of the history of the early breeding work was given by Valleau (1952).

*TMV Resistance From N. Glutinosa.*   The first transfer of genetic material from a *Nicotiana* species to *N. tabacum* was that of TMV resistance obtained from *N. glutinosa.* The original cross by Clausen and Goodspeed (1925) was *N. glutinosa* (12n) x *N. tabacum* (24n) to produce the fertile alloploid called *N. digluta.* Holmes (1938) interpreted this as an alien chromosome substitution. TMV resistance in current tobacco varieties is part of the tobacco genome and is inherited as a dominant monogenic character. Recent techniques for effecting this and other species crosses were discussed earlier in this chapter (*see* Interspecific Hybridization). Resistance to TMV from *N. glutinosa* (*N.* factor) is characterized by the well-known primary local lesion reaction, a hypersensitive response after inoculation (Holmes 1938). It results in development of small (2 mm diameter) localized areas of necrotic tissue in the immediate area of inoculation. Thus a resistant response appears to limit virus multiplication and prevent its usual systemic spread. Virus introduced into the vascular system may, under certain conditions, cause rapid death of plants that possess the *N.* factor.

Resistance to mosaic has also been found in the tobacco variety Ambalema, native to Colombia, South America. Breeders' experience with Ambalema has indicated that resistance is governed by two recessive genes linked with undesirable quality characteristics. Experimental evidence indicates that it is unsuitable for variety development. Different levels of TMV symptom expression among the tobacco introductions have recently been reported.

*Testing for Resistance to TMV.*    Screening for the hypersensitive form of resistance is accomplished easily in the greenhouse, preferably with small plants about 4 inches high. The necrotic reaction develops on plants of any age provided they are growing vigorously. Inoculum is prepared by mascerating fresh infected leaves in a blender. The residue is diluted with water in equal volume and strained through cheesecloth. A cheesecloth swab is then moistened with the liquid extract and used to rub gently the upper surface of individual leaves. A method used by Stokes (1959) consists of spraying young seedlings (15 psi) with suspension of inoculum mixed with No. 600 carborundum grit (1/2 teaspoon per 100 ml of inoculum). Populations are scored after 7 days by recording the number of plants with the *N.* factor that are killed by systemic necrosis. After homozygous genotypes are identified, additional plants from remnant seed lots provide material for agronomic and plant type selections.

## Bacterial Wilt of Tobacco

*Nature and Importance.*    The bacterial wilt disease or "Granville Wilt," incited by *Pseudomonas solanacearum*, E. F. Smith, was first found in the U.S.A. about 1880. It has now spread throughout most of the flue-cured area but is rare in the Burley area or the northern tobacco states. The disease occurs in nearly all countries where warm climates prevail, including the Dutch East Indies, Japan, and China. The pathogen has a wide host range among solanaceous and other plant families. In the United States bacterial wilt is not usually epiphytotic, but locally it has had devastating effects. Until 1940 in Granville County, North Carolina, losses reached 40-100% at a cost of nearly 40 million dollars annually. Symptoms are sudden wilting, at first on one side: plants then die promptly under conditions favorable for the disease. Internal symptoms are dark brown or black streaks in the vascular system, sometimes extending the length of the stalk. Infected roots show black discoloration, visible at first in one or a few of the main laterals. The pathogen is spread locally by surface water, infested soil, or diseased plant fragments. Infected transplants from seed beds spread the pathogen to the field. The bacterium is not known to be seed-borne. Survival of the pathogen in the soil is erratic: in some locations it may persist for many years in the host's absence, while in others it may decline in the presence of the host. The disease is described in detail by Lucas (1965), and a comprehensive literature review and bibliography was prepared by Kelman (1953).

Wilt-resistant varieties are widely grown in the southeastern United

States, and in some areas they have provided the only means for continued production.

*Source of Resistance and Breeding Work.*    Breeding work with bacterial wilt was reviewed by Valleau (1952) and by Kelman (1953). Resistance from *N. tabacum* was derived from T.I. 448A, the original source for all resistant American varieties. Clayton and Foster (1940) found that T.I. 448A was the only resistant selection among more than 1,100 T.I. lines tested. Several resistant Japanese varieties have been described. Resistance was not found in other *Nicotiana* species. The first resistant variety, Oxford 26, was released in 1945 and served as a progenitor for many successful flue-cured varieties. This impressive achievement is a tribute to the early work of Smith and Clayton (1948). The inheritance of wilt resistance from T.I. 448A was described as recessive and polygenic. Resistant plants were recovered in the $F_2$ or later generations, thus confirming recessive inheritance. Because of a general coexistence of black shank and bacterial wilt, it was desirable to combine resistance to both diseases in one variety. Since inheritance of the Fla. 301 type of black shank resistance and that of Granville wilt is polygenic, combining the two was considered highly improbable. However, resistance to both diseases was recovered in progenies of crosses between the two genotypes and from much smaller populations than expected. Thus, evidence of linkage between resistance to the two diseases was indicated. Most current varieties now possess resistance to both diseases.

*Variability of the Pathogen.*    Studies of pathogenic and physiologic responses of *P. solanacearum* indicate that it is a complex organism. Kelman (1953) found that the pathogen in laboratory cultures has a high mutation rate and rapidly loses virulence. He described a method to isolate and maintain virulent isolates with differential media. Okabe and Goto (1963) described pathogenic races on a basis of host specificity. In Georgia bacterial wilt is infrequent on tobacco, although it is common on tomato and peanuts. In Europe, Africa, and other countries, isolates of *P. solanacearum* were pathogenic on various solanaceous hosts, but not on tobacco. Tobacco breeders generally have not attached importance to pathogenic races of the wilt organism.

*Testing for Resistance.*    Naturally infested field nurseries are used by tobacco breeders in screening for resistance. Many of the practices used to maintain high levels of infestation for testing resistance to other soil-borne diseases also apply to bacterial wilt. Artificial inoculation methods are used at transplanting time in the field and greenhouse. Detailed procedures are discussed by Kelman (1953) and Moore et al. (1962).

## Multiple Resistance in Tobacco

Genes for resistance to several diseases have been introduced into established disease-resistant genotypes of tobacco. As a result, some varieties now possess resistance to 3 or 4 diseases. A variety developed by Moore et al. (1952) contained polygenically inherited resistance to bacterial wilt and black shank, derived from separate breeding lines despite problems with linkage. Where monogenic dominant factors control the inheritance of resistance, as in root knot, wildfire, and TMV, they may be added to established genotypes without serious difficulty by means of the backcross method. Ogden (1963) transferred wildfire resistance to the Wisconsin variety Havana 501, previously resistant to mosaic and black root rot. The varieties Burley 11A and 11B, which were resistant to Fusarium wilt and black root rot, were the basis of improved lines when resistance to black shank was added. The flue-cured variety NC95 resulted from the introduction of resistance to root knot and black shank into varieties that already possessed high resistance to Fusarium wilt and bacterial wilt (Moore et al., 1962). The variety Maryland 10, released in 1969, resulted from the addition of mosaic resistance to breeding lines that had resistance to Fusarium wilt and black root rot. Many other varieties released since 1960, representing all major types, have combined resistance to two or more diseases.

## Other Tobacco Diseases

Because of space limitations only four major diseases of tobacco were selected to illustrate the progress and problems in breeding for resistance. Omission of other diseases does not imply that they are less important or that they have received less attention from plant pathologists or breeders. On the contrary, significant advances in disease resistance breeding have been made in the control of black root rot, blue mold, wild fire, powdery mildew, and others. Table 13-1 presents a brief outline of progress with some major diseases not covered in this chapter.

**TABLE 13-1  SUMMARY OF SELECTED MAJOR TOBACCO DISEASES IN RELATION TO BREEDING FOR RESISTANCE.**

| DISEASE: | Blue Mold | Black Root Rot | Fusarium Wilt | Wildfire | Brown Spot | Powdery Mildew |
|---|---|---|---|---|---|---|
| PATHOGEN: | *Peronospora tabacina* Adam | *Thielaviopsis basicola* (Berk & Br) Ferr | *Fusarium oxysporum* (Schlecht) Wr. var. *nicotiana* Johnson | *Pseudomonas tabaci* (Wolf & Foster) F. L. Stevens | *Alternaria tenuis* (Ell. and Ev.) | *Erysiphe cichoracearum* D.C. |
| DISTRIBUTION: | Worldwide | Worldwide | Worldwide, warm zones | Worldwide | Sub-tropic, worldwide | Worldwide except North America |
| MEANS OF SPREAD: | Wind-borne spores | Soil, wild hosts | Soil-borne | Soil-borne, wind, rain | Wind-borne spores | Wind-borne spores |
| RESISTANT VARIETIES DEVELOPED: | Fla. shade (F₁ hybrids) European Australian | Burley, Conn. & Wisc. cigar, Canadian, New Zealand flue-cured | Flue-cured Burley, Maryland & Conn. cigar | Burley, Wisc. cigar, Penn. cigar | None | Chinese Japanese Rhodesian, Russian |
| TYPES AFFECTED: | All | All | Flue-cured Burley, Fla. Shade | Burley, Wisc. & Conn. Shade | All | All |
| STAGE AFFECTED: | Young and mature plants | All | All | Young and mature plants | Mature plants | Mature plants |
| CONDITIONS FAVORING DEVELOPMENT: | Cool, high humidity | Cool, nonacid soil | Warm soil | Rain soaked leaves | High humidity 26 C and above | High humidity, high altitudes |
| SOURCE OF RESISTANCE: | *N. debneyi* species | *N. tabacum, N. debneyi,* other species | *N. tabacum,* Turkish aromatic | *N. longiflora,* other species | *N. tabacum* (Beinhart 1000-1) | *N. glutinosa, N. tabacum,* other *N.* species |
| INHERITANCE: | Oligo- or polygenic | Polygenic, *N. tabacum* Monogenic (*N. debneyi*) | Polygenic intermediate | Monogenic dominant | Monogenic | Monogenic (*N. glutinosa*) recessive (*N. tabacum*) |
| REFERENCE: | Clayton (1967) | Valleau (1952) | Valleau (1952) | Ogden (1963) | Chaplin and Graham (1963) | Wan (1962) |

## Unresolved Problems

The few noncultivated *Nicotiana* species used in interspecific hybridization with *N. tabacum* have provided valuable sources of disease resistance. However, the greater portion of this store of germ plasm remains unexplored and unexploited. This genetic material offers an important source for further investigation in cytogenetic studies and for developing disease resistance. However, evidence suggests that problems may be involved in the narrow gene base of simply inherited resistance from alien species. Such genotypes may not be well buffered against new pathogenic races. However, our experience is limited with these relatively new genotypes, so a generalization would not apply in all circumstances. It is safe to conclude that continuing efforts will be required to keep pace with changing disease problems and with other stumbling blocks likely to result from alien gene transfer.

The unique method of breaking linkage between leaf size and root knot resistance with the alloploid 4n (*N. sylvestris* x *N. tomentosiformis*) was also found to be successful in breaking the linkage between poor leaf type and blue mold resistance transferred from *N. debneyi* (Clayton 1967). Clayton advocates this approach in resolving other unwanted linkage problems whether of intra- or interspecific origin.

The effect of modifying factors contributed by susceptible parents to enhance levels of resistance obtained from wild types of *N. tabacum* has been observed in different breeding studies. The significance of modifier genes in augmenting resistance merits further studies on methods to isolate and use them systematically. Research has shown that a major resistant gene placed in different genetic backgrounds is affected by different modifiers and, therefore, it behaves differently. If this reasoning is valid a wide range of desirable genetic combinations between *N. tabacum* and alien species should offer many possibilities.

Insufficient attention has been given to pathogenic races of disease producing organisms on tobacco. Some of the unresolved questions with black shank and other diseases relate to the capacity of the parasites to produce new races through genetic variability. Abundant evidence indicates that such variability in plant pathogens is stimulated in the presence of resistant hosts. This apparently occurred with the black shank fungus and the root knot nematode, two entirely unrelated pathogens. Basic studies on the physiology and genetics of tobacco pathogens have been neglected. More attention in this area should reveal a better understanding of unresolved questions.

Further exploration of the utility of F₁ hybrids is needed. Advantages of F₁ hybrids for disease control have been illustrated, and some problems were discussed. An intensified search should be made for new sources of disease resistance within the broad resources of the T.I. lines, tobacco varieties and types, and possibly certain of the wild *Nicotiana* species. Properly selected parental lines should show improved yields and quality or other desirable characteristics not yet discovered.

Much more information is needed for control of diseases not included in this review, for example, leaf spots, virus diseases, vascular wilts, and stalk and root rots. Some already are successfully controlled through resistant germ plasm; however, many others require more attention. Inheritance and transfer of resistance to insect parasites is now in a preliminary stage of investigation in the United States, and early evidence seems to offer some encouragement.

The high labor costs related to all aspects of tobacco culture and harvesting in the United States have brought about increased studies on mechanization of farm operations. Tobacco varieties adapted to mechanical harvesting and which are disease resistant and high yielding represent the current challenge to plant breeders.

## *Acknowledgments*

The authors gratefully acknowledge the helpful suggestions and criticisms offered by the following people while this chapter was being prepared: our associates in the Tobacco and Sugar Crops Branch of Plant Science Research Division, U.S. Department of Agriculture—J. F. Chaplin, Richard Gwynn, and J. Rennie Stavely; North Carolina State University, Department of Plant Pathology—G. B. Lucas, J. L. Apple, and N. T. Powell; Department of Genetics—T. J. Mann; and Department of Crop Science—D. U. Gerstel; Virginia Polytechnic Institute, Department of Plant Pathology—R. G. Henderson.

## *References*

Apple, J. L. 1962*a*. Transfer of resistance to black shank (*Phytophthora parasitica* var. *Nicotianae*) from *Nicotiana plumbaginifolia* to *N. tabacum*. Phytopathology 52:1. (Abstr.)

————. 1962*b*. Physiological specialization within *Phytophthora parasitica* var. *nicotianae*. Phytopathology 52:351-354.

Burk, L. G. 1967*b*. An interspecific bridge cross. *Nicotiana repanda* through *N. sylvestris* to *N. tabacum*. J. Hered. 58:215-218.

————, and V. H. Dropkin. 1961. Response of *Nicotiana repanda, N. sylvestris* and their amphidiploid hybrid to root knot nematodes. Plant Dis. Reporter 45:734-735.

————, and H. E. Heggestad, 1966. The genus *Nicotiana:* a source of resistance to diseases of cultivated tobacco. Econ. Bot. 20:76-88.

Chaplin, J. F. 1962. Transfer of black shank resistance from *Nicotiana plumbaginifolia* to flue-cured *N. tabacum*. Tobacco Sci. 6:182-187.

————. 1966. Comparison of black shank resistance from four sources. Tobacco Sci. 10:55-58.

————, and T. W. Graham. 1963. Brown spot resistance in *Nicotiana tabacum*. Tobacco Sci. 8:59-62.

————, and T. J. Mann. 1961. Interspecific hybridization, gene transfer and chromosomal substitution in *Nicotiana*. North Carolina Agr. Exp. Sta. Tech. Bull. 145:1-31.

————, T. J. Mann, and J. L. Apple. 1961. Some effects of the *Nicotiana glutinosa* type of mosaic resistance on agronomic characters of flue cured tobacco. Tobacco Sci. 5:80-83.

Clausen, R. E., and T. H. Goodspeed. 1925. Interspecific hybridization in *Nicotiana*, II. A tetraploid *glutinosa-tabacum* hybrid, an experimental verification of Winge's hypothesis. Genetics 10: 278-284.

Clayton, E. E. 1954. Identifying disease resistance suited to interspecific transfer. J. Hered. 45:273-277.

————. 1967. The transfer of blue mold resistance to tobacco from *Nicotiana debneyi*. Part III. Development of a blue mold resistant cigar wrapper variety. Tobacco Sci. 11:107-110.

————, and H. H. Foster. 1940. Disease resistance in the genus *Nicotiana*. Phytopathology 30. (Abstr.)

————, T. W. Graham, F. A. Todd, J. G. Gaines, and F. A. Clark. 1958. Resistance to the root knot disease of tobacco. Tobacco Sci. 2:53-63.

————, and T. E. Smith. 1942. Resistance of tobacco to bacterial wilt (*Bacterium solanacearum*). J. Agr. Res. 65:547-554.

Daulton, R. A. C. 1962. The behavior and control of the root knot nematode *Meloidogyne javanica* in tobacco as influenced by crop rotation and soil fumigation practices. Wed. Tobacco Sci. Congr. 3. Salisbury 181-190.

Dean, C. E., H. E. Heggestad, and J. J. Grosso. 1968. Transfer of blue mold resistance into F₁ tobacco hybrids. Crop Sci. 8:93-96.

Dukes, P. D., and J. L. Apple. 1968. Inoculum potential of *Phytophthora parasitica* var. *nicotiana* as related to factors of the soil. Tobacco Sci. 12:200-207.

Goodspeed, T. H. 1954. The genus *Nicotiana*. Chronica Botanica, Waltham, Mass.

————, and R. E. Clausen. 1928. Interspecific hybridization in *Nicotiana* VIII. The *sylvestris-tomentosa-tabacum* hybrid triangle and its bearing on the origin of tabacum. Univ. Calif. Pub. Bot. 11:245-256.

Graham, T. W. 1969. A new pathogenic race of *Meloidogyne incognita* on flue-cured tobacco. Tobacco Sci. 43-44.

Holmes, F. O. 1938. Inheritance of resistance to tobacco mosaic disease in tobacco. Phytopathology 28:553-561.

Kelman, A. 1953. The bacterial wilt caused by *Pseudomonas solanacearum*. North Carolina Agr. Exp. Sta. Tech. Bull. 99.

Lucas, G. B. 1965. Diseases of tobacco. Scarecrow Press, Inc. New York.

Moore, E. L., G. B. Lucas, and E. E. Clayton. 1952. Granville wilt and black shank resistant tobacco varieties. North Carolina State Exp. Sta. Bull. 378. 19 p.

———, and N. T. Powell. 1959. Dominant and modifying genes for resistance to black shank of tobacco. Proc. Assoc. of Southern Agr. Workers 211.

——— ———, G. L. Jones, and G. R. Gwynn. 1962. Flue-cured tobacco variety NC95 resistant to root knot, black shank and the wilt dieseases. North Carolina Agr. Exp. Sta. Bull. 419.

Ogden, W. B. 1963. Breeding wildfire-resistant tobacco Wisconsin Havana 501. Wisc. Agr. Exp. Sta. Bull. 562.

Okabe, N., and M. Goto. 1963. Bacteriophages of plant pathogens. Ann Rev. Phytopathol. 1:397-418.

Powell, N. T., and C. J. Nusbaum. 1960. The black shank root-knot complex on flue-cured tobacco. Phytopathology 899-906.

Schweppenhauser, M. A. 1968. Recent advances in breeding tobacco resistant to *Meloidogyne javanica*. Coresta Inf. Bull. (1) 9-20.

Smith, T. E., and E. E. Clayton. 1948. Resistance to bacterial wilt and black shank in flue-cured tobacco. Phytopathology 38:227-229.

———, and W. E. Earley. 1964. McNair's tobacco seed catalog. Laurinburg, N. C. p. 10-11.

Stokes, G. W. 1959. A rapid technique for indexing tobacco breeding lines for the necrotic reaction to TMV. Tobacco Sci. 3:129-130.

Troutman, J. L., and J. L. LaPrade. 1962. Effect of pH on black shank disease of tobacco. Va. Agr. Exp. Sta. Tech. Bull. 158.

Valleau, W. D. 1952. Breeding tobacco for disease resistance. Economic Botany 6, no. 1:69-102.

———, G. W. Stokes, and E. M. Johnson. 1960. Nine years experience with the *Nicotiana longiflora* factor for resistance to *Phytophthora parasitica* var. nicotianae in the control of black shank. Tobacco Sci. 4:92-94.

Wan, H. 1962. Inheritance of resistance to powdery mildew in *Nicotiana tabacum L.* Tobacco Sci. 6:180-183.

# 14 Sugarcane
## Ernest V. Abbott[1]

## *Origin, History, and Economic Importance*

Sugarcane is native to the tropical and subtropical regions of Asia, particularly India, China, and the islands of the southeastern Pacific. How, when, or where man first discovered the sweet juice of sugarcane and recognized its food value is not known, but primitive man carried sugarcane stalks on many of his early voyages. In this way sugarcane was transported from India and China to the Philippines, and by the Polynesians to many of the Pacific Islands. Overland it was carried from India to Arabia and thence to the countries bordering the Mediterranean. Columbus introduced it to the New World on his second voyage. It is now grown in some 50 countries or political units lying between approximately 40° north and 32° south latitude, in many of which it is a major item in the economy. It is an important crop for thousands of small farmers, in addition to being the basis of large plantation enterprises. World production of raw sugar in 1968 totaled 41,266,000 tons from 16,213,000 acres.

1. Plant Pathologist (Retired), Southern Region Crop Research Division, Agriculture Research Service, U.S. Department of Agriculture, Houma, Louisiana.

# Breeding and Seedling Testing Procedures

## Sources of Breeding Material

The basic material from which present-day commercial sugarcane varieties developed comprises a number of original forms from the *Saccharum* species *officinarum, spontaneum, sinense* (including *barberi*) and *robustum* (Arceneaux, 1965). Two World Collections of sugarcane germ plasm, composed of approximately 1,500 representatives of these species and related grasses, are available to sugarcane breeders. One is maintained by the United States Department of Agriculture at Canal Point, Florida, and the other by the Indian Sugarcane Breeding Institute at Coimbatore, India. Relatively few representatives of the species of *Saccharum* have been used extensively in breeding, and much remains to be done in systematically surveying the material available in the World Collections to identify new sources of disease resistance.

Each commercial sugarcane variety is a single clone that has been propagated vegetatively. The ease with which sugarcane may be propagated asexually permits the breeder rapidly to exploit any outstanding genotype (Warner, 1953; Burton, 1959), since increase and maintenance of the new genotype present no problem. Excellence as a sugar producer, however, is not always a reliable indication of the potential value of a sugarcane clone as a parent (Stevenson, 1965); but the best parents are generally those that most strongly exhibit the desired traits (Hebert and Henderson, 1959).

Sugarcane breeding follows the pattern of pre-Mendelian breeding: highly heterogeneous parents, often of the same ancestry, are crossed to give highly heterogeneous offspring. A single desirable individual may be propagated vegetatively to become a commercial variety, an empirical method that has yielded satisfactory practical results. Because of parental heterogeneity, the ability of a cane to transmit a particular trait can only be determined experimentally; thus, it is the custom to list parents that pass on particular traits rather than to list genetic information.

## Crossing

The entire sugarcane inflorescence is used as a unit in crossing. The flowers are not emasculated. Varieties that do not produce fertile pollen are used as females, and those with an abundance of pollen as males. Pollen fertility, not a constant characteristic, is influenced by climate or locality (Brandes and Sartoris, 1936).

In the tropics, sugarcane crosses are generally set up outdoors, since temperatures are favorable for pollen production, seed setting, and maturation. In subtropical areas, such as Florida, Louisiana, and South Africa, most of the crossing is done in plastic or glass-covered enclosures equipped for temperature control. For outdoor crossing the female variety may be grown in an isolated place to avoid contamination by undesired pollen, and the cut stalk with male tassel brought to the female, where it is placed in a sulfurous acid solution to maintain its viability during pollination.

For indoor crossing, stalks with female tassels are commonly airlayered or marcotted in the field several weeks in advance of flowering so that they may be transferred to the crossing enclosure on the roots formed at the nodes. The male tassels are then brought to the female at the proper time and placed in the acid solution (Dunckelman, 1959; Stevenson, 1965). Setting up the crosses in separate cubicles prevents pollen contamination. Parent canes may also be grown in large cans to facilitate their movement to photoperiod chambers or crossing enclosures. They may also be laid flat on the ground if necessary to avoid damage from high winds (Dunckelman, 1959).

In addition to bi-parental crosses of only one female and one male parent, the melting-pot cross, in which a number of elite males are interspersed among the tassels of selected female varieties, is used in Hawaii. In those crosses only the female parent is known with certainty (Mangelsdorf, 1953).

Photoperiod techniques are sometimes used to induce flowering of varieties that ordinarily do not flower, and for coordinating the blooming of desired male and female parents that do not normally flower at the same time (Dunckelman, 1959).

## Seedling Propagation

The number of seeds produced by a given cross will depend on compatibility of the parents, production and fertility of the pollen, and environmental conditions (particularly temperature) that influence seed setting and maturation. The number of seeds obtained per tassel may vary from zero to 2,000 or more, and can only be determined by trial.

The dried "fuzz" containing the seed is stripped from the tassel and planted in the greenhouse as soon as possible after harvesting. If planting must be delayed, the seeds may be stored for up to a year or more in a refrigerator, if frozen at approximately 0 F, though with some of viability.

The soil mixture used for the seedling flats should be sterilized to eliminate weed seeds and damping-off fungi. The "fuzz" is spread in a thin layer over the surface of the flats containing 3-4 inches of the soil mixture and watered thoroughly.

When of proper size the seedlings are transplanted to flats singly or as bunches of 5-15 seedlings (Warner, 1953), or to small pots singly or in bunches of 3-5 seedlings each (Hebert et al., 1962; Skinner, 1965). Procedures for transplanting to the field and spacing of the plants vary greatly at different breeding stations. Selections in the seedling nursery are made during the first year at most breeding stations, but are deferred until the stubble or ratoon year at others. The first selection is based primarily on agronomic type, including stalk diameter and height, sometimes on Brix as determined with a hand refractometer in comparison with control varieties, and on freedom from diseases. A minimum of 500-1,000 seedlings is desirable for evaluating a cross.

## Agronomic Testing of Selections

Agronomic testing of the selections begins in line tests established from the seedling nurseries, usually single rows 3-6 ft long. Stalk characters, vigor, plant habit, Brix, and freedom from diseases are the principal bases for selection in first line tests. Stand of cane is an added criterion in second line tests. Replicated plantings may begin with the second line test. Replication may be deferred until the third selection stage, which usually consists of 3-row plots. Plot weights and sucrose content based on 5- or 10-stalk samples of mature cane are usually obtained at this stage. Commonly 5-15% of the plants are selected in the original nursery; 5-15% of these may be selected for second line tests, and about 2% from second line tests are advanced to replicated plots.

As agronomic testing of the continually reduced number of selections progresses, plot size is increased and replications are usually not less than 4 or 5. Final selections are usually planted at several localities representing the soil and climatic conditions of the commercial area. From 7 to 12 years elapse between the breeding year and release of a variety to growers.

# Breeding for Resistance to Major Diseases

## Mosaic

Mosaic is one of the most widely distributed sugarcane diseases. Only Guiyana and Mauritius, among major producers of this crop, lack authentic reports of its occurrence. The disease causes varying degrees of destruction of the chlorophyll, and the general symptom is islands of normal green on a background of pale green or chlorotic areas. The principal effect is reduction of growth of the plant, the degree of which varies greatly with varieties and the strain of the virus present.

*Sources of Resistance.*    Clones of *Saccharum spontaneum* are the principal original sources of mosaic resistance in breeding (Dunckelman and Breaux, 1969). The Chinese canes (*S. sinense*) are highly resistant, but their hybrids have so many undesirable characteristics that this species can not be used to advantage in breeding. The noble canes (*S. officinarum*), the Indian canes (*S. barberi*), and the large wild cane of New Guinea (*S. robustum*), are generally susceptible (Abbott, 1961). Among commercial hybrids that have been used as sources of resistance in breeding are P.O.J. 2878, Cl. 41-223, and C.P. numbers 36-13, 47-193, 46-115, 52-1, 61-37, 61-39, 61-84, and 65-350 (Breaux and Dunckelman, 1969).

In a study of inheritance of resistance to mosaic, Breaux and Fanguy (1965) concluded that it was governed by the operation of multiple factors.

*The Pathogen.*    Ten strains of sugarcane mosaic virus (SCMV) have been formally described in the United States (Summers et al., 1948; Abbott and Tippett, 1966; Tippett and Abbott, 1968). Some of these have been identified from other countries (Abbott and Stokes, 1966).

Variability of the virus presents a formidable problem to the sugarcane breeder, who is continually confronted with new mutants that are virulent toward previously resistant commercial and breeding varieties. The frequent appearance of new virulent strains in Louisiana has seriously upset progress made in breeding for resistance and has necessitated continuous revamping of the breeding program (Breaux and Dunckelman, 1969).

*Techniques of Inoculation.*    Inoculum for use in resistance tests can be maintained on living stock culture plants. These may be grown from cuttings collected from infected plants in the field or established by inoculating young healthy plants with virus so collected. The stock cultures provide inoculum for transfer to differential hosts for strain identification, as well as for seedling inoculation. Inoculum is prepared by passing the youngest leaves from an infected plant through a food or other type of grinder, adding approximately 2 volumes of .01 M sodium sulfite, and straining the mass through cheesecloth. The inoculum can be stored in vitro at − 35 C for at least 12 months without loss of infectivity (Todd, 1961).

Two types of inoculation techniques may be used, the choice depending on the degree of severity of the test desired: (1) Sein's method (Sein, 1930), in which a strip of infected leaf is wrapped tightly around the spindle of the test plant and a fine needle used to prick through the tissues of infected leaf and host plant several times, or Matz's modification (Matz, 1933), in which a drop of infective juice is placed in the spindle and tissues pricked as above; and (2) the air brush (Bird, 1961) or air blast (Dean, 1960) methods, in which infective juice is forced into the tissues under pressure.

The pinprick methods are less severe and generally do not produce infection of the more resistant clones. However, they are laborious and not well adapted to testing large seedling populations. The pressure methods are preferred for large-scale testing, and where elimination of all except the very highly resistant individuals is desired. However, these methods are so severe that some clones that are resistant when exposed to natural spread in the field may become infected. Some susceptible clones may escape infection in greenhouse tests.

*Resistance Ratings.*    If immunity in parent varieties is the breeder's goal, he will rate them as resistant if there is no infection, or as susceptible if any plants become infected. However, the pressure inoculation techniques are so severe that some compromise may be necessary and clones classed as highly resistant if only a low percentage of plants becomes infected. Still further relaxation of standards may be in order when rating the clones selected from seedling progenies for agronomic testing. Moderate susceptibility may be acceptable in clones that excel in yields, or which possess some other especially desirable characteristic.

The standards of mosaic resistance must be geared to the commercial needs of the area. Resistance to mosaic infection under natural conditions may be given greater weight than the results of artificial inoculation. Field resistance is determined in replicated small plots of test varieties, and of controls which represent a range of resistance and susceptibility to mosaic. Source of inoculum is provided by bordering the test plots with mosaic-infected cane in alternate rows or every third row, and by inserting a plot of infected cane at intervals in the rows of test varieties. Plots of maize or sorgo may also be interplanted for multiplication of vectors and as an added source of mosaic inoculum.

## Red Rot

Red rot occurs in all sugarcane-producing countries and is a major disease in several countries in the subtropics. In Louisiana its principal damage is to planted seed cane. In Florida, Hawaii, India, Mauritius, South Africa, Taiwan, and Queensland, mill cane losses are of primary importance. The principal symptom is dull red discoloration of internal stalk tissues, interrupted by occasional whitish patches elongated at right angles to the long axis of the stalk. Sucrose content of mill cane is markedly reduced, and germination of seed cuttings is adversely affected.

*Sources of Resistance.*    Some forms of *S. spontaneum* are resistant to red rot and are important original sources of resistance in breeding (Abbott, 1938). In breeding commercial type canes, resistant hybrid varieties that have been used as parents include B. 43-337, C.P. 36-13, C.P. 47-193, C.P. 48-103, C.P. 52-68, Co. 213, Co. 281, H. 32-8560, H. 37-1933, Pindar, and Trojan.

*Nature of Resistance.*    Two types of resistance are recognized: physiological, or the ability of the plant to prevent or suppress development of the parasite within the host tissues; and morphological, in which infection by, or spread of, the parasite within the tissues is mechanically prevented or retarded. Physiological resistance is the more important, since the fungus does not develop rapidly within the tissues after infection has occurred. Phenolic compounds in the tissues of resistant plants may be involved in this type of resistance (Abbott, 1938).

Morphological resistance depends on the arrangement of components of plant tissues that prevent infection, or obstruction of vascular bundles that retard migration of the fungus spores through the stalks (Atkinson, 1938). Some varieties that are susceptible physiologically do not become infected readily. This type of resistance appears to be related to thickness of the epidermis and cuticle, thickness of rind, relative abundance of vascular bundles underneath the rind, character of bundle sheath, and thickness of bud scales (Edgerton, 1955).

*The Pathogen.*    The imperfect stage of the red rot fungus is *Collectotrichum falcatum* Went. The perfect stage is *Physalospora tucumanensis* Speg. Both may occur in lesions on leaf midribs or blades, sheaths, and bud scales, and conidia may be produced abundantly in stalk tissues (Carvajal and Edgerton, 1944).

Two cultural races are recognized: a "light" race which produces light gray, cottony, floccose mycelium on oatmeal decoction agar; and a "dark" race which produces compact, velvety, dark gray mycelium. Some isolates are intermediate (Abbott, 1938).

The fungus is highly variable in pathogenicity. Although specific pathogenic races have not been described, a comparison of several isolates of the fungus on a range of sugarcane varieties usually demonstrates differences in virulence. A common experience in areas where red rot is a major disease is that a variety resistant when it is released may become susceptible after several years of commercial culture (Abbott, 1938; Chona and Padwick, 1942). This indicates development of specialized races of the fungus.

*Inoculation Techniques and Resistance Ratings.*   The fungus is easily isolated from infected plant material by standard laboratory techniques. Oatmeal decoction agar is excellent for isolation, for maintenance of stock cultures, and for production of inoculum for resistance tests. Inoculum which consists of a spore suspension in water or weak oatmeal decoction is prepared by washing the surface of a sporulating culture, or by macerating the culture mechanically, as with a rotary blender, and filtering the mass through cheesecloth. For comparable results in different inoculation tests it is desirable to standardize the spore load in the inoculum, for which approximately 25 spores per low power microscope field are satisfactory. Cultures kept in vitro for any length of time may lose their sporulating ability and virulence (Abbott et al., 1965; Steindl, 1965).

The type of resistance test will depend on the nature of the disease in the area. In India, Hawaii, and Taiwan, for example, where the disease is primarily one of standing cane, mature stalks are inoculated in the field by introducing the spore suspension into the stalk with a hypodermic syringe. Usually 10-20 stalks are inoculated. In Louisiana, where the disease primarily affects seed cuttings, 5-10 stalks of cane such as are used for planting are inoculated by introducing the spore suspension into small holes bored into each of two internodes per stalk. In either case, the test varieties are rated on susceptibility after an incubation period by splitting the stalks longitudinally and observing or measuring the extent of rotting. Control varieties representing a range of resistance and susceptibility are included for comparison. Abbott et al. (1965) described five classes of resistance-susceptibility.

Srinivasan (1962) described a method of determining resistance in the seedling stage, in which susceptible seedlings may be eliminated on the basis of injury induced by spraying the seedling flat with a spore suspension of the fungus. Elimination of susceptible seedlings at this stage would save the expense of subsequent agronomic testing of them.

## Pythium Root Rot

Root rot, sometimes without specific identification of the cause, has been reported from practically every sugarcane-producing country. At one time a major disease in subtropical climates where sugarcane is grown, particularly on heavy, poorly drained soils, it has receded somewhat in importance with the development of resistant varieties and improved cultural practices. Nevertheless, it continues to be an important consideration in breeding for disease resistance in subtropical areas.

*Sources of Resistance.*    Most forms of *S. spontaneum* that have been tested are resistant to Pythium root rot, and those of *S. barberi* are slightly less so. These species are the principal sources of resistance in breeding. Most forms of *S. officinarum* are susceptible. The resistant commercial canes of the southern United States are mostly tri-specific hybrids of the three species (Rands and Dopp, 1938). The Chinese canes (*S. sinense*) are resistant, but hybrids derived from them are generally of poor economic quality. Hybrid canes that have been of value in breeding for resistance to root rot in India include Kassoer, P.O.J. 2878, and N.Co. 339 (Srinivasan and Narashimhan, 1963).

*The Pathogen.*    Although sugarcane roots may be infected by several species of *Pythium, P. arrhenomanes* Drechsl. is the principal causal agent. It is considered synonymous with *P. graminicolum* by some pathologists, but Drechsler (1936) concluded that the two are distinct and that the cane pathogen should be designated *P. arrhenomanes.*

Rands and Dopp (1938) demonstrated physiologic specialization and varietal adaptation. They found significant differences in virulence between localities and in multiplication of more virulent biotypes following general adoption of more resistant varieties.

The pathogen can be isolated in culture by thoroughly washing fragments of rotted rootlets in sterile water, drying with sterile filter paper, and plating on water agar. After incubation the plates are inverted and hyphal tips transferred to cornmeal agar, which is a suitable medium for maintaining stock cultures.

*Inoculation Techniques and Resistance Ratings.*    Koike (1965a) described two techniques for resistance tests. In one, single-eye cuttings of test varieties are planted in 2-gal cans of soil fumigated with methyl bromide, with 3 replications each inoculated and not inoculated. After 3-4 months, dry weights of roots and tops and the extent of rootlet rotting are determined. Results are expressed as per cent reduction in plant weight from inoculation. In the second technique, for testing seedling progenies, a thin layer of 7- to 10-day-old sand-cornmeal culture of *Pythium* is placed between two layers of fumigated soil of equal depth in a seedling flat. "Fuzz" of the sugarcane cross to be tested is planted on the surface. As the seedlings grow, their roots come in contact with the layer of inoculum, and the highly susceptible ones are killed. Seedlings showing little damage are transplanted to the field, and clones selected from them are further screened for root rot by planting in gallon pots of fumigated soil, with replications for inoculation and controls. After 2 months the clones are graded into 5 classes of resistance-susceptibility on the basis of vigor, height and diameter of stalks, and number of secondary shoots.

## Smut

Sugarcane smut is widely distributed in the Eastern Hemisphere, where it is a major disease in several countries. It has been reported only from Argentina and Brazil in the Western Hemisphere (Antoine, 1961). It is recognized by the long black smut "whips," which arise from the apex of affected plants.

*Sources of Resistance.*    With some exceptions, the noble canes (*S. officinarum*) are highly resistant to smut, although Srinivasan (1966) found some 80 forms to be susceptible. *S. barberi* and the Indian forms of *S. spontaneum* are generally highly susceptible and are principally responsible for the low degree of resistance of the Coimbatore (Co.) varieties. Forms of *S. spontaneum* from Java are much more resistant. However, Srinivasan and Chenulu (1956) rated 246 of 273 forms tested as resistant, and concluded that it should not be difficult to select variants of *S. spontaneum* clones that do not transmit smut susceptibility to their progeny. McMartin (1948) found in Natal that the most resistant canes were derived from crosses between *S. officinarum* and *S. spontaneum*. A difficulty in breeding for smut resistance in subtropical areas is that varieties derived from *S. barberi*, which are best adapted to subtropical conditions, are susceptible to smut.

Smut resistant hybrid varieties include P.O.J. 2878, N.Co. 334, No.Co. 339, Co. 449, Co. 527, M. 134/32, B. 41-227, B. 54-142, and Tuc. 2645 (Martin, 1965).

*The Pathogen.*    *Ustilago scitaminea* Sydow is a parasite of young meristematic tissue, entering through the lower part of the bud below the scales. Each diseased bud represents a separate and independent infection originating from an outside source.

The chlamydospores, which are produced abundantly in the smut whips, germinate readily under moist conditions, each giving rise to a promycelium, from each cell of which sporidia bud out. If kept dry, the chlamydospores may retain their vitality for months (Antoine, 1961).

Chona (1956) found no evidence of physiologic specialization among isolates of the fungus from different sources.

*Collecting and Preparing Inoculum.*   Chlamydospores are collected, dried in the shade, and stored in paper envelopes over $CaCl_2$. Only material showing at least 95% germination is used for inoculum. A heavy spore suspension is prepared by stirring 250 g of spores into 5 liters of water (Srinivasan, 1969).

*Inoculating and Rating.*   Twenty 3-bud seed cuttings of test varieties are dipped in the spore suspension for 1/2 hr., drained, and planted immediately. Resistance ratings are based on percentage of plants showing smutted whips. The whips are counted and removed as they appear.

## Downy Mildew

Downy mildew of sugarcane is confined to the western Pacific region and India (Hughes and Robinson, 1961).

*Sources of Resistance.*   Resistant or highly resistant clones occur in all five species of the genus *Saccharum*. These include Badila (*S. officinarum*), 28 N.G. 251 (*S. robustum*), and several forms of *S. spontaneum*. Hybrid varieties rated as resistant include Pindar, Ragnar, and Luna (Australia); F. 137, F. 138, and numerous P.T. varieties (Taiwan); H. 37-1933 and H. 32-8560 (Hawaii); and the worldwide commercial cane, N.Co. 310 (Chu, Liu and Lo, 1957; Robinson and Martin, 1956; Husain, Daniels and Krishnamurthi, 1968).

*The Pathogen.*   *Sclerospora sacchari* Miy. occurs in sugarcane in both the asexual and sexual forms. The asexual stage is the more common and forms the characteristic down under favorable moisture and temperature conditions. Infected seed pieces and conidia are the most important agencies in transmission. Conidia, produced during the night, are freed from the conidiophores as soon as they mature, and they germinate shortly afterwards, provided free moisture is present. They are carried by wind or rain to buds and young tissues, which are susceptible to infection. Factors affecting production of the sexual stage are not well understood. Oogonia are formed in the tissues of splitting leaves, but the role of oospores in transmitting the disease is undetermined. There is no evidence of physiological specialization.

*Resistance Tests and Ratings.*    Resistance tests depend on natural transmission in the field. Test varieties are planted in replicated plots of 20 sets and a 2 m plot of diseased cane is planted alternately with each variety plot. Later, a row of a susceptible variety of maize is planted beside each row of cane. These plants become infected quickly and act as sources of infection. During the growth of the crop the numbers of diseased stalks and stools are counted (Steindl, 1965; Leu, 1968), and resistance ratings are based on infection percentages.

## Leaf Scald

Leaf scald, a bacterial vascular disease, is widely distributed in both Eastern and Western Hemispheres. One of the troublesome aspects of the disease is that many cane varieties may be infected without exhibiting any symptoms, or the symptoms are so inconspicuous as to escape detection. Thus, the disease may be passed through quarantine and introduced into a country, where it suddenly may appear in epidemic proportions if a susceptible variety is being grown (Martin and Robinson, 1961).

The disease has two distinct phases: (1) chronic, manifested by narrow whitish stripes on leaves and sheaths, or stunted stalks with blanched leaves, and profuse development of side shoots; and (2) acute, in which affected shoots suddenly wilt and die as though killed by drought, but which are generally otherwise symptomless. In badly diseased stalks bacteria may accumulate in lysigenous cavities. The disease is transmitted in cuttings and by knives.

*Sources of Resistance.*    The species of *Saccharum* have not been systematically surveyed for sources of resistance to leaf scald. Some varieties of *S. officinarum* and *S. spontaneum* are resistant, while those of *S. robustum* are susceptible (Stevenson, 1965). Hybrid varieties rated resistant in the countries where they originated include: B. (Barbados) 37161 and 37172; Co. (Coimbatore) 241 and 419; M. (Mauritius) 134/32, 147/44, 31/45, 202/46, 93/48, 253/48, and Ebene 1-37; H. (Hawaii) 32-8560; and Pindar, Q. 50, and 47, bred in Queensland (Martin, 1965).

*The Pathogen.*    *Xanthomonas albilineans* (Ashby) Dowson is a short, slender rod, motile by a single polar flagellum. Wilbrink's agar

(Martin and Robinson, 1961) is satisfactory for isolation of the organism from sugarcane tissues and for maintenance of stock cultures. Marked differences in varietal reaction to the disease in different countries suggest the existence of strains of the pathogen, but experimental evidence is lacking. Varieties must be tested against the pathogen present in each cane-growing area.

*Resistance Tests and Ratings.*    Koike (1965*b*) in Hawaii prepared inoculum as a tap water suspension of a 7-10-day-old culture of the pathogen, which was squirted on the cut surfaces of primary shoots 20-30 cm tall, cut a few cm above the shoot apex. The inoculated shoots were capped with aluminum foil. The inoculated shoots and side shoots produced following inoculation were examined periodically for leaf scald symptoms as a basis for resistance ratings of test varieties, in comparison with controls of known reaction to the disease, which were interspersed through the plots and similarly inoculated.

In Queensland, Egan (1969) found the aluminum cap method a rapid and efficient means of segregating resistant and highly susceptible varieties, although it did not separate intermediate and susceptible groups.

## Ratoon Stunting Disease

Ratoon stunting disease (RSD) was not identified as an infectious disease of sugarcane until 1945 because it lacks specific external symptoms and manifests itself principally by general stunting and unthriftiness of the plant. A wide range of unfavorable environmental conditions may cause similar symptoms. Internal symptoms consist of reddish-orange discoloration of the vascular bundles at the node. During the past 20 years RSD has been reported from most of the cane-growing countries of the world. It is generally regarded as one of the principal causes of gradual yield decline of sugarcane varieties (Steindl, 1961).

*Sources of Resistance.*    Extensive surveys of parent material for sources of resistance to RSD have not been made because of the lack of methods of testing for varietal resistance other than comparison of field-plot yields from healthy and RSD-infected cane. Few if any varieties are highly resistant or immune to infection, but some show marked tolerance and their growth is not greatly affected. Among highly tolerant hybrid varieties are: Q. 44, Comus, Eros, and Pindar (Australia); Co. 301 (India); R. 337 and R. 397 (Réunion; and C.P. 29-116 and C.P. 52-68 (U.S.A.).

The pathogen is a spherical virus (Gillaspie et al., 1966). There is no conclusive evidence of the existence of strains.

*Inoculation Technique and Resistance Ratings.*    The virus is readily transmitted by mechanical inoculation with juice extracted from infected plants. For varietal resistance tests, progeny of heat-treated seed cane (disease free) are planted in replicated plots in the field, along with progeny of inoculated (diseased) cane. Use of progeny as a source of seed cane avoids any adverse effect of direct inoculation or stimulatory effect of heat treatment. All inoculated replications are planted in one row, and all uninoculated cane in an adjoining row, to prevent mechanical transmission of the virus along the row during cultivation and harvesting operations. Varieties of known reaction to the disease are planted as standards for comparison. Relative tolerance is based on percentage of yield reduction caused by disease in test varieties in comparison with the controls (Abbott et al., 1965; Steindl, 1961).

## Problems for Future Research

It is evident from the foregoing discussions that much remains to be done in surveying sugarcane species to discover new sources of disease resistance in breeding. Few studies of the inheritance of disease resistance in sugarcane have been made. Such research would aid breeders to select more intelligently breeding material and parental combinations. A challenge awaits pathologists to contribute to better understanding of the nature of disease resistance in sugarcane and the physiology of host-pathogen relationships. Breeding for disease resistance would also be advanced by artificial inoculation techniques that would more closely correlate resistance ratings with disease reaction of sugarcane varieties in the field.

## References

Abbott, E. V. 1938. Red rot of sugarcane. U.S. Dept. Agr. Tech. Bull. 641. 96 p.

————. 1961. Mosaic, p. 406-450. *In* J. P. Martin, E. V. Abbott, and C. G. Hughes [ ed.] Sugar-cane diseases of the world. Vol. I. Elsevier Publishing Co., Amsterdam.

————, and I. E. Stokes. 1966. A world survey of sugar cane mosaic virus strains. Sugar y Azucar 61(2):27-29.

————, and R. L. Tippett. 1966. Strains of sugarcane mosaic virus. U.S. Dept. Agr. Tech. Bull. 1340. 25p.

————, N. Zummo, and R. L. Tippett. 1965. Methods of testing sugarcane varieties for disease resistance at the U.S. Sugarcane Field Station, Houma, La. Int. Soc. Sugar Cane Technol. Proc.. 12:1138-1142.

Antoine, R. 1961. Smut, p. 326-354. *In* J. P. Martin, E. V. Abbott, and C. G. Hughes [ed.] Sugar-cane diseases of the world. Vol. #I. Elsevier Publishing Co., Amsterdam.

Arceneaux, G. 1965. Cultivated sugarcanes of the world and their botanical derivation. Int. Soc. Sugar Cane Technol. Proc. 12:844-854.

Atkinson, R. E. 1938. On the nature of resistance to red rot. Int. Soc. Sugar Cane Technol. Proc. 6:684-692.

Bird, J. 1961. Inoculation of sugarcane plants with the mosaic virus using the airbrush. J. Agr. Univ. Puerto Rico 45:1-7.

Brandes, E. W., and G. W. Sartoris. 1936. Sugarcane: its origin and improvement. U.S. Dept. Agr. Yearbook 1936:561-623.

Breaux, R. D., and P. H. Dunckelman. 1969. Breeding for resistance to sugarcane mosaic with interspecific hybrids. Int. Soc. Sugar Cane Technol. Proc. 13: 927-932.

————, and H. P. Fanguy. 1965. Breeding behavior of resistance to mosaic in sugarcane progenies and its association with some agronomic characters. Int. Soc. Sugar Cane Technol. Proc. 12:773-778.

Burton, G. W. 1959. Principles of breeding vegetatively propagated plants. Int. Soc. Sugar Cane Technol. Proc. 10:661-669.

Carvajal, F., and C. W. Edgerton. 1944. The perfect stage of *Colletotrichum falcatum*. Phytopathology 34:206-213.

Chona, B. L. 1956. Presidential address to Pathology Section. Int. Soc. Sugar Cane Technol. Proc. 9:975-986.

————, and G. W. Padwick. 1942. More light on the red rot epidemic. Indian Farming 3(2):70-73.

Chu, T. L., H. P. Liu, and H. C. Lo. 1957. Downy mildew resistance tests for new sugarcane varieties in Taiwan. Taiwan Sugar Exp. Sta. Rep. 16:1-21.

Dean, J. L. 1960. A spray method for inoculating sugarcane seedlings with the mosaic virus. U.S. Agr. Res. Service. Plant Dis. Reporter 44:874-875.

Drechsler, C. 1936. *Pythium graminicolum* and *P. arrhenomanes*. Phytopathology 26:676-684.

Dunckelman, P. H. 1959. Transition: effectiveness and implications of breeding sugarcane in an indoor environment. Sugar Bull. 74:229-252.

————, and R. D. Breaux. 1969. Screening for mosaic resistance in *Saccharum spontaneum* at Houma, Louisiana, 1964-68. Sugar y Azucar 64(10):16, 18.

Edgerton, C. W. 1955. Sugarcane and its diseases. La. State Univ. Press, Baton Rouge, La. 290 p.

Egan, B. T. 1969. Evaluation of the aluminum cap method for leaf-scald disease resistance testing in Queensland. Int. Soc. Sugar Cane Technol. Proc. 13:1153-1158.

Gillaspie, A. G., J. E. Irvine, and R. L. Steere. 1966. Ratoon stunting disease virus. Assay technique and partial purification. Phytopathology 56:1426-1427.

Hebert, L. P., R. D. Breaux, and H. P. Fanguy. 1962. Bunch-planting experiments with sugarcane seedlings at the U.S. Sugar Cane Field Station, Houma, La. Int. Soc. Sugar Cane Technol. Proc. 11:553-560.

———, and M. T. Henderson. 1959. Breeding behavior of certain agronomic characters in progenies of sugarcane crosses. U.S. Dept. Agr. Tech. Bull. 1194. 54 p.

Hughes, C. G., and P. E. Robinson. 1961. Downy mildew disease, p. 141-162. In Martin, Abbott, and Hughes [ed.] Sugar-cane diseases of the world, Vol. I. Elsevier Publishing Co., Amsterdam.

Husain, A. A., J. Daniels, and M. Krishnamurthi. 1968. The inheritance of resistance to downy mildew disease. Sugarcane Pathol. Newsletter 1:42-44.

Koike, H. 1965a. Methods for testing sugarcane for resistance to Pythium root rot. Int. Soc. Sugar Cane Technol. Proc. 12:1183-1187.

———. 1965b. The aluminum-cap method for testing sugarcane varieties against leaf scald disease. Phytopathology 55:317-319.

Leu, L. S. 1968. Methods for testing the resistance of sugarcane to disease (4) Downy mildew disease. Sugarcane Pathol. Newsletter 1:38-41.

McMartin, A. 1948. Sugar-cane smut. A report on visits to the sugar estates of Southern Rhodesia and Portuguese East Africa, with general observations on the disease. S. Afr. Sugar J. 32:737-749.

Mangelsdorf, A. J. 1953. Sugar cane breeding in Hawaii. Part II, 1921-1952. Hawaiian Planters' Rec. 54:101-137.

Martin, J. P. 1965. The commercial sugarcane varieties of the world and their resistance and susceptibility to the major diseases. Int. Soc. Sugar Cane Technol. Proc. 12:1213-1225.

———, and P. E. Robinson. 1961. Leaf scald, p. 78-107. In J. P. Martin, E. V. Abbott, and C. G. Hughes [ed.] Sugar-cane diseases of the world, Vol. I. Elsevier Publishing Co., Amsterdam.

Matz, J. 1933. Artificial transmission of the sugarcane mosaic. J. Agr. Res. 46:821-840.

Rands, R. D., and E. Dopp. 1938. Pythium root rot of sugarcane. U.S. Dept. Agr. Tech. Bull. 666. 96 p.

Robinson, P. E., and J. P. Martin. 1956. Testing sugar cane varieties against Fiji disease and downy mildew in Fiji. Int. Soc. Sugar Cane Technol. Proc. 9:986-1011.

Sein, F. 1930. A new mechanical method for artificially transmitting sugar-cane mosaic. J. Dept. Agr. Puerto Rico. 14:49-68.

Skinner, J. C. 1965. Sugar cane selection experiments. Int. Soc. Sugar Cane Technol. Proc. 11:561-567.

Srinivasan, K. V. 1962. A technique for the elimination of red rot susceptible seedlings at an early stage. Current Sci. 31:112-113.

———. 1966. Ann. Rep. Sugarcane Breeding Institute, Coimbatore. 1965-66:114.

———. 1969. Methods for testing the resistance of sugarcane to disease. (5) Sugarcane smut. Sugarcane Pathol. Newsletter. 2:7.

———, and V. V. Chenulu. 1956. A preliminary study of the reaction of Saccharum spontaneum to red rot, smut, rust and mosaic. Int. Soc. Sugar Cane Technol. Proc. 9:1097-1107.

————, and R. Narashimhan. 1963. Evaluating varieties for resistance of Pythium root rot. Indian Sugar Res. and Dev. 7:273-275.

Steindl, D. R. L. 1961. Ratoon stunting disease, p. 433-459. *In* J. P. Martin. E. V Abbott, and C. E. Hughes [ ed.] Sugar-cane diseases of the world. Vol. I. Elsevier Publishing Co., Amsterdam.

————. 1965. Testing sugarcane varieties for disease resistance at the Bureau Pathology Farm, Queensland. Int. Soc. Sugar Cane Technol. Proc. 12:1133-1137.

Stevenson, G. C. 1965. Genetics and breeding of sugarcane. Longmans, London. 284 p. Illus.

Summers, E. M., E. W. Brandes, and R. D. Rands. 1948. Mosaic of sugar cane in the United States, with special reference to strains of the virus. U.S. Dept. Agr. Tech. Bull. 955. 124 p.

Tippett, R. L., and E. V. Abbott. 1968. A new strain of sugarcane mosaic virus in Louisiana. U.S. Agr. Res. Serv. Plant Dis. Reporter 52:449-451.

Todd, E. H. 1961. Long term storage of the sugarcane mosaic virus. U.S. Agr. Res. Serv. Plant Dis. Reporter 45:178-179.

Warner, J.N. 1953. The evolution of a philosophy of sugar cane breeding in Hawaii. Hawaiian Planters' Rec. 54:139-162.

# 15 Barley
## C. W. Roane[1]

## *Introduction*

Barley is a widely adapted, genetically variable, dependable feed crop that has been studied extensively by geneticists, cytologists, plant pathologists, physiologists, entomologists, and soil scientists. Its taxonomy has been debated by many, and it has been the subject of study by numerous plant geographers and collectors. An introduction to the vast literature on barley is available in the publication entitled *Barley: Origin, Botany, Culture, Winterhardiness, Genetics, Utilization, Pests,* U.S. Dept. Agr. Handbook 338, 1968.

*Hordeum vulgare* L. is the primary cultivated barley, but *H. distichum* L. emend. Lam. is also grown throughout the world. These species may be hybridized easily. Cultivated barley is self-pollinated, and the techniques for breeding are well known.

1. Professor of Plant Pathology, Virginia Polytechnic Institute.

## Hybridizing Barley

Barley can be crossed most successfully in glasshouses where sturdy plants with prolific inflorescences are produced when grown in soil in 6-inch clay pots under conditions of 15-20 C, supplemental light, and intermittent fertilization. Plants should be started in a warm glasshouse, where for winter barley types the temperature can be reduced gradually to 2-5 C and held there for a period of 6 weeks to assure vernalization and adequate tillering. The temperature then is increased gradually to 15-20 C and held there for the final growing period. When feasible, lights may be turned on at midnight for 1 or 2 hr to shorten the time needed for flowering. Robust spikes usually are produced if the soil has a pH of near 6.0 and is maintained at a high fertility level. On 6-rowed cultivars, lateral florets are removed and central florets are clipped and emasculated one or two days before anthesis. Following emasculation and pollination, the spikes should be protected from stray pollen by glassine envelopes or tubes until the caryopses enlarge. A seed set of 90% or more can be expected if adequate pollen is available.

## Leaf Rust Resistance

Barley is attacked by *Puccinia graminis* Pers., *P. striiformis* West., *P. hordei* Otth., and *Uromyces* spp. Each of these rusts damages barley in some production center of the world, but *P. hordei* will be used to illustrate the principles of breeding rust-resistant cultivars.

### Sources of Resistance

There are numerous leaf-rust resistant barley cultivars (Mains and Martini, 1932; Moseman, 1956a; Nover and Mansfeld, 1959), but few have been analyzed genetically. The unified differential cultivars have been partially analyzed, and relationships have been established among genes in these and several other cultivars (Roane and Starling, 1967; 1970). Genes conditioning leaf rust reactions have been found at six loci. They are distributed among the following cultivars, names of which are accompanied by their U.S.D.A. cereal investigation numbers (CI nos.): $Pa$ in Kwan (1016), Oderbrucker* (940), Speciale* (7536), and Sudan* (6489); $Pa_2$ in Alagon (3530-2), Barley 305 (6015), Batna (3391), Carre 180 (3390), Juliaca (1114), Modia (2483), Morocco (4975), Peruvian (2441), Purple Nepal (1373), Reka 1* (5051),

Ricardo (6306), and Weider = No. 22 from N.S.W. (1021); $Pa_3$ in Estate (3510); $Pa_4$ in Franger (8811), Gold* (1145), and Lechtaler* (6488); $Pa_5$ in Cebada Capa (6193); $Pa_2$ and $Pa_5$ in Quinn* (1024); and $Pa_2$ and $Pa_6$ in Bolivia* (1257). Cultivars marked with asterisks are used along with Egypt 4* (6481) for differentiating races. $Pa_4$ is linked with the powdery mildew conditioning gene $Ml_a$ on chromosome 5 (Nilan, 1964). Most of the genes conditioning leaf rust reactions are dominant or partially dominant and are easy to work with in breeding programs. Additional genetic studies should be conducted to determine relationships of genes among several hundred cultivars for which resistance has been observed. This may lead to an increase in the number of genes available for breeding leaf-rust resistant cultivars.

Cultivars with resistance to other rusts of barley are cited by Moseman (1963, 1971a).

## Physiologic Specialization

Levine and Cherewick (1952) combined the previously described races of *P. hordei* into 52 "unified races." Although the differentials for the unified races have been found to possess 5 genes as shown above, other race differentiating genes may be present but undetected. A sixth gene, $Pa_3$, occurs in Estate (Roane and Starling, 1967). This variety should be included as a differential in future race studies.

## Manipulation of Rust Inoculum

Rust cultures characterized for their virulence traits should be preserved. They should be increased from single spores or at least from single pustules and propagated on susceptible seedlings. Spores may be shaken onto plastic or paper sheets or glass panes, passed through a 100-mesh screen, and air dried for 24 hours. Then they may be stored in stoppered vials at 2-5 C, or in sealed tubes for storage in liquid nitrogen refrigerators (Loegering, Harmon, and Clark, 1966) or, more simply, in vacuum tubes (Sharp and Smith, 1957). The latter method has been used by the author to maintain viable spores of several species of rust since 1954. Spores rehydrated in a saturated atmosphere for 24 hours will germinate adequately.

For creating rust epidemics in areas where dews may be expected, spores diluted with talc can be dusted onto spreader rows late in the day. Where dews are not expected, spores may be suspended in light oils and sprayed onto spreader rows (Rowell and Hayden, 1956) or they may be injected into young tillers with the aid of hypodermic syringes. For barley leaf rust, field inoculum must be applied to tillering or jointing plants to assure adequate infection in time for selection.

## Breeding

Seedling and adult plant or field resistance to leaf rust are known, but field resistance is required. Cultivars showing the best field resistance and adaptability should be used as parental material for the breeding program. It is essential to determine the virulence genes existing in the rust population. Representative rust cultures with particular virulence genes should be maintained permanently for use in seedling tests. Indexing for resistance may be done with $F_2$ and $F_3$ seedlings in glasshouses, but nursery epidemics are essential to assure selection of plants with field resistance from segregating populations.

The use of multiline varieties has not been exploited for any disease of barley, but this approach should not be overlooked.

# Scald Resistance

*Rhynchosporium secalis* (Oud.) J. J. Davis, the cause of scald, is destructive in many cool, humid, winter barley production areas of the world. Resistance has been reported in many cultivars (Reed, 1957) but breeding scald-resistant cultivars is complicated by the existence of several virulence genes in the pathogen (Schein, 1960). It is necessary to breed for resistance to all virulence genes occurring in *R. secalis* in a region. Fortunately, the spores of *R. secalis* seem not to move very far in nature; therefore, one needs to breed only against locally occurring races. Imported seed should be treated with mercurial fungicides to prevent the introduction of new races.

## Genetics and Sources of Resistance

Nine genes are reported to condition reaction to scald, although the number of loci involved is not known (Starling, Roane, and Chi, 1971). *Rh, Rh3,* and *Rh4* appear to be alleles or are closely linked and form a complex locus which is independent of *Rh2* and *Rh5*. The relationship of these genes to *rh6, rh7, rh8,* and *Rh9* is unknown. Cultivars grouped according to known genes conditioning resistance, as reported by various workers, are as follows: *Rh* in Brier (7157); *Rh2* in Atlas (4118) and Atlas 46 (7323); *Rh3* in Atlas 46, Bey (5581), Osiris (1622), Rivale (C.A.N. 258), Turk (5611-2), Unnamed (3515), and (8256); *Rh4* in LaMesita (7565), Modoc (7566), Osiris, and Trebi (936); *Rh5* in Turk; *rh6* and *rh7* in Jet (967) and Steudelli (2226); *rh8* in Nigrinudum (2222); *Rh9* in Abyssinian (668) and Kitchin (1296).

Hudson (8067) is monogenic at the complex *Rh-Rh3-Rh4* locus. The genetics of scald resistance is in a state of flux. A revision of the gene symbols may be forthcoming.

## Inoculation Procedures

*R. secalis* is a slow growing fungus; therefore, rapid increases of inoculum are best obtained by streaking plates of acidified (*p*H 5.0) lima-bean agar with spore suspensions and incubating them at 15-20 C for 10-14 days. Inoculum may be prepared by adding water to the plates and scraping the colonies to produce a spore suspension, or by blending the agar and fungus in a high-speed chopping device. Spore concentration is dictated by experience, but a minimum of 30,000/ml is suggested.

Seedlings in the 2-leaf stage are misted for 1 hr prior to inoculation, sprayed with a spore suspension, misted for an additional 8 hr, and dried slowly for 12 hr. An incubation period of 12-16 days at 16-20 C is appropriate. The scale of 0 to 4, described by Dyck and Schaller (1961), is useful for classifying reaction types, although additional classes may sometimes be necessary. Plants with 0 to 2 reactions are resistant, and those with 3 and 4 reactions are susceptible, although the partitioning of host responses into resistant and susceptible classes is arbitrary. Furthermore, higher temperature may alter the responses obtained at 16-20 C.

## Breeding

Selection of agronomically desirable, scald-resistant plants is best done in the field. However, scald may be of sporadic occurrence in nurseries if one depends only upon natural spread of *R. secalis*. To assure uniform distribution of scald, nurseries must be inoculated with laboratory grown spores or with scald-bearing residue from a previous crop. The latter source of inoculum is economical but may complicate selection of scald-resistant lines by introducing a high level of net and spot blotch inoculum. A breeder may consider this disease pattern either an advantage or disadvantage.

# Powdery Mildew Resistance

The fungus causing powdery mildew of barley, *Erysiphe graminis* (DC.) Mérat f. sp. *hordei* Marchal, is universally destructive in the absence of resistant cultivars. The genetics of this host-pathogen system has

TABLE 15-1 SOME BARLEY CULTIVARS WITH KNOWN GENES CONDITIONING REACTION TO *ERYSIPHE GRAMINIS HORDEI.*

| LOCUS[a] | CULTIVAR AND CI NUMBER[b] |
|---|---|
| **A. Monogenic cultivars.[c]** | |
| $Ml_a(=JMl_{sn}$ | Cebada Capa (6193), Heil's Hanna 3[d] (682), Marocaine 079 (8334), Multan (3401), Rabat (4979) |
| $Ml_a2 (= Ml_b, JMl^{ibr}_{sn})$ | Black Russian (2202) |
| $Ml_a3$ | Ricardo (6306) |
| $Ml_a4$ | No. 22 (= Weider, 1021) |
| $Ml_a5$ | Gopal (1091), Purple Nudum (2250) |
| $Ml_a6$ | Maris Badger, Maris Concord, HB 279/5/1/2, Franger (8811) |
| $Ml_{at} (=JMl_{r12})$ | Atlas (4118), Batna (3391), Colsess (11545), Modia (2483), Peruvian[d] (935), Russian 12 (11563) |
| $Ml_b (= Ml_a2, JMl^{br}_{sn})$ | (See above) |
| $Ml_d$ | Sel. dd (11640), Sel. 175 |
| $Ml_g (= JMl_g)$ | Atlas 46 (7323), Erie (8080), Goldfoil[d] (928), Menelik (5862), Stephan (8051), Weihenstephan I (11548) |
| $Ml_h (= JMl_h)$ | Chevron[d] (1111), Hanna (906) |
| $Ml_k (= JMl_k)$ | Kwan (1016) |
| $Ml_{na} (=JMl^{mc}_{sn})$ | Engledow India, Grandpa |
| $Ml_n$ | Nepal (595) |
| $Ml_p$ | Psaknon (6305), Tradak (5645) |
| $Ml_w$ | West China |
| $Ml_y$ | (Only in polygenic cultivars) |
| **B. Polygenic cultivars.[c]** | |
| $Ml_a + Ml_{at}$ | Algerian (1179) |
| $Ml_a + Ml_g$ | Palmella Blue (3609) |
| $Ml_a + Ml_m$ | Monte Cristo (1017) |
| $Ml_d + Ml_h + Ml_p$ | Duplex (2433) |
| $Ml_p + Ml_y$ | Arlington Awnless (702), Chinerme (1079), Nigrate (2444) |

C. Cultivars having genes with Japanese symbols not yet related to international gene symbols.[e]

| | |
|---|---|
| $JMl_{kb}$ | Kairyobuzu-mugi |
| $JMl_{nn}$ | Nigrinudum |
| $JMl_{nz}$ | Nakaizumi-zairai |
| $JMl_{r74}$ | Russian No. 74 |
| $JMl_{r81}$ | Russian No. 81 |
| $JMl^{h4}_{sn}$ | Hakata No. 2 |
| $JMl^{al}_{sn}$ | A.222 |

been thoroughly analyzed by Moseman and his co-workers and has been reviewed recently (Moseman, 1966). On the basis of this analysis, resistance factors may be astutely manipulated by breeders.

## Sources of Resistance

Many barley cultivars are reported to have genes conditioning resistance to powdery mildew (Hiura, 1960; Moseman, 1955; 1968). Cultivars with labeled genes for resistance are listed in Table 15-1. Several of the genes conditioning mildew reactions are on chromosome 5, and the chromosomal affiliations of others are known (Nilan, 1964).

## Physiologic Specialization

Eighteen genes conditioning reaction in barley to powdery mildew occur at 13 loci, with six alleles at one locus. There probably are many more. By using several groups of differentiating cultivars, more than 50 races have been identified in Europe, 24 in North America, 11 in Japan, and lesser numbers elsewhere (Moseman, 1966).

A precise relationship exists between genes conditioning powdery mildew resistance in barley and those conditioning virulence in the pathogen. Moseman (1971b) has discussed the co-evolution of host-pathogen genetic systems that has led to this relationship. It is apparent that for each gene in barley conditioning reaction to *E. graminis hordei*, there exists in the pathogen a corresponding gene for virulence. Since *E. graminis* has an active sexual mechanism for recombining the virulence genes, an enormous number of races can exist and it would be useless to catalog them. However, it is significant that all virulence genes, like all resistance genes, are not universally distributed, and this should facilitate on a regional basis the breeding of cultivars resistant to powdery mildew.

## Manipulation of the Fungus

The isolation and maintenance of cultures of *E. graminis hordei* with certain specific virulence genes is necessary to expedite the breeding

---

[a] | Loci are identified by the subscript letters immediately following $Ml$, the international designation for genes conditioning reaction to *E. graminis hordei*. $Ml_a$—$Ml_a6$ indicates 6 alleles have been identified at the $Ml_a$ locus; there may be others.

[b] CI numbers are cereal investigation numbers of the United States Department of Agriculture. CI numbers are provided if they are known.

[c] Information from various sources cited by Moseman (1966).

[d] North American differential cultivars.

[e] From Hiura (1960); additional sources are listed by him.

program. Any newly occurring race found on cultivars that previously have been resistant should be isolated.

Methods of maintaining isolates of *E. graminis hordei* have been described by Hiura (1960), Moseman (1956*b*), and Newton and Cherewick (1947). Isolates may be maintained on plants growing in large test tubes and stored in illuminated refrigerators at 2-4 C or on plants in clay pots under lamp chimneys or in plexiglass tubing plugged with cotton. Since it is difficult to prevent the spread of mildew in glasshouses, test plants should be grown in mildew-proof containers, growth chambers, or in houses where sulfur may be placed on the heating units at weekly intervals. Inoculation is facilitated by simply shaking plants with sporulating colonies over the test plants. Generally, five infection responses are recognized; these are described by Newton and Cherewick (1947) and beautifully illustrated by Hiura (1960).

### Breeding

The problems of breeding for mildew resistance are only slightly different from those described for leaf rust. In seedling testing, particularly when several mildew isolates must be considered, special precautions should be taken to avoid contamination by mildew of unknown origin. Variations of the basic procedures are described by Hiura (1960). Natural or artificial epiphytotics of mildew in field plots are enhanced by application of adequate amounts of nitrogen fertilizer.

## Brown Loose (Nuda) Smut Resistance

Unlike scald, leaf rust, and powdery mildew pathogens, the smut fungus, *Ustilago nuda* (Jens.) Rostr., cannot become epiphytotic on a healthy crop in a single year. The incidence of smutted plants increases over a period of 2 to several years, depending on the susceptibility of the cultivar and environmental conditions which affect floral infection from year to year. The seed-borne mycelium is present in the embryo and nearby tissues. The practice of eliminating smut mycelium from infected seed by hot water, cold soak, or anaerobic treatments is well known, but the treatments are bulky and cause reduction in germinability. Consequently, much emphasis has been placed on controlling nuda smut with resistant varieties. The recent introduction of oxathiin derivatives (Vitavax, for example) for chemical control of smut by seed treatment may cause waning interest in genetic control of smut.

## Sources of Resistance

Many cultivars have been reported to have resistance to nuda smut (Cloninger and Poehlman, 1954; Moseman and Metcalfe, 1969; Niemann, 1961; Pedersen, 1962). Cultivars which have labeled genes for resistance to nuda smut are grouped according to genes conditioning their reactions as follows: *Un* in Trebi (936) and Valentine (7242); *Un2* in Missouri Early Beardless (6051); *Un3* in Jet (967) and Nigrinudum (2222); *Un4* in Dorsett (4821); *Un5* in Wisconsin Sel. X173-10-5-6-1; *Un6* in Jet and Nigrinudum; *Un7* in Anoidium (7269); *Un8* in PR 28, and Milton (4966); and probably *Un3* plus *Un6* in Abyssinian (668), Bifarb (3951-3), and Kitchin (1296). Genes *Un, Un3, Un4, Un7,* and *Un8* appear to convey the most effective resistance. The numerous studies on inheritance of reaction to *U. nuda* through 1962 are cited by Smith (1951), Nilan (1964), and Metcalfe (1966).

## Physiologic Specialization

From 4 to 13 races of *Ustilago nuda* have been described (Halisky, 1965; Nilan, 1964). Since different workers employed various sets of differential cultivars, inconsistent numbers of races have been reported. Merely to classify a culture as to race is of little value. Useful analysis of the genetics of host-pathogen relations may be made without identification of a culture as to race (Moseman and Metcalfe, 1969).

## Manipulation of the Fungus

A collection of *U. nuda* is truly a collection. If taken from a susceptible cultivar, it may be a mixture of races or, more specifically, genotypes, and considerable time may be spent isolating lines homozygous for virulence from the collection. The isolation of homozygous lines is necessary for precise genetic studies of host-parasite interactions, but for breeding cultivars resistant to prevailing virulence genes, mixtures may be time saving. Cultivars listed above may be useful for screening out various smut genotypes.

Inoculations must be made at anthesis. The barley flower may not open, and the natural introduction of teliospores into the floral chamber is somewhat accidental. Several techniques have been devised for artificial inoculation of barley florets, but the number that can be inoculated is restricted by the brief period of susceptibility.

Both dry and wet spores have been used to inoculate barley spikes (Moseman and Metcalfe, 1969). A rubber bulb fitted with a hypodermic needle may be filled with either dry or water-suspended teliospores; the lemma is punctured with the needle, and the spores may be either puffed or squirted into the floral chamber. Each floret is inoculated. If help is plentiful, this basic technique is perhaps the most efficient. The dry-spore method is least damaging to the florets. One may inoculate with a single collection or race, or he may use several bulbs and inoculate different spikes with different races.

If one chooses to work with field mixtures of smut, all the florets on a spike may be clipped, and the spike may be dusted by knocking a few smutted spikes against it or by using a puff-gun fitted with a tube to slip over and enclose the spike. If smutted spikes are abundant in the inoculation area, only the clipping is necessary; the spores will arrive.

The classification of progenies is a problem. If one crosses a smut-resistant with a susceptible parent and inoculates the female floret, the resistance or susceptibility of the ovary wall of the maternal parent may control the response in the $F_1$. The smut mycelium must pass through a homozygous resistant or susceptible ovary wall to reach a heterozygous embryo. If resistance is dominant, inoculation of florets on $F_1$ plants may not give a true picture of genetic constitution in the $F_2$. Consequently, inoculation in the $F_2$ and later generations is essential to determining the genetic constitution of genetic or breeding material.

## Breeding

Resistance to nuda smut may be bred by the following procedure. Having obtained cultivars reported to have genes conditioning resistance, one should expose these cultivars to natural infection by nuda smut and should artificially inoculate them with locally collected spores. If satisfactory resistance is observed, crosses with locally adapted cultivars should be advanced to the $F_2$ and $F_3$ generations. Head selections from these generations should be seeded into "head rows" 1 m long. Rows should be selected for plant type and resistance to other diseases, and random spikes from these selected rows should be inoculated. In subsequent generations, the inoculated spikes should be planted in rows of 1 m or less and should be accompanied by a bulk seeding from the selected row in which inoculations were made. Smut-free rows from the inoculated material should be candidate rows for additional reinoculations. Additional inoculations may also be made in the bulk rows. After two or three successive generations of selecting smut-free rows, it may be assumed that smut-resistant lines have been established. Before any lines are released, they should be retested for smut reaction.

# Barley Yellow Dwarf Virus (BYDV) Resistance

Several viruses attack barley, but BYDV is more cosmopolitan and damaging than the others. Furthermore, the use of resistant cultivars is the only practical method of reducing losses caused by BYDV. Two comprehensive reviews on BYDV have been published (Bruehl, 1961; Rochow, 1961). The virus has a wide range of gramineous hosts, including several that are perennial and are hosts of various aphid species which serve as vectors. The virus is also persistent in aphids, and some species of these overwinter as adults; thus, as a result of overwintering in both perennial grasses and viruliferous aphids, reservoirs for early spring dissemination of BYDV are assured.

## Sources of Resistance

Many barley cultivars have been classified for their reaction to BYDV; most of the highly resistant cultivars are of Ethiopian origin or derive their resistance from Ethiopian cultivars (Arny and Jedlinski, 1966; Catherall and Hayes, 1966). Among those rated consistently as highly resistant were unnamed cultivars with the following CI numbers: 1113, 1227, 1240, 1429, 2230, 2376, 3208-2, 3208-4, 3737, 3906-1, 3906-1, 3908-1, 3926-3, 5809, 6471, 8279, 9588, 9622, 9654, 9704, 9730, and 9794. The gene Yd2 on chromosome 3 conditions the reaction to BYDV in 14 varieties of Ethiopian origin (Damsteegt and Bruehl, 1964; Schaller, Qualset, and Rutger, 1964).

## Variation in the Virus

"Since barley yellow dwarf involves an interaction among three biological systems, plant, virus, and aphid, it is not surprising that much variation has been encountered in the disease" (Rochow, 1969). Variations in pathogenicity, host range, and vector specificity have been observed. Some cultivars are resistant in one area but susceptible in others (Arny and Jedlinski, 1966; Bruehl, 1961), but fortunately some cultivars have been universally resistant.

## Manipulation of the Virus and Breeding

BYDV is transmitted only by aphids. Nine species of aphids have been shown to be vectors of one or more strains of BYDV. The most common vectors include the English grain aphid, *Macrosiphum granarium*

(Kirby) = *M. avenae* (Fab.); apple grain aphid, *Rhopalosiphum fitchii* (Sanderson) = *R. padi* (L.); corn leaf aphid, *R. maidis* (Fitch); and greenbug, *Toxoptera graminum* (Rondani) = *Schizaphis graminum* (Rond.). *M. granarium* transmits most isolates of BYDV that have been studied and has been used to induce artificial epiphytotics of BYDV in the field. Some virus isolates are not transmitted by *M. granarium* but are readily transmitted by *R. fitchii;* thus, vector specificity compels one to establish cultures of regionally prevalent aphids capable of transmitting the virus strains that are causing damage. Both the virus and aphid cultures may be maintained in screen cages. Hudson (CI 8067), a winter barley, is a suitable host for maintaining the virus and vector (Brown and Poehlman, 1962). Viruliferous aphids may be transferred directly to test plants in greenhouse or field: after 3 days greenhouse plants are usually freed of aphids with a suitable insecticide. Field plants may be inoculated from virus-susceptible aphid trap crops such as Club Mariout (CI 261), Bonneville (CI 7248), or spring-sown winter types such as Hudson. Spread of the virus from trap crops may not occur sufficiently early in the spring to assure natural inoculation of winter-type barley, because winter barley grows rapidly before the virus can be spread uniformly from trap plots. The method has been employed for some spring-type barley nurseries. For winter-type barley, viruliferous aphids must be increased in greenhouses and transferred directly to test plants in the field. The hill-plot planting technique may facilitate more efficient inoculation and selection of resistant lines in early generations of breeding material of winter barley. Field and greenhouse techniques are summarized by Bruehl (1961).

## Resistance to Miscellaneous Diseases

### Covered Smut

Resistance to covered smut, *Ustilago hordei* (Pers.) Lagerh., is conditioned by four genes. The genes and cultivars in which they occur are *Uh* in Titan, OAC21, Ogalitsu, and Anoidium; *Uh2* in Anoidium; *uh3* in Ogalitsu; and *uh4* in Jet (Nilan, 1964). Inoculations with *U. hordei* may be made by peeling the lemma away from the region of the embryo and mixing batches of such dehulled seed with spores of selected cultures, or by using the partial vacuum technique. Shallow planting of smutted seeds is recommended for field studies.

## Stem Rust

Barley is attacked by *Puccinia graminis* f. sp. *tritici* Eriks. and Henn., and by *P. graminis* f. sp. *secalis* Eriks. Genes *T* and *T2* condition resistance to some races of *P. graminis tritici,* but not to *P. graminis secalis.* Gene *T* is found in the cultivars Chevron (CI 1111), Peatland (CI 5267), and Valentine (CI 7242); gene *T2* occurs in Hietpas 5 (CI 7124). Blackhulless (CI 666), Vaughan (CI 1367), Hispont (CI 8828), Gospick (CI 9094), and Lopac (CI 9095) are resistant to some races of *P. graminis tritici,* and it is suspected that these cultivars have the *T* gene (Moseman, 1963). Gene *T* is linked with *un7* on chromosome 1 (Andrews, 1956).

## Spot and Net Blotch Complexes

Helminthosporium leaf blotches, spot blotch [ *Cochliobolus sativus* (Ito and Kurib.) Drechsl.], and net blotch [ *Pyrenophora teres* (Died.) Drechsl.] are frequently found together, but each is a damaging disease of barley foliage. *C. sativus* also causes foot rot and seedling blight. It has been difficult to breed varieties with good resistance to these diseases, probably because the fungi are highly variable. The mode of inheritance of resistance has not been completely unraveled. By work with specific isolates of the fungi, resistance has been found to be conditioned by single gene pairs (Nilan, 1964).

One of the difficulties of breeding helminthosporium-resistant cultivars has been in devising adequate seedling tests. Methods devised by Cohen, Helgason, and McDonald (1969) may lead to increased efficiency of breeding blotch-resistant varieties. Resistance to *C. sativus* is found in Anoidium, Br 3962-4, Lenta, Opal B, CI 8873, CI 8969, and others (Cohen, Helgason and McDonald, 1969). Resistance to *P. teres* is conditioned by the genes *Pt* in Tifang (CI 4407-1); *Pt2* in Harbin (CI 4929), Manchurian (CI 739), and Ming (CI 4797); and *Pt3* in Canadian Lake Shore (CI 2750) and CI 4922 (Mode and Schaller, 1958). There is conflicting evidence on the distribution and linkage relations of these genes in certain cultivars (Khan and Boyd, 1969).

Two cultivars recently released by the Indiana Experiment Station, Harrison (CI 10667) and Jefferson (CI 11902), have outstanding resistance to both net and spot blotch.

# Acknowledgments

I am grateful to the late Dr. C. C. Wernham for giving me the opportunity to contribute a chapter to this book.

My friends and colleagues, Drs. T. M. Starling, Professor of Agronomy, Virginia Polytechnic Institute; J. G. Moseman, Project Leader, Barley, U.S.D.A., A.R.S.; and R. G. Henderson, Professor of Plant Pathology, V.P.I., have reviewed the manuscript and made many helpful suggestions.

## *References*

Andrews, J. E. 1956. Inheritance of reaction to loose smut. *Ustilago nuda,* and to stem rust. *Puccinia graminis tritici,* in barley. Can. J. Agr. Sci. 36:356-370.

Arny, D. C., and H. Jedlinski. 1966. Resistance to the yellow dwarf virus in selected barley varieties. Plant Dis. Reporter 50:380-381.

Brown, G. E., and J. M. Poehlman. 1962. Heritability of resistance to barley yellow dwarf virus in oats. Crop Sci. 2:259-262.

Bruehl, G. W. 1961. Barley yellow dwarf. Amer. Phytopathol. Soc. Monogr. No. 1. 52 p.

Catherall, P. L., and J. D. Hayes. 1966. Assessment of varietal reaction and breeding for resistance to the yellow-dwarf virus in barley. Euphytica 15:39-51.

Cloninger, C. K., and J. M. Poehlman. 1954. Resistance of winter barley to *Ustilago nuda* (Jens.) Rostr. Missouri Agr. Exp. Sta. Res. Bull. 560. 35 p.

Cohen, E., S. B. Helgason, and W. C. McDonald. 1969. A study of factors influencing the genetics of reaction of barley to root rot caused by *Helminthosporium sativum.* Can. J. Bot. 47:429-443.

Damsteegt, V. D., and G. W. Bruehl. 1964. Inheritance of resistance in barley to barley yellow dwarf. Phytopathology 54:219-22.

Dyck, P. L., and C. W. Schaller. 1961. Inheritance of resistance in barley to several physiologic races of the scald fungus. Can. J. Genet. Cytol. 3:153-164.

Halisky, P. M. 1965. Physiologic specialization and genetics of the smut fungi. III. Bot. Rev. 31:114-150.

Hiura, U. 1960. Studies on the disease-resistance in barley. IV. Genetics of the resistance to powdery mildew. Ber. Ohara Inst. Landwirt Biol. 11:235-300.

Khan, T. N. and W. J. R. Boyd. 1969. Inheritance of resistance to net blotch in barley. II. Genes conditioning resistance against race W.A.-2. Can. J. Genet. Cytol. 11:592-597.

Levine, M. N., and W. J. Cherewick. 1952. Studies on dwarf leaf rust of barley. U.S.D.A. Tech. Bull. 1056. 17 p.

Loegering, W. Q., D. L. Harmon, and W. A. Clark. 1966. Storage of urediospores of *Puccinia graminis tritici* in liquid nitrogen. Plant Dis. Reporter 50:502-506.

Mains, E. B., and M. L. Martini. 1932. Susceptibility of barley to leaf rust (*Puccinia anomala*) and to powdery mildew (*Erysiphe graminis hordei*). U.S.D.A. Tech. Bull. 295. 34 p.

Metcalfe, D. R. 1966. Inheritance of loose smut resistance. III. Relationships between the "Russian" and "Jet" genes for resistance and genes in 10 barley varieties of diverse origin. Can. J. Plant Sci. 46:487-495.

Mode, C. J., and C. W. Schaller. 1958. Two additional factors for resistance to net blotch in barley. Agron. J. 50:15-18.

Moseman, J. G. 1955. Sources of resistance to powdery mildew of barley. Plant Dis. Reporter 39:967-972.

————— 1956a. Evaluation of varieties and selections of barley for disease resistance and winter hardiness in southeastern United States. U.S.D.A. Tech. Bull. 1152. 33 p.

—————. 1956b. Physiological races of Erysiphe graminis f. sp. hordei in North America. Phytopathology 46:318-322.

—————. 1963. Present status of plant pathological research on barley in the United States. Barley Genetics. I. Proc. 1st Int. Barley Genet. Symp., Wageningen. p. 250-258.

—————. 1966. Genetics of powdery mildews. Ann. Rev. Phytopathol. 4:269-290.

—————. 1968. Reactions of barley to Erysiphe graminis f. sp. hordei from North America, England, Ireland, and Japan. Plant Dis. Reporter 52:463-467.

—————. 1971a. Studies of inheritance of resistance in barley to pathogenic organisms, 1963-1969. Proc. 2nd Int. Barley Genet. Symp. Pullman, Wash. 1969:535-541.

—————. 1971b. Co-evolution of host resistance and pathogen virulence. Proc. 2nd Int. Barley Genet. Symp. Pullman, Wash. 1969:450-456.

—————, and D. R. Metcalfe. 1969. Identification of resistance genes in barley by reactions to Ustilago nuda. Can. J. Plant Sci. 49:447-451.

Newton, M., and W. J. Cherewick. 1947. Erysiphe graminis in Canada. Can. J. Res. C. 25:73-93.

Niemann, E. 1961. Flugbrandresistente Gersten sorten. Z. Pflanzenzücht 45:8-16.

Nilan, R. A. 1964. The cytology and genetics of barley 1951-1962. Wash. State Univ. Res. Studies, Monogr. Suppl. 3. 278 p.

Nover, I., and R. Mansfeld. 1959. Resistenzeigenschaften im Gersten- und Weizensortiment Gatersleben. 3. Prüfung von Gersten auf ihr Verhalten gegen Zwergrost (Puccinia hordei Otth.) Kulturpflanze 7:29-36.

Pedersen, P. N. 1962. Testing varieties of barley for physiological resistance to loose smut, Ustilago nuda (Jens.) Rostr. Royal Vet. and Agr. Yearbook, Copenhagen. p. 124-143.

Reed, H. E. 1957. Studies on barley scald. Tenn. Agr. Exp. Sta. Bull. 268. 43 p.

Roane, C. W., and T. M. Starling. 1967. Inheritance of reaction of Puccinia hordei in barley. II. Gene symbols for loci in differential cultivars. Phytopathology 57:66-68.

————— —————. 1970. Inheritance of reaction to Puccinia hordei in barley. III. Genes in the cultivars Cebada Capa and Franger. Phytopathology 60:788-790.

Rochow, W. F. 1961. The barley yellow dwarf virus disease of small grains. Adv. Agron. 13:217-248.

—————. 1969. Biological properties of four isolates of barley yellow dwarf virus. Phytopathology 59:1580-1589.

Rowell, J. R., and E. B. Hayden. 1956. Mineral oils as carriers of urediospores of the stem rust fungus for inoculating field-grown wheat. Phytopathology 46:267-268.

Schaller, C. W., C. O. Qualset, and J. N. Rutger. 1964. Inheritance and linkage of the Yd2 gene conditioning resistance to the barley yellow dwarf virus disease in barley. Crop Sci. 4:544-548.

Schein, R. D. 1960. Physiologic and pathogenic specialization of *Rhynchosporium secalis*. Penn. Exp. Sta. Bull. 664. 29 p.

Sharp, E. L., and F. G. Smith. 1957. Further study of the preservation of Puccinia uredospores. Phytopathology 47:423-429.

Smith, L. 1951. Cytology and genetics of barley. Bot. Rev. 17:1-51; 133-135.

Starling, T. M., C. W. Roane, and K.-R. Chi. 1971. Inheritance of reaction to *Rhynchosporium secalis* in winter barley cultivars. Proc. 2nd Int. Barley Genet. Symp. Pullman, Wash. 1969:513-519.

# 16 Beans
## M. J. Silbernagel[1]
## W. J. Zaumeyer[1]

## *Origin, History, and Value of Crop*

Remains of the common bean (*Phaseolus vulgaris* L.), radiocarbon dated at 4000 B.C., have been found in caves in Ocampo, Mexico, and vessels containing beans have been found in pre-Inca tombs in Peru. Beans were mentioned by nearly all of the European explorers of the Americas. They were eaten by the natives, who generally grew them between maize plants, as they still do in much of Central and South America. They were brought from America to Europe in the sixteenth century. Their cultivation was soon extended throughout Europe, and today beans are grown in all parts of the world.

Vavilov (1935) was of the opinion that *P. vulgaris* originated in the tropical southern part of Mexico, Guatemala, Honduras, and a part of Costa Rica. Beans spread from this center of origin to North and South America long before the arrival of European explorers.

1. Plant Pathologists, Crops Research Division, Agricultural Research Service, U. S. Department of Agriculture, Prosser, Washington and Beltsville, Maryland (retired), respectively.

The farm value of dry beans (including snap beans grown for seed) in the United States in 1970 was approximately $150 million; snap beans for fresh market were valued at $40 million; and snap beans for processing, $54 million.

In 1970, about 1.5 million acres produced about 18 million 100-pound bags of dry beans in the U.S. Approximately 86,000 acres of snap beans were harvested for fresh market, and 228,000 acres for processing (which produced 311 million pounds and over 1.25 billion pounds of beans, respectively).

The dry bean is the principal edible legume grown and eaten in most countries of Central and South America. In some of these countries it is the second most important crop grown; in many it is the principal source of human dietary protein. Brazil, which produces 33 million 100-pound bags of dry beans annually on 5.5 million acres, leads the world in acreage and total production, followed by Mexico and the United States, with acreages of about 3 and 1.5 million acres, respectively. Per capita consumption is about 60 pounds per year in Brazil, compared to 7 pounds in the United States.

# Sources of Information, Disease Resistance, and Genetic Variability

Recent literature reviews on breeding for disease resistance in beans include those by Yarnell (1965), Zaumeyer and Thomas (1957), Hubbeling (1957), and Walker (1965).

Through the Bean Improvement Cooperative (B.I.C.) Annual Reports[2] and their biennial meetings, a breeder is able to contact virtually every bean worker in the world and to obtain literature reprints, current research results, and breeding stocks.

The National Seed Storage Laboratory in Fort Collins, Colorado, maintains basic stocks of commercial varieties, and many older genetic stocks. The USDA Plant Introduction Station at Pullman, Washington, maintains a world *P. vulgaris* collection of over 8,000 items, as well as several thousand accessions of related *Phaseolus* species, many of which can be hybridized with *P. vulgaris*. Computerized information on these lines enables a breeder to quickly scan the collection for desired characteristics.

2. Contact Dr. D. P. Coyne, B.I.C. Chairman and Editor, Department of Horticulture and Forestry, University of Nebraska, Lincoln, Nebraska 68503.

Samples of all released varieties and other useful germ plasm should be preserved in a permanent depository, such as the National Seed Storage Laboratory or the P.I. collection, to insure its safe preservation for future generations.

Breeders should be cautious in acquiring new accessions. There are many seed-borne diseases, such as anthracnose, bean common mosaic virus, and all of the bacterial diseases. Present quarantine regulations are not adequate to preclude spread of these dangerous pathogens into or within this country. Unless all new accessions are isolated until they are verified to be pathogen free, there is real danger of infesting breeding lines or the commercial bean production in a given area.

If specific desired characteristics are not found in commercial varieties, wild *P. vulgaris* collections, or related species, artificially induced mutations are another source of resistance or genetic variability, although the probability of inducing and recovering a specific useful characteristic is extremely low.

## Breeding Methods

A simple intraspecific cross, followed by sufficient backcrossing, selfing, and screening for several generations, is usually all that is needed to establish a new disease-resistant variety. Sometimes characteristics of more than one commercial variety are desired, and additional outcrosses are required.

Sib intercrossing is very effective in wide crosses where disease resistance, or any other desired characteristic, results from the additive effects of polygenic factors, as in the case of root-rot resistance in beans.

By intercrossing partially tolerant hybrid sibling lines derived from different crosses between susceptible commercial *P. vulgaris* parents, and tolerant wild parents, one can simultaneously accumulate genes for disease tolerance and commercial plant type. Backcrossing the same partially tolerant hybrids to either parent sacrifices progress towards resistance or type, depending on the parent.

Interspecific hybridization is usually the most difficult way to obtain factors for resistance, as it is often hindered by high levels of incompatibility. The *P. vulgaris* x *P. coccineus* L. cross is usually successful only if *P. vulgaris* is the female parent. Ovule abortion and low hybrid fertility often occur with interspecific crosses, sometimes necessitating embryo culture, as in the case of the *P. vulgaris* x *P. acutifolius* var. *latifolius* Freeman (tepary bean) cross (Honma, 1956).

However, interspecific hybrids can be a unique source of disease resistance and wide genetic variability. If a breeder does not succeed in transferring all the desired genes for resistance in the first interspecific attempt, hybrids obtained can usually be used as bridging lines that are more compatible than the susceptible parent for further backcrosses to the resistance source.

## Hybridization Techniques

Hybridization is best accomplished in the spring, under greenhouse conditions, where high humidity can be maintained. In summer, high temperatures and bright sunshine may cause flowers to dry up. In winter, growing conditions are generally poor, and pollen production is low. An illustrated review of crossing techniques was published by Buishand (1956).

The proper age for emasculation and hybridization of a young unopened bud is about one day before anthesis. Fresh anthers that have recently matured and ruptured or which are on the verge of rupturing are the best pollen source. Some prefer to use the pollen-laden stigma from the male parent to transfer pollen to the female plant stigma.

It is best to limit a female parent to 3 or 4 pods, preferably using the first flowers which form on a young plant, rather than those on an old plant.

About 30-80% successful pollinations can normally be expected, depending upon the skill of the operator and environmental conditions. To insure a desired cross, as many as 10-12 reciprocal pollinations may be needed.

## The Disease Screening-Type Selection Cycle

Most disease screening can be done either in the field or greenhouse. Duplication of known field disease reactions in the greenhouse usually establishes the validity of greenhouse disease screening tests.

While the desirability of simultaneous testing with several pathogens is apparent, it should be established that the host-pathogen relationships observed in single-pathogen tests are not affected by the presence of other pathogens.

Greenhouse and growth chamber facilities are best suited for the critical establishment of environmental parameters of disease resistance and for studying the differential effects of related pathogenic strains. Plant development under greenhouse conditions is not comparable to development under field conditions. Therefore, concomitant selection for horticultural type and disease resistance can only be done under field conditions. To what extent and how early in the program this can be done depends upon the homozygosity of the hybrid population and the availability of resistance. Resistance to the curly top virus in beans is readily available in commercial varieties, easily transferred and stabilized; therefore, severe selection pressure for horticultural type can begin in early generations. Conversely, *Fusarium* root-rot resistance is not clearly defined, exists only as degrees of tolerance in wild types, and seems to be the result of the cumulative effects of many factors. Early selection for horticultural type in these populations may eliminate what little root-rot tolerance is present. So, in the case of bean root rots, first priority is given to recovering resistance, with selection for horticultural type being done in later generations.

## Selection Methods

In the case of genetically diverse crosses, as in interspecific hybrids or crosses between drastically different *P. vulgaris* types, several early generation cycles of single plant selection are often necessary to sort the radically different types into manageable groups.

A pedigree, single plant selection system is also desirable in early generations ($F_1$ and $F_2$) to help eliminate selfs and to facilitate $F_2$ and $F_3$ family selection.

Modified mass selection techniques are useful from $F_3$ through $F_8$ in genetically narrow crosses between similar parents, to maintain maximum genetic diversity while stabilization of homozygosity takes place. Thus, a select bulk of 6-12 individuals might be kept from a promising line which still has several undesirable traits; while a rogued bulk might be kept from a line approaching homozygosity, which has only a few off-type individuals to be removed. As a line approaches homozygosity for disease resistance and horticultural type ($F_6$ to $F_{12}$) and begins to show promise as a varietal candidate, it is subjected to increasingly critical and detailed single plant selection. At about $F_6$ to $F_8$ 100-200 single plant selections are made from each of the more promising lines. These selections are increased for further evaluation during the next two years while the bulk is used for testing commercial potential.

# Evaluation

After successful performances in local trials, a promising line is distributed in 1 or 2 lb. lots for observation by regional, national, and/or international cooperators to determine environmental adaptability, phenotypic stability, uniformity, relative maturity, yield, quality, quality, and appearance of the new lines in comparison with standard and appearance of the new lines in comparison with standard varieties.

After one or two seasons of such trials the breeder should have enough information to enter a line in replicated variety trials, and from there to commercial trials.

A variety release based on an artificially reconstructed bulk made up of 50-100 essentially identical single plant selections provides maximum genotypic flexibility within necessary phenotypic limitations. However, some breeders prefer to base a variety release on one late generation, single plant selection increase.

# Seed Storage

Germination, vigor, storage life, yield, uniformity, and disease resistance are a few of the factors affected by seed quality. Seed quality is dependent upon proper harvesting, handling, and storage (Pollock and Roos, 1971). Seeds should be harvested gently at about 15% moisture content to minimize injury.

For seed to be replanted the following spring, seed moisture content equilibrates at about 10-11% in a seed room kept at 50 F and a relative humidity of 50%. Seed for long storage equilibrates at 7-8% moisture content at a relative humidity of 30%.

When seeds are stored under very dry conditions, hard seededness can be a problem in some varieties and may necessitate holding the seed at 70% relative humidity for a week or two before planting (Harrington, 1964).

# Breeding for Resistance to Specific Diseases

## Virus Diseases

*Bean Common Mosaic.*   Once a major disease, BCMV is now almost nonexistent in the U.S. because most varieties developed since 1940 are resistant to both the type strain and the New York 15 strain. Several other strains have been reported, but none has become serious (Silbernagel, 1969). Resistance in most U.S. snap bean varieties is derived from Corbett Refugee (Zaumeyer, 1969). Ali (1950) has shown that Corbett Refugee resistance is based upon a dominant inhibitor gene *I* epistatic to a dominant gene *A* required for virus infection. Corbett Refugee resistance confers resistance to all strains of BCMV known in the U.S., and it is an excellent example of Van der Plank's (1963) horizontal resistance.

Resistance in most dry bean varieties (Red Mexican, Great Northern, Pinto) seems to be dependent upon a different pair of recessive alleles specific for each strain of the virus (Silbernagel, *unpublished data*). However, some of the Michigan navy bean varieties, such as Seaway and Saginaw, may have obtained NY-15 resistance from the Corbett Refugee source.

BCMV is seed-borne and remains active as long as the seed is viable. It does not retain its activity for long periods in frozen juice, although leaf tissue will retain infectivity in a frozen desiccated condition for about two years.

Inoculum is prepared by grinding fresh, young infected plant material in a mortar-pestle or electric blender, straining through cheesecloth, and diluting the juice about 1:50 with water. The leaves to be inoculated are lightly dusted with 600 mesh carborundum powder and gently rubbed with a soft pad saturated with the inoculum. Atomizing seedlings with dilute infective juice containing 1% carborundum at about 30 psi also results in good infection and is much quicker under field conditions. Symptoms develop in about 10 days at 27 C.

*Bean Yellow Mosaic.*   Strains of BYMV occur wherever beans are grown, and they occasionally cause serious crop losses. Dickson and Natti (1968) reviewed the literature on inheritance of resistance and reported single factor dominant resistance derived from *P. coccineus.* Resistance from *P. coccineus* has also been reported to be inherited in a recessive manner conditioned by two or three major genes, with additional modifiers affecting the variation in symptom expression. *P. coccineus* is resistant to more strains of BYMV than any known variety of *P. vulgaris*. Some of the Great Northern varieties resist one or more strains of the virus.

Resistance to a pea isolate of BYMV was found to result from a single dominant factor in a Red Kidney variety, which was susceptible to every other known strain of BYMV. Resistance to two severe pod-distorting strains of BYMV in Great Northern U.I. 31 was found to be conditioned by three major recessive genes with additional modifiers.

Hagel et al. (1972) reported aphid resistance in the Black Turtle Soup variety to be related to field resistance to BYMV. This kind of general resistance should be effective against all strains of all viruses carried by this vector.

Inoculum preservation and inoculation techniques are the same as for BCMV, except BYMV is not seed transmitted.

*Curly Top.*   The feasibility of breeding for resistance to the CTV disease was first demonstrated by Mackie and Esau in 1932, when they showed that resistance could be recovered from segregating populations of crosses between resistant and susceptible varieties. The University of Idaho and the U.S. Department of Agriculture have had bean breeding programs for curly top resistance for over 40 years, located in the desert regions of Washington and Idaho where the sugar-beet leafhoppers *Circulifer tenellus* (Baker), the only known vector of the disease, is found in great abundance.

Resistance to CTV in present varieties was obtained from California Pink, Red Mexican, and Burtner Blightless (Danna, 1934). The first two parents were used extensively in the development of such curly top-resistant dry beans as Bigbend Red Mexican, Pinto U.I. 114, Royal Red Kidney, and Great Northern U.I. 31.

Burtner Blightless provided the curly top resistance for the bush snap bean varieties Apollo, Idelight, Jackpot, Rodeo, and Wondergreen. According to Schultz and Dean (1947), resistance is dominant and conditioned by two factors in dominant and recessive epistasis.

Greenhouse work with this disease is difficult, since it entails rearing and working with the vector. The virus cannot be mechanically trans-mitted. Field exposure of segregating populations is facilitated by planting rows of susceptible sugarbeets about 20 ft apart a month before the beans. The beets attract the migratory vector and become a virus reservoir. Leafhopper activity on emerging bean seedlings is increased by cutting off the beet tops, forcing the leafhoppers to feed on the less desirable bean host. Inoculation during the seedling stage is preferable, since plants become more tolerant with age.

Strains of the virus have long been recognized, and differences between varieties in degrees of susceptibility are common (Silbernagel,
1965). However, even the best resistance can be overcome under
greenhouse conditions of unfavorable host development and high
inoculum levels.

That CTV resistance has remained stable under commercial conditions for over 40 years in the western U.S. seems to lend support to
Ballantyne's (1970) suspicion that resistance to CTV may be a general
or horizontal type of resistance in the sense of Van der Plank (1963).
She found that 56 curly-top resistant bean varieties from the U.S. were
resistant to Australian Summer Death virus, another yellows type,
phloem-restricted, leafhopper-transmitted disease. Moreover, many of
the curly top-summer death resistant varieties also seem to have a high
degree of tolerance to the subterranean clover stunt virus, a yellows
type, phloem-restricted, aphid-transmitted disease.

*Southern Bean Mosaic.*    SBMV produces systemic mottle symptoms
on some varieties and local lesions on others. Both symptoms are
never produced on the same variety.

Zaumeyer and Harter (1943) studied the inheritance of symptom expression to infection in bean crosses involving nine parents. A single
gene was found to govern local lesion development, which was
dominant to systemic mottling.

*Pod Mottle.*    Although not related to SBMV, PMV, also produces
local and systemic mottle symptoms, but never both on the same
variety. The local lesions differ in appearance from those
produced by SBMV, and the mottle symptoms are somewhat milder
than those produced by the former virus. The mottling of the pods cannot be distinguished from that produced by SBMV. Thomas and
Zaumeyer (1950) found that plants carrying the single dominant factor
are susceptible to the local-lesion type of infection, while the recessive
plants are susceptible to the systemic-mottle type of infection.

## Fungus Diseases

*Bean Anthracnose.*    Bean anthracnose is presently a minor disease in
the U.S. because of disease-free seed production in the western
deserts. Most major bean breeding programs in Europe, Africa, and
Central and South America consider anthracnose resistance one of their
primary objectives.

Emerson No. 847, an old Cornell University breeding line, is resistant to alpha, beta, and gamma races of *Colletotrichum lin-demuthianum* (Sacc. & Magn.) Scrib. Wells Red Kidney was reported to be resistant to alpha and beta races. In Michigan, Saginaw and Gratiot are resistant to alpha and beta, and Seafarer resists all three races. Manitou, a recent Red Kidney release from Michigan, also resists three races of the organism. P.I. 304,110 was found to resist beta, gamma, and delta races, but it is susceptible to alpha. Cornell 49-242 is resistant (*ARE* gene) to all four races, but it has undesirable genetic linkages.

The organism can readily be grown in culture, although it loses its ability to sporulate if conidial transfers are not made. Sporulation also occurs when the fungus is reisolated from fresh diseased material. Sporulation occurs most readily on bean pod agar or sterilized bean pods.

Seedlings are readily infected by atomizing them with a spore suspension of the organism and holding them in a humidity chamber for 24-48 hours at 18 C. Infection on the stems and leaves is observed in about a week. Susceptible seedlings often die if inoculated with a heavy spore suspension. The detection of disease-resistant plants can be made in the seedling stage in the greenhouse.

The first contribution on the inheritance of resistance to any bean disease was made by Burkholder (1918), who investigated anthracnose resistance. Crosses between Wells Red Kidney, which was resistant to alpha and beta races, and Perry White Marrow, resistant to only one of them, when inoculated in the $F_2$ generation with a single race of the organism, indicated that resistance was dominant and governed by a single gene.

Schreiber (1933) supplemented these studies using 37 isolates of the organism, which he divided into 3 main groups corresponding to alpha, beta, and gamma races. Reciprocal crosses between resistant Dry Shell No. 22 and Konserva, and between Dry Shell and a German variety, Wachs Best von Allen, showed a 3:1 ratio with resistance dominant when inoculated in the $F_2$ with one race of the organism.

*Bean Rust.*    Bean rust, caused by *Uromyces phaseoli* var. *typica* Arth., has been reported from almost every part of the world. The disease has been of considerable importance in certain parts of the western mountain states like Colorado, but currently it is significant in the U.S. only along the Atlantic Seaboard.

No varieties are known to resist all 34 races of rust, although the varieties Pinto 5 and 14 resist more races than any others. These were developed in 1945 by the U.S. Department of Agriculture but never became commercially popular because of their viny growth habit. The most resistant snap bean variety is Tenderwhite; others showing considerable resistance are Custer, Earligreen, Florigreen, Olympia, and Seminole.

Zaumeyer and Harter (1941) extended the study of inheritance of resistance to six races of rust. Their results showed that resistance to races 1 and 2 was governed by a single factor, but that more than one factor was involved in resistance to races 6, 11, 12, and 17.

Bean rust spores stored under refrigeration will maintain their viability for about a year.

Inoculation is readily accomplished under greenhouse conditions. A drop of a dispersing agent is used in making a spore suspension with water. Seedlings are sprayed with the spore suspension and placed in a humidity chamber for 12-24 hours at 20 C. Symptoms are observed within 5 days, and lesions reach their maximum size in 12-14 days.

Disease resistance is based on the size of the lesions. Ratings from 0-5 or 0-10 are used, with zero denoting immunity or no lesions. Intermediate grades denote degrees of field resistance.

*Bean Root Rots.*    Long a problem in the U.S., *Fusarium* root rot is the most serious root rot, followed by *Rhizoctonia, Pythium,* and *Thielaviopsis* root rots, any of which can be very destructive. A detectable level of tolerance has been found against all of these pathogens, and some differential resistance and/or susceptibility between varieties is noted.

Breeding for resistance to root rot caused by *Fusarium solani* (Mart.) Appel & Wr. F. *phaseoli* (Burk.) Synd. & Hans. has been difficult (Wallace and Wilkinson, 1965). Numerous cultural and environmental factors affect severity of disease expression (Burke, 1966). Many attempts have been made to control the disease by chemical treatments and organic soil amendments. However, no completely effective chemical, cultural, or genetic control has yet been developed. Burke et al. (1972) recently reported that subsoiling significantly increased yields in compacted root rot soils. More frequent watering is another cultural practice which seems to counteract the effects of *fusarium* root rot, and shallow planting (for rapid emergence) seems to reduce the severity of *Rhizoctonia* root rot. Breeding for resistance is further hampered by the lack of a high level of resistance. A degree of tolerance is available in P.I. 203958, collected in Mexico by Oliver Norvell, and in P.I.'s 165426 and 16535. The latter two lines have also been reported resistant to *Rhizoctonia* (McClean et al., 1968) and to root knot nematodes (Fassuliotus et al., 1967). Other lines have also been reported resistant to *Rhizoctonia* (Prasad and Weigle, 1970). Resistance in the scarlet runner bean *P. coccineus* is superior to that found in *P. vulgaris*. No commercial varieties have been released with root-rot resistance derived from any of the above mentioned sources; however, several dry bean breeders in this country reportedly have measurable root-rot tolerance in advanced breeding lines.

Bravo et al. (1969) concluded that resistance, whether derived from P.I. 203958 or scarlet runner, is incompletely dominant. Additive gene effects are larger than dominant gene effects. Estimates of the number of genes controlling resistance ranged from three to seven, and the effects of individual genes could not be distinguished.

Adequate controls in field or greenhouse screening tests are necessary because of the extreme variability in severity of disease expression. Greenhouse plants grown at 21 C are rated 3-4 weeks after planting on a 0-4 scale, with zero indicating clean roots and hypocotyl, and 4, dead or dying from root rot. Populations are given a disease index rating (0-100) to take into account both the severity of symptoms and the number of plants in each disease category. Tolerant selections (0-1 rating) from populations with a low disease index are repotted for seed production, followed by a cycle of field testing. Greatest emphasis should be placed upon field tests, since greenhouse tests only estimate resistance over a short period of time. Field grown plants are usually rated at or near maturity so seed of resistant selections can be saved and compared for yield.

The ability to germinate under cold, wet (cold tolerance) conditions, or to develop a large, vigorous root system, and the presence of inhibitory substances in seed coat and hypocotyl may enhance the level of root-rot tolerance. Naturally infested field soil is usually satisfactory for greenhouse *Fusarium* screening tests, although macroconidial suspensions may be used to insure uniform inoculation.

Resistance to *Pythium* blight caused by five species of *Pythium* was recently reported in P.I. 203958 and in Bush Green Pod (Adegbola and Hagedorn, 1970).

Hassan et al. (1971*a*) reported that P.I. 203958 and New York root-rot breeding line 2114-12 possess the same genes for resistance to *Thielaviopsis basicola* (Berk.) Ferr., and that resistance is partially recessive and controlled by 3 genes. They also noted that the genes controlling resistance to *Fusarium* and *Thielaviopsis* are different and not linked (Hassan et al., 1971*b*).

## Bacterial Diseases

Halo Blight.    The halo blight *Pseudomonas phaseolicola* (Burk.) Dows. epidemics of 1963 to 1967 in the U.S. stimulated a great deal of research on many aspects of bean bacterial diseases, which was reviewed by Frazier (1970). Schuster crossed the resistant varieties Red Mexican and Ankara Yellow to susceptible U.S. 5 Refugee and reported resistance caused by a single major recessive gene. In the cross Red Mexican with Asgrow Stringless, two recessive genes were involved. Dickson and Natti, using mixed inoculum (races 1 and 2), found resistance in P.I. 181954 to be due to one or two recessive genes with modifiers that can increase resistance levels. Walker and Patel reported hypersensitive resistance against race 1 in Red Mexican U.I. 3 to be controlled by one dominant gene, while tolerance to races 1 and 2 in a selection out of P.I. 150414 was conditioned by one recessive gene. Coyne, Schuster, and Shaughnessy reported tolerance to halo blight race 1 in Great Northern Nebraska No. 1 sel 27, when crossed with White Seeded Tendergreen, to be conditioned primarily by a major dominant gene. Other known sources of resistance include BO-19, Cornell 49-242, Great Northern 1140, Great Northern U.I. 59, Pinto Scout, Redkote Red Kidney, California Small White, Red Mexican U.I. 34, P.I. 203958, OSU-190, OSU-10183, Wisconsin HBR-40, and Calpac M-77. Coyne et al. (1967) and Baggett and Frazier (1967) summarize the sources of resistance in beans and their reactions to races 1 and 2 of halo blight.

Inoculum may be kept as infected seed, dried plant parts, agar cultures, or under sterile water. Inoculation can be effected by: (1) sprinkling dried ground infected plant tissue, or pouring liquid cultures over the seed at planting and allowing the seed to emerge through 4 inches of damp vermiculite; (2) infiltrating dry seed with a liquid culture under vacuum; (3) spraying tender plant parts with a liquid culture suspension under pressure; or (4) tissue puncture with an infected needle. Use of a moist chamber generally facilitates infection. A temperature of about 21 C is best for halo development.

*Common Blight, Fuscans Blight, Brown Spot and Bacterial Wilt.*    The Great Northern varieties Tara and Jules were developed (Coyne and Schuster, 1969a) using Great Northern Nebraska No. 1 selection 27 as a source of resistance to common blight *Xanthomonas phaseoli* (E.F. Sm.) Dows. and fuscans blight *X. phaseoli* var. *fuscans* Freeman (Burk.) Starr & burk. Tara has moderate tolerance, and Jules has high tolerance to both bacteria.

In the cross G.N. 1140 x P.I. 165078, Coyne et al. (1966) concluded that resistance to bacterial wilt *Corynebacterium flaccumfaciens* (Hedges) Dows. was inherited quantitatively. In crosses involving P.I. 165078 by G.N. Neb. 1 sel 27, susceptibility appeared to be conditioned by two complementary dominant genes, with absence of either or both genes resulting in tolerance. Inoculation consisted of inserting an inoculum-laden needle into the stem at the point of cotyledon attachment.

Resistance to brown spot incited by *Pseudomonas syringae,* Van Hall, was reported in G.N. 1140 and Tempo (Coyne and Schuster, 1969*b*). Truegreen was tolerant to one but susceptible to another strain of the bacterium. Tempo, Sanilac, Saginaw, and Truegreen were tolerant to brown spot in Wisconsin (Frazier, 1970). Inoculum is handled in the same way as for halo blight.

Resistance to fuscans blight is also found in the tepary bean, which is extremely difficult to hybridize with *P. vulgaris* (Honma, 1956).

Inoculation with common or fuscans blight is accomplished by spraying bacterial suspensions on plants at about 150 psi. High relative humidity and warm temperatures (26-32 C) favor infection by these pathogens.

## Challenges for the Future

Beans are naturally rich in protein, averaging about 24% with a range of 15-31%. Many people in developing nations (for example, in Central and South America and parts of Africa) are largely dependent upon beans to supply most of their daily dietary protein needs. Although beans are deficient in certain essential amino acids, such as methionine and tryptophan, recent work indicates that high methionine-containing lines are available for hybridization to high protein lines (Kelly et al., 1971). Hopefully, plant breeders will improve beans nutritionally for a hungry world in which protein is one of the greatest needs. However, nutritionally improved varieties must also carry sufficient disease resistance to be produced in the areas where they are needed most.

Breeders in the future should provide resistance based on horizontal resistance (Van der Plank, 1963) or on at least two or more vertical factors, since chances that a double mutation in a pathogen will overcome resistance in the host are many times smaller than the probability that a pathogen mutation will overcome single-factor vertical resistance.

Varieties must be developed that complement and make greater progress possible in the way of technological breakthroughs, such as the high-density harvesters for snap beans. The new harvesters require a plant with better anchorage than the current root-rot susceptible varieties with diseased roots.

Finally, the breeder must recognize that most genetic factors (genes) for disease resistance are "perishable" entities, which in most instances can be protected in many ways to preserve the longevity of their usefulness. For example, resistance to such seed-borne pathogens as anthracnose and the bacteria is needed under many Midwestern and Eastern growing conditions. When halo blight-resistant varieties such as Redkote are developed, special care should be taken to be sure seed stocks are continually being replaced by western desert-grown seed to reduce the possibility of the survival and perpetuation of a mutant strain of the bacterium, which may overcome the resistance in Redkote.

Likewise, if tolerance to *Rhizoctonia solani* is ever successfully transferred from its wild state to commercially acceptable varieties, the use of a chemical seed protectant such as Demosan should be continued for two reasons. First, the genetic tolerance to *Rhizoctonia* is not strong enough by itself. Second, even if it were strong, it would be wise to continue the use of the chemical seed treatment barrier to reduce the opportunity of the more resistant host screening out more virulent strains of the pathogen.

As much effort on the part of the breeder should be put into devising means of protecting the usefulness of genes for resistance as is spent in finding the resistance and developing a resistant variety in the first place.

# *References*

Adegbola, M. O. K., and D. J. Hagedorn. 1970. Host resistance and pathogen virulence in *Pythium* blight of bean. Phytopathology 60:1477-1479.

Ali, M. A. 1950. Genetics of resistance to the common bean mosaic virus (bean virus 1) in the bean (*Phaseolus vulgaris*). Phytopathology 40:69-79.

Baggett, J. R., and W. A. Frazier. 1967. Sources of resistance to halo blight in *Phaseolus vulgaris*. Plant Dis. Reporter 51:661-665.

Ballantyne, B. 1970. Field reaction of bean varieties to summer death in 1970. Plant Dis. Reporter 54:903-905.

Bravo, A., D. H. Wallace, and R. E. Wilkinson. 1969. Inheritance of resistance to *Fusarium* root rot of beans. Phytopathology 59:1930-1933.

Buishand, T. J. 1956. The crossing of beans (*Phaseolus* spp.). Euphytica 5:41-50.

Burke, D. W. 1966. Predisposition of bean plants to *Fusarium* root rot. Phytopathology 56:872.

———, D. E. Miller, L. D. Holmes, and A. W. Barker. 1972. Counteracting bean root rot by loosening the soil. Phytopathology 62:306-9.

Burkholder, W. H. 1918. The production of an anthracnose-resistant white marrow bean. Phytopathology 8:353-359.

Coyne, D. P., and M. L. Schuster. 1969a. "Tara," a new Great Northern dry bean variety tolerant to common blight bacterial disease. Univ. Nebraska Agr. Exp. Sta. Bull. 506:1-10.

————— —————. 1969b. Moderate tolerance of bean varieties to brown spot bacterium (Psuedomonas syringae). Plant Dis. Reporter 53:677-680.

————— —————, and L. W. Estes. 1966. Effect of maturity and environment on the genetic control of reaction to bacterial wilt in Phaseolus vulgaris L. crosses. Proc. Amer. Soc. Hort. Sci. 88:393-410.

————— —————, and R. Fast. 1967. Sources of tolerance and reaction of beans to races and strains of halo blight bacteria. Plant Dis. Reporter 51:20-24.

Danna, B. F. 1934. Progress in investigations of curly top of vegetables. Proc. 49th Ann. Meet. Oreg. State Hort. Soc.

Dickson, M. H., and J. J. Natti. 1968. Inheritance of resistance of Phaseolus vulgaris to bean yellow mosaic virus. Phytopathology 58:1450.

Fassuliotus, G., J. C. Hoffman, and J. R. Deakin. 1967. A new source of root-knot nematode resistance in snapbeans. Nematologia 13:141.

Frazier, W. A. 1970. Breeding beans tolerant to several bacterial diseases. Bean Improvement Cooperative Report No. 13:12-19.

Hagel, G. T., M. J. Silbernagel, and D. W. Burke. 1972. Resistance to aphids, mites, and thrips in field beans to infection by aphid relative-borne viruses. U.S. Dept. of Agr., ARS Bull. 33-139. 4 p.

Harrington, J. F. 1964. Mechanics of seed preservation. Seed World, July 24. p. 4-6.

Hassan, A. A., R. E. Wilkinson, and D. H. Wallace. 1971a. Genetics and heritability of resistance to Thielaviopsis basicola in beans. Proc. Amer. Soc. Hort. Sci. 96:623-627.

————— ————— —————. 1971b. Relationship between genes controlling resistance to Fusarium and Thielaviopsis root rots in beans. J. Amer. Soc. Hort. Sci. 96:631-632.

Honma, S. 1956. A bean interspecific hybrid. J. Hered. 47(5):217-220.

Hubbeling, N. 1957. New aspects of breeding for disease resistance in beans (Phaseolus vulgaris L.). Euphytica 6:111-141.

Kelly, J. F., A. Firman, and H. L. Adams. 1971. Microbiological methods for the estimation of methionine content of beans. Tenth Dry Bean Res. Conf. Rep. USDA-ARS 74-56:84-90.

McClean, D. M., J. C. Hoffman, and G. B. Brown. 1968. Greenhouse studies on resistance of snap beans to Rhizoctonia solani. Plant Dis. Reporter 52:486-488.

Pollock, B. M., and E. E. Roos. 1971. Chapter on seed and seedling vigor. In T. Kozlowski [ed.], Seed biology. Academic Press, New York. (In press.)

Prasad, K., and J. L. Weigle. 1970. Screening for resistance to Rhizoctonia solani in Phaseolus vulgaris. Plant Dis. Reporter 54:40-44.

Schreiber, F. 1933. Resistenz-zuchtung bei buschbohnen. Kuhn-Arch. 38:287-292.

Schultz, H. K., and L. L. Dean. 1947. Inheritance of curly top resistance reaction in the bean. Amer. Soc. Agron. J. 39:47-51.

Silbernagel, M. J. 1965. Differential tolerance to curly top in some snap bean varieties. Plant Dis. Reporter 49:475-477.

————. 1969. Mexican strain of bean common mosaic virus. Phytopathology 59:1809-1812.

Thomas, H. R., and W. J. Zaumeyer. 1950. Inheritance of symptom expression of pod mottle virus. Phytopathology 40:1007-1010.

Van der Plank, J. E. 1963. Plant diseases; epidemics and control. New York, Academic Press. 349 p.

Vavilov, N. I. 1935. Origin, variation, immunity and breeding of cultivated plants. K. S. Chester Trans. Chron. Bot. 1951.

Walker, J. C. 1965. Disease resistance in vegetable crops. III. Bot. Rev. 31:331-380.

Wallace, D. H., and R. E. Wilkinson. 1965. Breeding for *Fusarium* root rot resistance in beans. Phytopathology 55:1227-1231.

Yarnell, S. H. 1965. Cytogenetics of the vegetable crops. IV Legumes (cont.). Bot. Rev. 31:247-330.

Zaumeyer, W. J. 1969. The origin of resistance to common bean mosaic in snap beans. Seed World 105(4):8-9.

————, and L. L. Harter. 1941. Inheritance of resistance to six physiologic races of bean rust. J. Agr. Res. 63:599-622.

———— ————. 1943. Inheritance of symptom expression of bean mosaic virus 4. J. Agr. Res. 67:295-300.

————, and H. R. Thomas. 1957. A monographic study of bean diseases and methods for their control. U.S. Dept. of Agr. Tech. Bull. No. 868. 255 p.

# 17 Soybeans
## John Dunleavy[1]

## Introduction

### Origin

The soybean is believed to have originated in eastern Asia. The wild soybean, *Glycine ussuriensis*, is generally considered to be the progenitor of the cultivated soybean, *G. max* (L.) Merrill (Nagata, 1960). Nagata believed that *G. max* was brought to Japan from Korea, and that the species was brought to Korea from north China between 200 B.C. and 300 A.D. The soybean was in Ceylon and Vietnam in the eighteenth century, and was cultivated in Indonesia by 1855. The first record of soybeans in the Western Hemisphere was in the writings of E. Kaempfer in 1712. Soybean introductions from Manchuria, China, Korea, and Japan were important in the development of varieties now widely grown in the United States (Johnson and Bernard, 1962).

1. Research Plant Pathologist, Crops Research Division, Agricultural Research Service, U.S. Department of Agriculture; and Professor of Plant Pathology, Iowa State University.

## Geographical Distribution

Soybeans are cultivated on all continents and in such island centers as Taiwan, the East Indies, and Japan. The United States produced 73% of the world production in 1967 and is the major producer in North America. Brazil produces about 90% of the soybeans in South America. Romania and Yugoslavia are the principal European producers. Tanzania, Rhodesia, and Nigeria are the only significant soybean-producing countries in Africa, but even this production is limited. China is the largest soybean producer in Asia and ranks second to the United States in world production; Indonesia, Japan, and Korea are the other main Asian producers. Soybean production in the U.S.S.R. is relatively small compared to that in China and the United States. In 1968 the U.S.S.R. produced 0.6 million metric tons, whereas China produced 6.5, and the United States 29.4 million tons of soybeans.

## Economic Importance

Soybeans are one of the world's leading sources of vegetable oil and processed vegetable protein. They contain about 20% oil and 40% protein, on a dry weight basis. Soybean oil is composed of about 80% unsaturated fatty acids consisting of 52% linoleic, 20% oleic, and 7-8% linolenic. Soybean protein has excellent nutritional quality. The amino acid distribution very closely approximates that required in animal diets.

In the United States most of the soybean oil is used for edible products. Edible refined oil is used primarily in shortenings, margarines, and cooking oils. Technical refined oil is used in the manufacture of many diverse products, a few of which are paint, soap, varnish, resins, linoleum, insecticides, and disinfectants. The fatty acids are used in many creams and lotions, in emulsifiers, plasticizers, waxes, polishes, and liquid soaps. Lecithin is used in the manufacture of a host of food products, pharmaceuticals, textiles, chemicals, and automotive products.

Soybean protein has its chief use in animal foods in the United States. In 1969 the United States produced 15.2 million tons of soybean cake and meal. Soybean protein has many industrial uses. Soy flour is utilized in many bakery products, and recently a process was developed for isolating spun soybean protein. This fiber has been used in the manufacture of food products simulating the texture and flavor of bacon, ham, and chicken.

# The Soybean Flower

Soybean flowers are small (6-7 mm long) and complex, which limits the ease with which hand pollinations can be made. The flower has a tubular calyx that terminates in five unequal lobes. The anterior lobe is the longest and is located near the front and center of the opened flower. The next two lobes are lateral and shorter, and the posterior lobes are shortest. The corolla is composed of five petals: two keel petals, two wing petals, and one large standard petal. The flower contains 10 stamens, 9 of which are supported by a basal ring that surrounds the ovary. The tenth stamen is posterior and separate. The single ovary narrows into a short, slender style with a single bulbous stigma (Johnson and Bernard, 1962).

Soybeans will remain vegetative almost indefinitely if the days are long enough, or they will flower in 3-4 weeks if the days are short. Because the length of the daily dark period is a function of latitude, soybean varieties are adapted to narrow belts of latitude.

It is not uncommon for soybean plants to abort up to 75% of the flowers. The cause of loss of substantial numbers of flowers is unknown. Aborted ovaries contain fertilized ovules, and abortion is not the result of a lack of viable pollen.

# Crossing

## Natural

Soybean flowers are normally self-pollinated. One per cent outcrossing for plants in close contact has been observed. This frequency increased four to six fold in plants from seeds subjected to irradiation with X-rays and thermal neutrons. As much as 16% outcrossing was observed in plants infected with the tobacco ringspot virus because of induced male sterility (Johnson and Bernard, 1962).

## Artificial

Soybean flowers are fragile and must be handled carefully. Partially opened flowers showing 2-3 mm of the corolla are used as females. Calyx lobes are removed individually with the forceps by pulling each lobe downward. The corolla is removed by grasping the tip with the forceps, slightly turning and pulling it; the anthers can then be removed with the points of the forceps. Johnson and Bernard (1962) recommend removal of the pistil and column of anthers from an older pollen flower and using this as a brush to apply pollen.

# Breeding

## Hybridization and Selection

In the early years of soybean improvement in the United States virtually all improvement came from selections made from soybean plants introduced from other countries. In contrast, nearly all of the improved varieties currently in production are the result of selection in segregating populations resulting from artificial crosses. Johnson and Bernard (1962) have reviewed the literature on this subject in detail.

The bulk and pedigree systems, with various combinations and modifications, and the backcross system are all used by soybean breeders, depending on the objectives in making a cross. Maturity, height, seed size and quality, and resistance to disease, lodging, and shattering are the traits that receive most attention in early generations prior to $F_4$. Yield and chemical composition of seed receive attention in later generations.

Preliminary evaluation of genotypes may be based on a replicated test at one well-chosen location (Johnson and Bernard, 1962). Early selection of plant material representing a wide range of phenotypes has merit; however, when the phenotypic range is narrow, greater progress results when selections are made in $F_4$ through $F_7$ (Hanson and Weber, 1961). Johnson and Bernard (1962) believed that advancing a large number of plants from an $F_2$ population through single-seed descent (selecting but a single seed from each plant for each generation) to about the $F_5$ generation before extracting lines, has considerable merit as a breeding procedure.

## Backcrossing

Backcrossing (the crossing of a hybrid to one of its parents) is an important soybean breeding method. Repeated backcrossing is particularly effective in transferring simply inherited characters such as disease resistance to nonresistant but otherwise valuable germ plasm. The technique is useful in the development of closely related lines differing in simply inherited characteristics.

# Bacterial Diseases

## Bacterial Blight

*Source and Type of Resistance.*   Resistance to *Pseudomonas glycinea* Coerper, the bacterium that causes bacterial blight, was first reported by Dunleavy et al. (1960). A soybean introduction, P.I. 68708, was resistant to bacterial blight under natural conditions and when inoculated. Later, Cross et al. (1966) reported 7 pathogenic races of *P. glycinea* and distinguished these races by the disease reaction on leaves of the varieties Acme, Chippewa, Flambeau, Harosoy, Lindarin, Merit, and Norchief. Chippewa was the most resistant variety tested: it was resistant to all races except race 4, for which it gave a reaction intermediate between resistance and susceptibility.

Resistance to race 1 of *P. glycinea* is simply inherited and attributed to a single dominant gene designated $Rpg^1$ (Mukherjee et al., 1966). Data from the same study indicate that the inheritance of resistance to race 2 involves more than one gene and requires additional study.

*Nature of the Pathogen.*   *P. glycinea* is one of the most widespread bacterial species that cause diseases of soybeans. Bacteria splashed from soil or nearby diseased plants gain entrance into the leaves through the stomata. The bacteria multiply rapidly in the substomatal chambers and intercellular spaces and produce small, angular, wet spots in about 7 days. The spots turn yellow and then brown as the tissue dies. Many small lesions sometimes coalesce; the large necrotic areas which result may fall out. Lower leaves may drop as a result of heavy infection. The optimum temperature for growth of *P. glycinea* is 24-26 C, and bacterial blight is therefore usually most severe during periods of cool weather when rain or dew occurs frequently. The disease is usually observed early in the season before plants flower.

Cells of *P. glycinea* are gram negative rods 1.2-1.5 $\mu$ by 2.3-3.0 $\mu$. They are motile with polar flagella and produce a green fluorescent pigment in culture. This bacterium is facultatively aerobic, and its maximum temperature for growth is 35 C. Colonies appear on beef-peptone agar in 24 hr and are circular, creamy white, smooth, shining, and convex. *P. glycinea* is pathogenic only to soybeans and is widespread throughout the soybean producing regions of the world (Elliott, 1951).

*Inoculum.*   *P. glycinea* can be maintained in the laboratory on trypticase soy agar or potato dextrose agar at *p*H 7.0. Small quantities of inoculum can be prepared by suspending the bacteria from 48-72 hr test tube slants in sterile water. Larger quantities can be prepared from slanted cultures in bottles, or simply by growing the bacteria in trypticase soy broth at 28 C in a shaker incubator for 24 hr. Other methods of preparing inoculum were described by Dunleavy et al. (1960) and Leben et al. (1968).

*Inoculation.* A rapid method of inoculation for greenhouse plants consists of dusting the leaves lightly with 600 mesh carborundum and gently rubbing the surface of the leaves with a cloth pad saturated with a suspension of *P. glycinea* ($10^7$ cells/ml) in sterile water.

Kennedy and Cross (1966) used an artist's airbrush for inoculation and reported that it was the most efficient and precise method tested for differentiating races of *P. glycinea.* Dunleavy et al. (1960) used a motor-driven sprayer operated at 80 psi to inoculate field plants. An improvement of this technique consists of using a motor-driven air compressor with a paint sprayer. Carborundum (600 mesh) is added to the inoculum (2 g/l). During inoculation the inoculum must be agitated periodically to keep the carborundum in suspension.

*Rating System.* A common rating system for bacterial diseases ranges from 1-5 as follows: 1, immune; 2, resistant; 3, slightly susceptible; 4, susceptible; 5, very susceptible (Dunleavy et al., 1960). Cross et al. (1966) used three categories in their rating system: resistant, intermediate, and susceptible. Inoculated plants are usually rated 10-14 days after inoculation.

*Literature.* Under conditions ideal for the development of bacterial blight, the disease has reduced yield by as much as 22%. Although these conditions are not often encountered in fields, severe losses from bacterial blight can result when they do occur. The bacteria are seed-borne and survive in dead leaves from one growing season to the next. The disease was first described by Coerper in 1919 (Elliott, 1951).

## Bacterial Pustule

*Source and Type of Resistance.* The CNS variety of soybeans is highly resistant to *Xanthomonas phaseoli* (Smith) Dows. var. *sojensis* (Hedges) Starr and Burkh. and is the source of all bacterial-pustule resistance incorporated in improved U.S. varieties. Resistance to bacterial pustule is recessive and conditioned by a single major gene pair with modifying genes which determine the degree of susceptibility. The designation for this gene is *rpx*. Resistance is generally expressed much more strongly in the field than in the greenhouse. Under some conditions it is possible to obtain infection of resistant varieties in the field when bacterial concentration of the inoculum is $10^{10}$ cells/ml or higher. Bacteria in lesions produced in this way will not spread and cause additional lesions under natural conditions, however. Only two varieties grown in the northern U.S. carry the *rpx* gene: Clark 63 and Wayne. Nearly all varieties grown in the southern U.S. are bacterial-pustule resistant: a few of these are Bragg, Hampton, Hardee, Hill, Hood, Lee, and Scott.

*Nature of the Pathogen.*    *X. phaseoli* var. *sojensis* is present to some extent in most soybean-growing regions of the world. Elliott (1951) lists *Phaseolus lunatus macrocarpus* and *P. vulgaris* as alternate hosts. The bacteria infect soybean leaves in the same way as that described for bacterial blight. Lesions caused by the bacteria are usually confined to the leaves. At first, they appear as small, yellowish-green areas with reddish-brown centers, more conspicuous on the upper surface of the leaf. A small, raised pustule usually develops at the center of the lesions, especially on the lower leaf surface.

Cells of *X. phaseoli* var. *sojensis* are gram negative rods 0.5-0.9 $\mu$ x 1.4-2.3 $\mu$, motile by a single polar flagellum. The bacterium is aerobic and produces a yellow, non-water soluble pigment on agar media. Colonies on trypticase soy agar are pale yellow and become deeper yellow with age, small, circular, smooth (occasionally rough), margins entire, and butyrous to slightly viscid. The optimum temperature for growth ranges from 30-33 C; the minimum is 10 C and the maximum 38 C (Elliott, 1951).

*Inoculum.*    Inoculum is prepared in the same ways described for *P. glycinea,* except that the cultures are incubated at 30 C. Jones and Hartwig (1959) described a method for preparing inoculum for field inoculations in which infected leaflets were stored in sealed glass jars and frozen for use the following season. Thawed leaves were passed through a food chopper. Ten chopped leaves were added to 300 ml of water. This mixture was then filtered through 2 layers of cheesecloth after 2 hours, the volume increased to 1 gal, and the suspension sprayed on field plants with a pressure sprayer.

*Inoculation.*    Inoculation procedures described for bacterial blight apply for bacterial pustule inoculations; however, pustule inoculations in the field are frequently unsuccessful unless the temperature is 30 C or higher for substantial periods during the daylight hours. Selection for resistance in segregating plant populations in the greenhouse is difficult and not recommended. If it is attempted, inoculum concentration must be rigidly controlled. Chamberlain (1962) was able to differentiate resistant CNS plants from susceptible Lincoln plants in the greenhouse on the basis of pustule formation at inoculum concentrations between 8 x $10^4$ and 8 x $10^5$ cells/ml. CNS developed necrotic spots on the leaves, but no pustules developed; whereas most of the lesions formed on Lincoln leaves contained pustules.

*Rating System.*    The rating system used for bacterial blight is also commonly used for rating bacterial pustule.

*Literature.*    Resistant selections yield as much as 11% more than susceptible selections, and susceptible plants produce seed 4% smaller than seed from resistant plants. *Brunnichia cirrhosa* (red vine), a weed which grows in the southern U.S., is a natural host for *X. phaseoli* var. *sojensis* and is involved in local spread of bacterial pustule in soybean fields (Jones, 1961). Hedges first described bacterial pustule in 1924.

Wildfire, caused by *Pseudomonas tabaci* (Wolf and Foster) F. L. Stevens, occurs only where bacterial pustule is found. *P. tabaci* alone does not produce a disease in soybeans. The bacteria invade only the parts of the leaf already parasitized by the pustule bacteria. The disease is not common, and all varieties resistant to bacterial pustule are resistant to wildfire.

# Fungus Diseases

## Phytophthora Rot

*Source and Type of Resistance.*    Bernard et al. (1957) studied the inheritance of resistance to Phytophthora rot of soybeans. They observed that the varieties A. K., Arksoy, Blackhawk, CNS, Dorman, Harly, Illini, Monroe, and Mukden are resistant. They reported that resistance is controlled by a single dominant gene, designated *Rps,* and that the resistance of A.K. and Mukden is due to the same gene. Hartwig et al. (1968) studied the genotype of soybeans having a resistant reaction to race 1 of the fungus and a susceptible reaction to race 2. They reported that *Rps* is dominant to *rps²*, and that *rps²* is dominant to *rps*. There is some evidence of linkage in the coupling phase between *Rps* and the allele for lateness in maturity. Paxton and Chamberlain

(1969) described two types of resistance: resistance in young plant tissue (0-2 weeks old), in which presence of a phytoalexin is important; and morphologic resistance characterized by woody tissue.

*Nature of the Pathogen.*   *Phytophthora megasperma* Drechs. var. *sojae* Hildeb. is the fungus that causes Phytophthora rot. Sporangia are obpyriform, slightly papillate, and average 52 x 35 $\mu$. They may germinate by germ tubes or sporangia. The oogonia are thin-walled, spherical, and average 36 $\mu$. Oospores are smooth-walled, yellow, and average 31 $\mu$ in width. Chlamydospores are irregular to spherical, and nearly as large as oogonia. Optimum temperature for growth is 25C.

Two races of the fungus have been identified. A second pathogenic race of the fungus was isolated from D60-9647, which is resistant to race 1. The varieties Harrel and Nansemond are also resistant to race 1 and susceptible to race 2, and they effectively distinguish between these races.

The fungus can attack the plant at any stage in its development. It causes preemergence damping-off, postemergence killing of seedlings, or a more gradual killing or reduction in vigor of plants throughout the growing season.

*Inoculum.*   The fungus grows well on cornmeal agar or lima bean agar. These media are generally used for production of inoculum for stem inoculations. Lima bean broth is an excellent medium for preparation of inoculum to be added to soil. All medium should be rinsed from the mycelium before the fungus is placed in the soil.

*Inoculation.*   The toothpick method of inoculation was used successfully by Kaufmann and Gerdemann (1958). They boiled toothpick tips in water for 1 hr, autoclaved the tips, added the tips to cornmeal agar plates, and inoculated the plates with the fungus. Plants were inoculated by inserting a toothpick tip into the hypocotyl of each seedling and sealing the wound with petroleum jelly. Several researchers use a modification of this method, in which mycelium scraped from the surface of a plate is inserted into the hypocotyl of a seedling with a scalpel or half-spear-point needle, and the wound sealed with petroleum jelly. Inoculum may be added to soil and to seed sown above infested soil. However, this method is quite variable, and it is less reliable than stem inoculation.

*Rating System.*   Bernard et al. (1957) rated field plants as either resistant or susceptible. Resistant plants showed no symptoms after inoculation, but over 90% of the susceptible plants were killed within 3-4 days, and the remaining plants showed some necrosis. Smith and Schmitthenner (1959) rated plants from 1 to 5 as follows: 1, healthy roots; 2, trace of root rot; 3, one-third of root rotted; 4, two-thids of root rotted; and 5, all roots rotted.

*Literature.* Phytophthora rot is especially noticeable in low, poorly drained areas of a field, but in wet seasons it also appears on higher ground. It is more severe on heavy clay soils than on the lighter soils. The disease occurs over most of the soybean-producing areas of the United States and Canada.

Chamberlain and Gerdemann (1966) reported that resistant Harosoy 63 soybeans became susceptible to *P. megasperma* var. *sojae* after 1 hr at 43-45 C, and also became susceptible to *P. cactorum* (lebert & Cohn) Schroet., a nonpathogen of soybeans. Harosoy soybean plants can be locally crossprotected with *P. cactorum* against subsequent infection by *P. megasperma* var. *sojae*. Production of a phytoalexin by soybeans is implicated in this crossprotection.

## Downy Mildew

*Source and Type of Resistance.* Dunleavy (1970) tested 72 varieties of soybeans for disease reaction to 14 of the 23 races (2, 5, and 7-18) of *Peronospora manshurica* (Naum.) Syd. ex Gaum., the downy mildew fungus. Only Mendota was immune to all races to which it was exposed. This variety has yellow seed with a yellow hilum, and is in maturity group I. Kanrich was infected by races 2 and 9. Kanro was infected by races 5 and 13 and segregated for immunity and high susceptibility to races 10 and 14. Both Kanrich and Kanro are large-seeded vegetable type soybeans in maturity group II. Only Bansei was rated highly susceptible to all races to which it was exposed.

No information has been published dealing with the inheritance of downy mildew resistance since Geeseman (1950) first investigated this subject. He proposed a system of three gene pairs ($Mi_1$ $mi_1$, $Mi_2$ $mi_2$, and $Mi_R$ $mi_R$) to explain the complex results he obtained.

*Nature of the Pathogen.* *P. manshurica* is an obligate parasite that infects only soybeans. Conidia germinate in water on the surface of expanding young leaves, and the hyphae penetrate epidermal cells. As the mycelium progresses through host tissue in the intercellular spaces, it penetrates cells at intervals and forms intracellular haustoria. Chlorotic or necrotic leaf lesions range in size from less than 1 mm up to lesions that cover the entire leaf. The fungus produces clusters of conidia borne on single conidiophores on the lower side of the leaf 7-10 days after infection. The lower surface of the leaf must be wet or the fungus will not sporulate. These spores are short lived and usually persist for only 1 day. The fungus grows within the plant, invades pods, and covers some seeds with a white crust of thick-walled oospores. Oospores also form in infected leaf tissues. These spores overwinter in the fallen leaves and probably provide a source of infection for the next season's crop. When oospore-encrusted seeds are planted, a small percentage of the seedlings are infected systemically. Such plants will usually produce conidia throughout the growing season whenever sufficient moisture is available. Twenty-three races of *P. manshurica* have been described (Dunleavy, in press).

*Inoculum.* Conidia are increased on susceptible varieties grown in the greenhouse. Leaves infected for 8-10 days are placed in a moist chamber overnight. During this period the fungus produces conidia on the lower surface of the infected areas of leaves. These conidia are washed into water containing 0.1% Tween-20 and sprayed over the seedling leaves with an air-pressure sprayer. Primary leaves are inoculated when they are about three-fourths unfolded (8-10 days after planting when grown at 24-28 C). Inoculated leaves must be kept moist in a moist chamber during infection. Good infection occurs in 16-24 hrs between 18 and 23 C.

*Inoculation.* Greenhouse-grown plants are inoculated by spraying the leaves with a suspension of freshly harvested conidia in water containing 0.1% Tween-20. The leaves should be sprayed until the surface is wet. Seedling plants with unifoliolate leaves expanding are most suitable for inoculation; although older plants may be used leaf lesions will be smaller. Inoculated plants must be placed in a moist chamber for 16-24 hr before the inoculum on the leaf surface dries.

Inoculation of plants in the field should be made after sundown, when dew formation is likely. If dew is unlikely, the soil near the inoculated plants should be saturated before inoculation. After inoculation, the plants must be covered with a plastic sheet and the edges covered with soil. The sheet must be removed shortly after sunrise on clear days to prevent excessive accumulation of heat under the sheet.

*Rating System.* The rating system for downy mildew ranges from 1 (immune) to 5 (very susceptible). This system should be used only for rating unifoliolate leaves, because lesions on older leaves are smaller and more difficult to rate.

*Literature.*    Downy mildew is one of the most common soybean diseases in the United States. The disease has world-wide distribution and occurs almost everywhere soybeans are grown. In Europe the disease occurs from Sweden to the Balkans. Downy mildew is an important disease in all of the soybean-producing regions of the Orient, and it also occurs in India and Australia. In South America the disease is serious in Ecuador and Brazil.

Lehman (1953) studied the effect of temperature on oospore germination. He observed that when oospore-encrusted seed was germinated at 13 C, 40% of the seedlings were systemically infested, whereas no infection occurred at 18 C or above. Dunleavy and Snyder (1962) observed that oospores germinated by means of a germ tube, and that oospores on agar plates inhibited germination of nearby conidia. They believed that this inhibitor might be the same as that produced by conidia.

# Future Goals

## Sources of Resistance

There is little doubt that breeding for disease resistance is the most direct means of achieving disease control in soybeans, *if* a source of resistance is available. There is no known adequate source of resistance to the following soybean diseases: Rhizoctonia root rot; Fusarium root rot; stem canker [ *Diaporthe phaseolorum* (Cke. & Ell.) Sacc. var. *caulivora*] Athow & Caldwell; Pythium rot; anthracnose (*Colletotrichum truncatum*); and bud blight (tobacco ringspot virus). Once resistance is incorporated in a variety, there is the threat that pathogenic races of an organism will become capable of attacking the resistant variety. Researchers recently reported a strain of the soybean cyst nematode (*Heterodera glycines*) in several U.S. states that can cause damage and complete its life cycle in resistant plants. Other soybean diseases for which pathogenic races have been found are: bacterial blight (Cross et al., 1966 , Phytophthora rot, downy mildew (Dunleavy, 1971), and frog eye leaf spot (*Cercospora sojina* Hara) (Athow et al., 1962).

## The Nature of Resistance: Role of Multiple Infections

*Cytoplasmic Inheritance.*    No observation in the field has impressed me more than specific soybean diseases apparently imparting host resistance to other diseases. Soybean mosaic-infected plants tend to be remarkably resistant to stem canker and brown stem rot (*Cephalosporium gregatum* Allington & Chamberlain).

Over a period of years I attempted to select stem-canker resistant plants from a cross of *Glycine ussuriensis* (resistant) x Ford (susceptible). Segregates commonly displayed symptoms of soybean mosaic virus (SMV), which is seed-transmitted in soybeans. These infected plants were invariably resistant to stem canker but were discarded because of poor agronomic characteristics and low yield. Resistance was never transferred to a line with acceptable agronomic traits. SMV can be masked under various conditions, and all plants of at least one soybean variety are SMV-infected (Quiniones and Dunleavy, 1970). SMV is seed-transmitted in *G. ussuriensis,* but symptoms in the field are masked. Ford has not expressed symptoms of SMV in our field plots or in the greenhouse (Dunleavy and Quiniones, *unpublished data*). These observations suggest that the original stem-canker resistance in *G. ussuriensis* may have been caused by interference with normal metabolism by SMV.

Kunkel (1965) reported that of the varieties tested, Midwest was most resistant to brown stem rot. Quiniones (1968) found that plants of this variety were also infected with SMV, and that sap from plants of Midwest reacted with antisera of two strains of SMV. This again raises the question: does SMV infection of soybean plants influence resistance to other diseases?

Dunleavy and Quiniones (*unpublished data*) recently completed a study in which SMV was transmitted mechanically from L65-461, a soybean isoline possessing a wavy leaf character, to Harosoy, the normal leaf counterpart of the isoline. The youngest leaves on each of 10 Harosoy plants showed the wavy leaf character 2 weeks after inoculation with sap from L65-461. The virus produced rugose symptoms on leaves of 25 virus-free Bansei soybean plants inoculated with sap from L65-461. The virus was identified by symptomatology on Bansei, and by local lesion production on 25 plants each of *Dolichos lablab* and *Cyamopsis tetragonoloba* (Guar.). The microprecipitin test was used to identify the virus serologically. Antiserum of the M isolate of SMV (SMVM) formed a precipitate when added to clarified sap from SMVM-infected Bansei plants, and also when added to clarified sap from L65-461 plants. No precipitate formed in clarified sap from Harosoy plants when the antiserum was added. In a field test, L65-461 yielded 32% less seed than Harosoy, and it produced seed that was 2g/100 seed lighter. There is little doubt that phenotypic expression of a wavy leaf character was mechanically transferred from one variety to another. If one wishes to cling to current genetic principles, then we must conclude that the gene which controls this wavy leaf character was transferred to Harosoy. Since SMV was also transferred, we are left with the idea that a gene may be a virus. This is not a new idea, but I know of no current researchers seriously pursuing it.

Luria (1953) defined viruses as submicroscopic entities capable of being introduced into specific living cells and of reproducing inside such cells only. Existence of both DNA and RNA viruses has long been established. Because SMV is an RNA virus, we are left with a postulated RNA "gene" in the case of the wavy leaf character transfer. Initially, this seems unlikely because of the preoccupation of geneticists with genes and DNA, but a search of the microbiological literature reveals that, besides most plant viruses, there are several RNA viruses (bacteriophages) capable of reproduction (Stent, 1963). Indeed, nothing illustrates the genetic capabilities of RNA more clearly than the production of mutants of tobacco mosaic virus through chemical alteration of the virus by Gierer and Mundy (1958). Once we are accustomed to the idea that extra-nuclear RNA, like DNA, is capable of reproduction in a cell and carries genetic capability, the idea of an RNA "gene" is more palatable. In fact, the alteration of soybean plant cells infected by SMV, and the genetic alteration of transduced bacteria through introduction of specific DNA into the bacterial cells are quite analogous. It is important to note that DNA used in this way fits Luria's definition of a virus, because the submicroscopic DNA that enters the bacterial cell replicates itself.

Dunleavy and Urs (1970) reported the occurrence of an 80-nm polyhedral-shaped DNA phage (SBX-1) and its *Xanthomonas* (bacterial) host in certain soybean varieties, but not in others. Plants of soybean varieties that contained no phage contained a species of Xanthomonas that was resistant to SBX-1. We successfully isolated the Xanthomonas from soybean pollen, and we believe that both the Xanthomonas and phage are cytoplasmically transmitted via the seed to the next generation.

*The Nuclear Genome vs. the Plastid and Mitochondrial Genomes.*   Until relatively recently geneticists were primarily interested in effects produced by nuclear DNA in plants, but there is now much interest in the double standard DNA in such cytoplasmic inclusions as the mitochondria and plastids. These inclusions are structurally independent components of the cytoplasm of all respiring plant cells. Mitochondria are the principal energy source of the cell, and plastids have much in common with mitochondria. Chloroplasts are the organelles in which photosynthesis occurs in plants. The DNA of these inclusions could add greatly to the genetic apparatus of the cell.

Germane to the question of whether a genetic apparatus exists separate from the chromosomal genes, are theoretical computations based on the information carried by the chromosomal genes (Goldman, 1953). Bremermann (1963) computed that the stimulus-response mechanism of immunity in man would require at least $5 \times 10^6$ items of information. Considering all of the additional genetic traits under chromosomal control, it becomes obvious that the chromosomal "memory bank" is shamefully overloaded. If chromosomal genetic information is in short supply, as appears to be the case, evolution could have placed a premium on nonchromosomal sources of information storage and transfer. These nonchromosomal genetic systems may be similar to the DNA-template apparatus (Watson and Crick, 1953), or the genetic information may be stored in three-dimensional patterns of cellular organization, as implied in Sonneborn's (1960) studies of *Paramecium aurelia*.

There is strong evidence that the nuclear genes have considerable control over the formation of mitochondria (Ephrussi and Hottinguer, 1951; Solonimski, 1953). At present we do not know whether the control is by passage of messenger RNA into the cytoplasm, or whether a mitochondrial precursor is synthesized in or near the nucleus. Hartman (1954) suggested that mitochondria arise from sub-microscopic granules at the nucleo-cytoplasmic interface, rather than by division of existing mitochondria.

In summary, plant genetics is in a state of flux. Some patterns of inheritance in plants and microorganisms contradict Mendelian theory and can only be explained if we are willing to assume that other parts of the cell, in addition to the chromosomes, contain genetic information. T. H. Morgan (1909), who formulated the chromosome theory of heredity, made this comment on Mendelian theory, which may still be considered timely: "The superior jugglery sometimes necessary to account for the results may blind us, if taken naively, to the commonplace that the results are so often excellently explained because the explanation was invented to explain them." Soybean pathologists and breeders are challenged by the complex problems that face us, but fortunately we have a unique tool to help us meet the challenge—the soybean.

# References

Athow, K. L., A. H. Probst, C. P. Kurtzman, and F. A. Laviolette. 1962. A newly identified physiological race of *Cercospora sojina* on soybean. Phytopathology 52:712-714.

Bernard, R. L., P. E. Smith, M. J. Kaufmann, and A. F. Schmitthenner. 1957. Inheritance of resistance to Phytophthora root and stem rot in the soybean. Agron. J. 49:391.

Bremermann, H. J. 1963. Limits on genetic control. IEEE Trans. on Military Electronics 7:200-205.

Chamberlain, D. W. 1962. Reaction of resistant and susceptible soybeans to *Xanthomonas phaseoli* var. *sojensis*. Plant Dis. Reporter 46:707-709.

————, and J. W. Gerdemann. 1966. Heat-induced susceptibility of soybeans to *Phytophthora megasperma* var. *sojae*, *P. cactorum* and *Helminthosporium sativum*. Phytopathology 56:70-73.

Cross, J. E., B. W. Kennedy, J. W. Lambert, and R. L. Cooper. 1966. Pathogenic races of the bacterial blight pathogen of soybeans, *Pseudomonas glycinea*. Plant Dis. Reporter 50:557-560.

Dunleavy, J. 1970. Sources of immunity and susceptibility to downy mildew of soybeans. Crop Sci. 10:507-509.

————. 1971. Races of *Peronospora manshurica* in the United States. Am. J. Bot. 58:209-211.

————, and G. Snyder. 1962. Inhibition of germination of oospores of *Peronospora manshurica*. Iowa Acad. Sci. 69:118-121.

————, and N. V. R. Urs. 1970. Occurrence of a phage and its Xanthomonas host in soybean varieties. Bacteriological proceedings, p. 39. Amer. Soc. Microbiol. Washington, D. C. 258 p.

————, C. R. Weber, and D. W. Chamberlain. 1960. A source of bacterial blight resistance for soybeans. Proc. Iowa Acad. Sci. 67:102-125.

Elliott, C. 1951. Manual of bacterial plant pathogens. Chronica Botanica Co., Waltham, Mass. 186 p.

Ephrussi, B., and H. Hottinguer. 1951. Cytoplasmic constituents of heredity. Cold Spring Harbor Symp. Quant. Biol. 16:75-85.

Geeseman, G. E. 1950. Physiologic races of *Peronospora manshurica* on soybeans. Agron. J. 42:257-258.

Gierer, A., and K. W. Mundy. 1958. Production of mutants of tobacco mosaic virus by chemical alteration of its ribonucleic acid in vitro. Nature 182:1457.

Goldman, S. 1953. Information theory. Prentice-Hall, New York. 371 p.

Hanson, W. D., and C. R. Weber. 1961. Resolution of genetic variability in self-pollinated species with an application to the soybean. Genetics 46:1425-1434.

Hartman, J. F. 1954. Electron microscopy of motor nerve cells following section of axones. Anat. Rec. 118:19-34.

Hartwig, E. E., B. L. Kelling, and C. J. Edwards, Jr. 1968. Inheritance of reaction to Phytophthora rot in the soybean. Crop Sci. 8:634-635.

Johnson, H. W., and R. L. Bernard. 1962. Soybean genetics and breeding. Advanc. Agron. 14:149-221.

Jones, J. P. 1961. A weed host of *Xanthomonas phaseoli* var. *sojense*. Phytopathology 51:206.

————, and E. E. Hartwig. 1959. A simplified method for field inoculation of soybeans with bacteria. Plant Dis. Reporter 43:946.

Kaufmann, M. J., and J. W. Gerdemann. 1958. Root and stem rot of soybean caused by *Phytophthora sojae* n. sp. Phytopathology 48:201-208.

Kennedy, B. W., and J. E. Cross. 1966. Inoculation procedures for comparing reaction of soybeans to bacterial blight. Plant Dis. Reporter 50:560-565.

Kunkel, J. F. 1965. Brown stem rot syndrome development in soybeans. Ph.D. Thesis. Iowa State Univ. Diss. Abstr. 25:6146.

Leben, C., G. C. Daft, and A. F. Schmitthenner. 1968. Bacterial blight of soybeans: population levels of *Pseudomonas glycinea* in relation to symptom development. Phytopathology 58:1143-1146.

Lehman, S. G. 1953. Systemic infection of soybean by *Peronospora manshurica* as affected by temperature. Elisha Mitchell Sci. Soc. J. 69:83.

Luria, S. E. 1953. General virology. John Wiley and Sons, New York. 427 p.

Morgan, T. H. 1909. What are "factors" in Mendelian explanations? Am. Genet. Ass. Ann. Rep. 47 p.

Mukherjee, D., J. W. Lambert, R. L. Cooper, and B. W. Kennedy. 1966. Inheritance of resistance to bacterial blight (*Pseudomonas glycinea* Coerper) in soybeans (Glycine max). Crop Sci. 6:324-326.

Nagata, T. 1960. Studies on the differentiation of soybeans in Japan and the world. Mem. Hyogo Univ. Agr. 3:63-102.

Paxton, J. D., and D. W. Chamberlain. 1969. Phytoalexin production and disease resistance in soybeans as affected by age. Phytopathology 59:775-777.

Quiniones, S. S. 1968. Soybean mosaic. Ph.D. Thesis. Iowa State Univ. Diss. Abstr. 29:3157B.

————, and J. M. Dunleavy. 1970. Identity of a soybean mosaic virus isolated from *Glycine max* variety Hood. Plant Dis. Reporter 54:301-305.

Smith, P. E., and A. F. Schmitthenner. 1959. Further investigations of the inheritance of resistance to Phytophthora rot in the soybean. Agron. J. 51:321-323.

Solonimski, P. P. 1953. Third symposium of the society for general microbiology. Cambridge University Press, Cambridge, England. 257 p.

Sonneborn, T. M. 1960. The gene and cell differentiation. Proc. Nat. Acad. Sci. U.S. 46:149.

Stent, G. S. 1963. Molecular biology of bacterial viruses. W. H. Freeman Co., San Francisco. 474 p.

Watson, J. D., and F. H. C. Crick. 1953. Genetical implications of the structure of deoxyribonucleic acid. Nature 171:164.

# 18 Cucurbits
## Wayne R. Sitterly[1]

## Introduction

The Cucurbitaceae are among the more important and widespread plant families that supply man with food and fiber. Resistant cucurbit parental material usually occurs in plants obtained from areas where a pathogen is endemic and where resistance has occurred by mutation and natural selection. Thus, one must be familiar enough with the host to understand the "normal" condition and not be confused by any deviation caused by temperature or other external factors.

Cucurbits exhibit both advantages and disadvantages in plant breeding. The disadvantages are that plants require much space, which makes large populations expensive; hand pollination is generally necessary for controlled genetic work; usually (except in the case of squash) pollination must be done before selection; chromosomes are not easily differentiated from cytoplasm in pollen mother cells; and

1. Professor of Plant Pathology, Clemson University Truck Experiment Station.

chromosomes are small and not well separated from each other. Advantages are that plants are easily grown by simple methods; flowers are relatively large and easily hand-pollinated; plants are indeterminate with flowers available over a long period of time; fruits are fairly durable; and most fruits yield many seeds. Cucurbits are an interesting group of crops.

# Host

## Origin, History, and Economic Importance

Cucurbits may not be as important on a world basis as the cereals and legumes, but from the tropics to the milder portions of the temperate zone they serve as sources of carbohydrates, as dessert and salad ingredients, and as pickles. Some cucurbits serve as pottery and baskets, insulation, and in oil filters. The cucurbits have needed man for survival, as bona fide specimens of the wild counterparts have apparently never been collected. There are 90 genera and 750 species in the family Cucurbitaceae.

DeCandolle (1882) presented data showing watermelon (*Citrullus vulgaris* Schrad.) to be indigenous to tropical Africa (the drier open areas on both sides of the equator), with perhaps a strong secondary center of genus diversification in India. Watermelon has been cultivated for centuries by people bordering the Mediterranean Sea.

DeCandolle (1882) also showed that cucumber (*Cucumis sativus* L.) was native to India, where it has been cultivated for over 3,000 years. Cucumber spread east to China, and west where it was enjoyed by the Greeks and Romans. The chromosome count of 7, and such morphological features as angular stems, separate it from others in this genus.

The place of origin of cantaloupe (*Cucumis melo* L.) has never been fully resolved (Whitaker and Davis, 1962), but it appears to be tropical and subtropical Africa. From there it spread into Asia, with well developed secondary centers in India, Persia, southern Russia, and China. When placed in a congenial environment it exploded into a number of subspecies.

The squashes and pumpkins (*Cucurbita* spp.) are indigenous to North America. The center of distribution seems to be central or southern Mexico, or the northern part of Central America. As northern migration occurred they developed both xerophytic and humid adaptive types. The Indians of this area utilized the genus for many of their needs.

## The Floral Organs and Fertilization

*Citrullus* flowers are smaller and less colorful than *Cucumis* flowers, and most are monoecious with pistillate flowers occurring every 7th node. *Cucumis melo* flowers may be either andromonoecious or monoecious. Hermaphroditic flowers are borne singly at first and second nodes of fruiting branches, pistillate flowers are also borne singly, and staminate flowers are borne in axillary clusters of 3-5. *Cucumis sativus* flowers are generally monoecious. Pistillate flowers occur singly at nodes, while staminate flowers appear at other leaf nodes in clusters of 5. *Cucurbita* flowers are monoecious, large, bright yellow, and occur singly in leaf axils. In vine cultivars, staminate flowers are located near the center of the vine on slender pedicels, while pistillate flowers are borne on short peduncles distal to the staminate ones.

The number of staminate flowers always exceeds the number of pistillate flowers. Staminate flowers have 3 stamens, with free filaments, united by their anthers. Pistillate flowers are epigynous, with a pistil having 1-5 carpels and a thick short style terminated by 3 papillate stigma. There may also be sterile rudimentary stamens.

The fruit is an inferior berry or pepo (fruit indehiscent with fleshy floral tube adnate to pericarp). In *Citrullus* and *Cucumis sativus* the greater portion of the fruit is edible flesh derived from the placenta. In *Cucumis melo* and *Cucurbita* it is derived from pericarp tissue.

Seaton and Kremer (1939) showed that the most important climatological influence on anthesis and anther dehiscence is temperature. Minimum temperature for pumpkin and squash is 10 C; for watermelon and cucumbers it is 15 C; and for cantaloupe it is 18 C. Mann and Robinson (1950) showed that pollen grains germinate in a few hours, and that the pollen tubes reach the ovary after 24 hours. Tiedjens (1928) demonstrated light-influenced fertilization of ovules, and that shape and growth of fruit can be affected by influencing sugar availability. He also showed light-influenced sex expression, with an abundance of light increasing the number of staminate flowers. Hibbard (1940) demonstrated the pronounced inhibitory effect of growing fruit on the subsequent setting of pistillate flowers.

# Pathogen

## Anthracnose — *Colletotrichum lagenarium* (Pass.) Ell. & Halst.

*Symptoms.*   On cucumber leaves the lesions start on a vein and expand into roughly circular 1-cm brown spots. Young leaves may be distorted. Stem lesions are shallow, elongate, and tan. Lesions appear on fruit as circular, sunken, watersoaked areas. In moist weather pinkish spore masses occur. On watermelons the leaf lesions are black, and coalescence of the lesions produces a scorched appearance in a severely infected field. Young fruits may have black sunken spots, resulting in malformation. On older fruits the lesions are elevated and flat initially, and later they become sunken.

*Nature of Pathogen.*   The anthracnose pathogen produces black stromata containing conidiophores. Conidia are budded off apically, disseminated by water, and penetrate by an infection peg from an appressorium. Humid or rainy weather is essential for sporulation and penetration. *C. lagenarium* is composed of several physiologic races. Goode (1958) demonstrated the presence of 3 races that were morphologically and culturally indistinguishable. Race 1 was virulent on all cucumber varieties tested: it produced slight infection on Charleston Gray, Congo, and Fairfax watermelon varieties, and it moderately attacked Butternut squash. Race 2 was virulent on all cucumber and watermelon varieties, and it moderately attacked Butternut squash. Race 3 was identical to race 1, except that Butternut squash was immune. Goode also showed that cucumber PI 163217 and PI 196289 were resistant to race 2. Four more races of anthracnose have more recently been identified. Jenkins et al. (1964) stated that race 4 was virulent on all hosts; race 5 was weakly virulent on their cucumber differentials and highly virulent on watermelons; race 6 was weakly virulent on cantaloupe and highly virulent on watermelons; race 7 was like race 3, except only weakly virulent on Pixie cucumber. There is some conflict concerning race 4, as Dutta et al. (1960) stated that Model cucumber and Chris Cross watermelon were moderately resistant, and that cucumber PI 163213 and Charleston Gray watermelon were resistant. Races 1 and 2 appear to be most common.

*Nature of Resistance.*   Since some varieties within a species are resistant and others susceptible, breeders have to work with resistance to specific pathogenic races. Barnes and Epps (1952) found two types of resistance among Plant Introduction cucumber accessions. One type involved an extremely high level of resistance controlled by several genes (PI 197087), while the other was a moderate degree of resistance controlled by a single dominant gene (PI 175111). The genetic ratio of mode of inheritance of the multigenic resistance has not been solved. Busch and Walker (1958) confirmed monogenic resistance in PI 175111, but stated that the lack of definite segregation ratios suggested the presence of modifiers. They also showed that the resistance

mechanism involved the parasite penetrating both resistant and susceptible cells, but that hyphal progress was much slower and cells collapsed
more slowly in resistant tissue.

Winstead et al. (1959) evaluated many watermelon items and found
that those resistant to race 1 were also resistant to race 3 but susceptible
to race 2. The mode of inheritance of resistance to races 1 and 3 was
governed by a single dominant gene. Winstead et al. also found African
Citron W-695 was resistant to race 2, but the mode of inheritance was
not determined.

*Sources and Range of Resistance in Varieties.*    Sources of cucumber anthracnose resistance are: race 1—PI 163213 (monogenic); race
2—PI 196289 (multigenic); and race 3—PI 197087 (multigenic). Actually, PI 197087 has resistance to all 3 races. A range of reaction
would be: highly resistant—PI 197087 and Poinsett; moderately
resistant—PI 175111; slightly tolerant—Ashley; and susceptible—
Marketer.

Resistance in watermelon to races 1 and 3 can be found in African 8,
and to race 2 in African Citron W-695. A range of reaction would be:
susceptible—Klondike; races 1 and 3 resistant—Charleston Gray and
Congo; race 2 resistant—African Citron W-695.

*Collecting and Preparing Inoculum.*    Anthracnose is maintained on
potato-dextrose agar until needed. Inoculum is grown on fresh or frozen
green snap-bean pods (the salt content of canned beans is too high for
growth) in test tubes for 3-8 days at 28 C. The snap-bean cultures are
then ground for 5 seconds in a Waring blendor, strained through
cheesecloth, and diluted with distilled water to a spore count of 50,000-
100,000 per ml. This preparation can be stored at 5 C for 72 hr
without loss of viability or infectivity (Littrell and Epps, 1965).

*Inoculation in Greenhouse and Field.*    Inoculum is applied as a
spray at 10 psi when plants are in the first true leaf stage. Plants in the
greenhouse are then placed in a moist chamber for 16-24 hr at 26 C.
Goode (1958) modifies this by placing his plants under the fluorescent
light of four 40-watt bulbs for 7-10 days of continuous light. Field
inoculation should be done in the late afternoon to take advantage of
low temperatures and the entire dew period. Field inoculations are not
successful at temperatures above 29 C or below 18 C. In both
greenhouse and field, symptoms occur on susceptible plants in 3-7 days,
on resistant plants in 5-10 days, and on highly resistant plants in 11-13
days. A scale for rating anthracnose susceptibility, devised by Goode
(1958), is: 1—highly resistant; 2—resistant with a few small lesions;

3—intermediate with a few well-formed stem or leaf lesions; 4 — susceptible with many large stem or leaf lesions; 5 — highly susceptible, plants dead.

## Powdery Mildew — *Erysiphe cichoracearum DC.*

*Symptoms.*    Tiny, white, superficial spots appear on leaves and stems of cantaloupe and cucumber, becoming powdery as they enlarge. Black pinpoint bodies rarely occur, but they are conspicuous when they do. Under ideal conditions premature defoliation may occur as fungus covers the leaf surface. The chief damage in cantaloupe is the reduced quality of fruits. Yield reduction occurs in proportion to time and severity of disease development.

*Nature of Pathogen.*    Conidia are continuous, elliptic, hyaline, and borne in chains on short unbranched conidiophores. Conidia are readily detached and borne by air currents. Penetration is confined to the epidermal cells, with most of the fungus remaining on the host surface. Cleistothecia are borne on the host surface as dark bodies with flexuous, indeterminate appendages. Cleistothecia contain about 15 asci, each having two hyaline ascospores.

Although powdery mildew appears to have many biotypes, the extent of specialization has not been determined. Some race investigation has been done in cantaloupe by Jagger et al. (1938). These investigators noted powdery mildew was attacking the previously resistant cantaloupe variety PMR 45 in some areas; in other areas susceptible lines were being attacked, but not PMR 45; and in the nursery some breeding lines were not attacked, but PMR 45 was attacked. On this basis they established the existence of races 1 and 2. Pryor and Whitaker (1942) established differentials for distinguishing the two races; race 1—virulent on Hales Best, Casaba, and Tip Top cantaloupe; race 2—virulent on the preceding varieties plus PMR 45 and PMR 8.

*Nature of Resistance.*    Using resistant cantaloupe material from India (PI 79374), Jagger and Scott (1937) found resistance to race 1 was controlled by a single dominant gene, referred to as $Pmr^1$. Bohn and Whitaker (1964), using race 2 resistant varieties, also developed from resistant Indian material, demonstrated race 2 resistance was controlled by a partly dominant gene, designated $Pm^2$. They stated that 2 modifier genes differentiated extreme resistance to race 2, and were epistatic to $Pm^2$ but hypostatic to $pm^2$. Harwood and Markarian (1966), using

Seminole as the resistant stock, demonstrated that resistance in their germ plasm was controlled by 2 gene pairs whose effects were unequal and partly additive. The major gene showed incomplete dominance, while the minor gene showed complete dominance. This resistance was different from that for $Pm^1$ (race 1), but was either identical, allelic, or closely linked to $Pm^2$.

Smith (1948) found Puerto Rico 37 cucumber was resistant to powdery mildew and, upon crossing it with Abundance, found the $F_1$ to be susceptible, but he found resistant individuals in the $F_2$. This suggested that resistance is polygenic. Barnes (*personal communication*) states that perhaps 3 genes are involved, but material with all 3 genes may have been discarded as a result of the close linkage with manganese deficiency proposed by Robinson (1960). Kooistra (1968) states that resistance of PI 1200818 is of the hypersensitive type, controlled by 2 genes, and is possibly recessive.

When cantaloupe breeding lines are inoculated with powdery mildew, according to Bohn and Whitaker (1964), the susceptible reaction is luxuriant fungus growth on all organs 6-8 days after inoculation. There is little apparent early tissue injury, but tissue is dead after 20 days. In the resistant reaction the fungus infects the plant but grows sparsely, with sporulation and growth very sensitive to environmental fluctuation. The $F_1$ reaction is intermediate, but it closely approaches the resistant parent.

Whitaker (1959) showed *Cucurbita lundelliana* to be resistant to powdery mildew. Rhodes (1964a,b), utilizing the interspecific cross technique, developed a common gene pool of divergent germ plasm in a series of crosses between *C. lundelliana, C. pepo, C. moschata, C. mixta,* and *C. maxima.* He demonstrated that resistance in squash is controlled by a single dominant gene which had been transferred from *C. lundelliana* to *C. moschata.* Sitterly (*personal communication*) obtained germ plasm from Rhodes and transferred resistance to bush type *C. pepo.* On susceptible squash plants powdery mildew grows on both upper and lower leaf surfaces and later spreads to petioles and stems. On resistant plants fungus growth is more restricted, and very small colonies grow only on upper leaf surfaces. An intermediate reaction consists of powdery mildew on mature leaves, but not on younger leaves.

*Sources and Range of Resistance in Varieties.*     Sources of cantaloupe resistance are PI 79374, PI 124111, and PI 134198. A range of resistance would be: susceptible—Hales Best and Honey Dew; moderately resistant—Georgia 47; races 1 resistant—PMR 45; race 2 resistant—Seminole; highly resistant to race 1 and 2—Planters Jumbo. Sources of cucumber resistance are PI 197087 and PI 1200815. A range of resistance would include: susceptible—Model; 1 resistant gene—Palmetto and Ashley; 2 resistant genes—Poinsett and Cherokee; 3 resistant genes—PI 197087. A source for squash powdery mildew resistance is *Cucurbita lundelliana.* No resistant varieties have been released.

*Collecting and Preparing Inoculum.*   If infected leaves are available, one has only to collect them and either wash off conidia and adjust spore concentration, or allow the leaves to remain on the plants and use indirect transfer. Powdery mildew cannot be grown in artificial culture, but susceptible hosts may be grown in the greenhouse and the pathogen propagated by continuous living host transfer. For storage, infected leaves may be dried at room temperature and kept for a year in paper bags.

*Inoculation in Greenhouse and Field.*   In the greenhouse plants are inoculated twice—at cotyledon expansion, and again at first true leaf expansion. Conidia are blown over plants either by human breath or by a small electric blower, or applied in a spray at 10 psi. It is not necessary to wet the plants, but moist chamber walls may be dampened to raise humidity. Severity can be increased by abundant inoculum, increased plant population, protection from free water, and the use of cheesecloth tents. Harwood and Markarian (1966) modified this procedure by growing plants at 24 C until first true leaves were 1/2 inch in diameter and then lowering the temperature to 15 C for the remainder of the test.

For field inoculation one has only to place an inoculum suspension in a knapsack sprayer and apply at 10 psi. The last hour of daylight is the best application time so as to utilize the night dew period. In both field and greenhouse, ratings may be made 16-20 days after inoculation with the following scale modified from Bohn and Whitaker (1964): 1—sparse fungus growth, no sporulating colonies, some early necrosis; 2—sparse fungus growth, few to numerous sporulating colonies, some early necrosis; 3—fungus growth luxuriant, few colonies, no early necrosis; 4—fungus growth luxuriant, numerous colonies, no early necrosis.

## Downy Mildew — *Pseudoperonospora cubensis (Berk. & Curt.) Ros.*

*Symptoms.*   Leaf lesions are normally angular in shape and yellow on the upper side, while the purplish mildew appears on the lower side.

Affected leaves may die as lesions coalesce, and plant injury may be severe enough to cause death or stunting.

*Nature of the Pathogen.*   Mycelium lives intercellularly in the host with haustoria being intracellular. Sporangiophores arise in groups through the stomata, are branched, and produce sporangia on the tips. Sporangia are grayish-purple, ovoid, and have a papilla at the distal end. Sporangia germinate by producing biciliate zoospores, which in turn produce germ tubes that penetrate through the stomata. Although downy mildew cannot overwinter in areas with temperatures below 0 C, it can thrive in both warm and cool temperatures. Thus humidity is the most important factor in establishment.

*Nature of Resistance.*   Cochran (1937) crossed the tolerant Indian cucumber cultivar Bangalore with commercial items and stated that resistance was apparently determined by several factors. J. M. Jenkins (1946) crossed the highly resistant cultivars Chinese Long and Puerto Rico 37 with commercial items and found the $F_1$ to be intermediate in resistance. Successive selection demonstrated this resistance was also multigenic. After stabilization of resistance this material was crossed by Barnes (1948) with Cubit and Marketer to produce Palmetto and Ashley. Barnes and Epps (1954) reported a second source of a high degree of resistance in PI 197087, which appeared multigenic, with inheritance controlled by 1 or 2 major genes and 1 or more minor genes. These investigators also reported a new symptom associated with the physiological response of PI 197087 to downy mildew. The mildew lesions were initially light brown and then turned dark brown without passing through a yellow stage. Resistance was an expression of hypersensitivity.

Cantaloupe downy mildew resistance genetics has not been fully investigated. Ivanoff (1944), using 4 tolerant West Indian varieties, found resistance to be partially dominant. By combing and selecting he was able to raise the resistance level. Ivanoff also observed an apparent close relationship of downy mildew resistance with aphid resistance.

*Sources and Range of Resistance in Varieties.*   Sources of downy mildew resistance in cucumber are Chinese Long and PI 197087. A range of resistance would include: susceptible—Marketer; intermediate resistance—Santee; resistant—Ashley; highly resistant—Poinsett. Sources of resistance in cantaloupe are Seminole and PI 124112. A range of resistance would include: susceptible—Hales Best; tolerant—Smiths Perfect; moderately resistant—Georgia 47; resistant—Edisto 47; immune—Seminole.

*Collecting and Preparing Inoculum.* The same procedure used to collect and prepare powdery mildew for spray inoculation may be used for downy mildew. More spores are obtained if harvested leaves are placed in a moist chamber at 21 C for 12-24 hours before washing. Decaying leaves should never be used, as they apparently produce a toxin that inhibits germination.

*Inoculation in Greenhouse and Field.* Inoculation is done as for powdery mildew, but only at the first true leaf expansion. If done in the greenhouse, plants must be placed in a moist chamber overnight. Susceptible cucumber leaves never collapse, but inoculum should be adjusted to 3-5 conidia per low power microscope field because a higher concentration causes even the leaves of resistant cantaloupe lines to collapse and die. In the greenhouse there is practically no sporulation on cantaloupe leaves, which is in direct contrast to cucumber. Rating of severity may be done on a simple 1-5 scale.

For interested individuals, Ivanoff (1944) developed a seedling technique for greenhouse testing in which he screened his plants for resistance to melon aphids. By utilizing the relationship between downy mildew resistance and aphid resistance he circumvented environmental influence on downy mildew field development.

## Scab—Cladosporium cucumerinum *Ell. & Arth.*

*Symptoms.* Scab appears on leaves as small, circular to angular, watersoaked, brownish spots difficult to distinguish from angular leaf spot. Under high humidity sporulation occurs on the leaf surface as an olivaceous mold. On young fruit watersoaked spots appear. A gummy brown exudate occurs on the surface, and as the lesion expands, the tissue sinks and the exudate dries to a brown bead. If fruit are infected after reaching full size, the host "corks off" the infected area and forms a tan, shallow scab. None of these fruit are marketable.

*Nature of Pathogen.* Mycelium is hyaline when young and turns olivaceous with age. Oblong, colored, continuous conidia are borne successively on short branched conidiophores. Intermediate cells between the conidiophore and conidia may also be detached and germinate. Infection is through stomata, and spread and development are favored by relatively low night temperatures and heavy dew and fogs. The optimum temperature for disease development (below 20 C) is lower than that for either the host or the pathogen (Walker, 1950). At temperatures above 21 C lesions are rapidly cicatrized by host reaction, and further disease development is stopped.

*Nature of Resistance.* Bailey and Burgess (1935) showed that some selections of cucumber cultivar Longfellow were resistant to scab. These were self-pollinated, and the segregating resistant and susceptible plants were crossed. Analysis of the $F_2$ showed resistance to be controlled by a single dominant gene. Walker (1950) confirmed this and showed that plants grown in a moist chamber at 17 C could be accurately rated for resistance. Local lesions developed on resistant lines without rapid stem invasion, while stems and leaves of susceptible items became water-soaked and died rapidly. Pierson and Walker (1954) demonstrated that at 17 C both resistant and susceptible hosts were invaded, but in the resistant hosts progress was stopped by a series of actions associated with cell wall thickening and cell necrosis. At temperatures above 21 C there is no damage to susceptible lines because of rapid cicatrization. Although scab is an important disease of squash, no genetic investigations have been completed on this crop.

*Sources and Range of Resistance in Varieties.* Sources of cucumber slicer and pickle resistance are Maine 2 and Highmoor. A range of resistance would include: susceptible—Chicago Pickling and Marketer; resistant—Highmoor and Wisconsin SMR 15.

*Collecting and Preparing Inoculum.* If covered with sterile mineral oil, the scab organism can be stored on potato-dextrose agar for a year in a refrigerator. To obtain inoculum, loops of a spore suspension are transferred from a nurse tube to test tubes, medicine bottles, and so forth, of potato-dextrose agar. Containers are held at room temperature for 48 hr, then incubated at 19-21 C for 14 days. After incubation, distilled water is placed in tubes, agitated vigorously, decanted, and the solution adjusted to the desired spore concentration. If spores are difficult to harvest, a small amount of a mild detergent is added to aid wetting.

*Inoculation in Greenhouse and Field.* Plants are inoculated in the second true leaf stage. In the greenhouse, plants are placed in a moist chamber, inoculated, and incubated at 17 C for 24 hr. In the field, inoculum can be applied via knapsack sprayer at 10 psi when the temperature is forecast for 17-20 C accompanied by a heavy dew. Ratings may be obtained in 5-7 days.

## Angular Leaf Spot — *Pseudomonas lachrymans* E. F. Sm. & Bryan

*Symptoms.* The disease first appears on cucumber plant organs as small watersoaked spots. On leaves these spots become tan on the upper surface and shiny on the lower surface, and they are angular in shape. Many of the spots loosen and fall out. On stems and fruits the water-soaked areas become covered with a crusty white exudate. As fruits mature, brown lesions form in the fleshy tissue beneath the rind, and the discoloration continues along the vascular system to the seeds.

*Nature of the Pathogen.*   The angular leaf spot pathogen is a rod-shaped bacterium that has 1-5 polar flagella and forms capsules. The organism overwinters on infected plant residue and is spread by surface water and rain. Penetration of the host is through stomata, and initial development is intercellular. In infected fruit the bacteria are found in the xylem elements of the mesocarp and in the tissue surrounding the developing seeds. Optimum disease development occurs at 24-28 C.

*Nature of Resistance.*   Chand and Walker (1964b), using resistant PI 169400 from Turkey, found that when crosses were made with Wisconsin SMR 15 and Wisconsin SMR 18, the segregation in the $F_2$ and backcross generations did not fit any ratio. This indicated that resistance is possibly multigenic. Resistance gradually increased with continuous selection of resistant individuals. Barnes (1961) demonstrated that PI 197087 from southern India had even more resistance than PI 169400, and that this resistance was also multigenic. Chand and Walker (1964a) found that mature leaves of susceptible plants were more resistant than young leaves. Young and old leaves reached maximum bacterial concentration 3-4 days after inoculation, but young leaves had 20-40 times as many bacteria. When inoculated leaves were compared, the multiplication was greatest in susceptible plants, least in resistant plants, and intermediate in a hybrid between resistant and susceptible plants.

*Sources and Range of Resistance in Varieties.*   PI 197087 is a source of cucumber angular leaf spot resistance. A range of resistance would include: highly resistant—PI 197087; resistant—Southern Cross; tolerant—Pixie; susceptible—Model and Wisconsin SMR 18.

*Collecting and Preparing Inoculum.*   The angular leaf spot pathogen may be maintained by continuous transfer on nutrient-dextrose agar. One may also collect infected leaves, dry them at room temperature, and store them in a cardboard box at 15-21 C (may be kept for a year). To prepare inoculum from tube cultures, a bacterial water suspension is transferred via a loop to slants of nutrient-dextrose agar in 200 ml bottles. These are kept at room temperature for 48 hr and the bacteria are then removed by flooding and scraping the surface. This suspension can be standardized at an optical density of 0.10 by measurement with a Bausch and Lomb Spectronic 20 colorimeter. In preparing inoculum from dried leaves, the leaves are crushed and placed in water to soak for an hour. The plant residue is strained and diluted to desired concentration.

*Inoculation in Greenhouse and Field.* In the greenhouse plants are inoculated at fifth leaf stage by spraying inoculum against the underside of the leaf with an atomizer at 10-20 psi until watersoaking of substomatal chambers is visible. Plants are placed in a moist chamber for 48 hr at 21 C and 98-100% relative humidity. For field inoculation, 15 cc of carborundum per gallon of diluted inoculum are placed in the tank of a sprayer and applied as with a virus inoculum at 30-60 psi in the evening. In both greenhouse and field, ratings are obtainable after 4-5 days using the following scale by Chand and Walker (1964*a*): 0—no infection; 1—very few small lesions; 2—few small lesions; 3—many small lesions; 4—many moderate-sized lesions and medium exudate; 5—very severe infection, large lesions, abundant exudate.

## Fusarium wilt. Watermelon: *Fusarium oxysporum f. niveum (E. F. SM.) Snyder and Hansen. Cantaloupe: Fusarium oxysporum f. melonis (Leach and Currence) Snyder and Hansen.*

*Symptoms.* In young seedlings the cotyledons and small leaves lose their green color, droop, and wither. The hypocotyl is girdled by watery, rotten tissue, and plants may become stunted. In older plants, leaves wilt during the day for several successive days, then wilt permanently. Cottony mycelium may be produced on the surface of dying vines. The vascular bundles become yellow or brown.

*Nature of the Pathogen.* *Fusarium* produces a septate mycelium in which chlamydospores are formed and which produces conidia. Infection occurs through the root-tip region, wounds, and through the peg of germinating seed. The fungus invades the root cortex and becomes established and moves within the xylem elements. Other tissues are not invaded until the plant dies. Seedling injury is severe when soil temperatures are 20-30 C. In older cantaloupe plants the optimum soil temperature is 27 C, and for watermelon plants it is 33 C. *Fusarium* does not develop above 33 C. There is a positive correlation between degree of soil infestation and rate of wilting. When soil becomes infested, a 10-year rotation period is required between melon crops.

Although there are many strains of the fungus, little investigation into physiologic specialization has occurred. Crall (1963) stated that 2 races exist in Florida. Race 1 was most common and produced 100% wilt in susceptible Florida Giant variety, and scattered wilt in Charleston Gray and Summit. Race 2 produced 100% wilt in susceptible varieties, and no wilt in Charleston Gray or Summit—not even in seedlings.

*Nature of Resistance.*    Little is known of the inheritance of resistance because of the numerous strains and because there is no precise method of measuring resistance. Orton (1911) crossed Stock Citron with the variety Eden and demonstrated that resistance could be transferred from one item to another by hybridization and selection. Porter and Melhus (1932) demonstrated that wilt-resistant individuals do occur in commercial varieties, thus avoiding the use of the inedible citron in a breeding program. They also showed that resistance is multigenic but not completely fixed. Braun (1942) attempted to uncover the physiological basis of resistance to wilt by demonstrating that acetic acid retards fungus growth, and that a similar chemical was found in the stems of watermelon. Perhaps this helps explain the rapid growth of the fungus in the roots and slower growth in the stem.

Mortensen (1959) devised an apparently successful technique for distinguishing resistant and susceptible cantaloupe genotypes. He showed the best separation could be made at a constant soil temperature of 30 C. Resistance was found to be dominant to susceptibility, with resistance being controlled by one principal dominant gene ($R$) plus 2 complementary dominant genes ($A$) ($B$). The hypothetical genotypes would be: resistant, *RAB, RAbb, Raab, Raabb,* and *rrBA;* susceptible, *rrAbb, rraaB,* and *rraabb.*

*Source and Range of Resistance in Varieties.*    Sources of watermelon wilt resistance are Stock Citron and Iowa Belle. A varietal range would include: resistant—Charleston Gray and Garrisonian; susceptible—Klondike and Florida Giant. A source of cantaloupe wilt resistance is Golden Gopher. A range of resistance would include: susceptible—Hales Best and Gulfstream; resistant—Golden Gopher; highly resistant—Spartan Rock.

*Collecting and Preparing Inoculum.*    The wilt organism may be isolated and stored via standard pathological procedures. The inoculum preparation technique is described by Wellman (1939), utilizing liquid culture procedures.

*Inoculation in Greenhouse and Field.*    Field inoculation is almost impossible on an annual basis because of environmental fluctuations. For field screening, melons are planted in heavily infested soil and the same field is continued in use for every crop. For greenhouse screening, the mycelial mats are macerated, a 1 inch-deep trench is cut beside the host, the inoculum suspension is poured in, and the trench is re-covered. Heat cables may be used to keep soil temperature at an optimum for fungus development.


## Virus, Cucumber Mosaic

*Symptoms.*    The yellow and green mosaic pattern, commonly seen on young leaves near the growing tip, often changes to an indistinguishable mottle as the leaves mature. A dwarfing of vine internodes becomes more conspicuous as leaf mottle symptoms recede. Infected young fruit reactions range from a mild mottle to extensive warty malformation accompanied by a lack of color ("white pickle"). After midseason, vines may show death of all leaves except for a few near the growing point, and are of no productive value.

*Nature of the Pathogen.*    Cucumber mosaic virus (CMV) has many variants which differ in host range and in the type of symptoms on specific hosts, but which are similar in physical properties. CMV is not seed-transmitted, but is transmitted by the green peach and melon aphids, and by mechanical agents (farm machinery). Perennial host plants (*Commelina nudiflora, Melothria* species, and so forth) are common between-crop virus reservoirs or overwintering hosts.

*Nature of Resistance.*    Porter (1928) found the cucumber varieties Chinese Long and Tokyo Long Green to be highly resistant to CMV. Upon crossing these varieties with susceptible material, Shifriss et al. (1942) demonstrated that, although resistance was dominant, it involved at least 3 major genes and possibly some modifiers. They maintained that resistance was associated with the presence (susceptible) or absence (resistant) of chlorosis in the cotyledonary stage, and that the degree of tolerance could be determined by the distance from the cotyledons to the leaf in which symptoms first appear. Ratios were constantly changing as the plants developed from the first true leaf stage. At this stage, several gene modifiers cause a low frequency of symptomless plants in a mature plant population. Thus it is possible to select plants with varying degrees of tolerance determined by the number of resistant genes involved. Wasuwat and Walker (1961), crossing resistant Wisconsin SMR 15 x National Pickling, demonstrated that in their specific source of resistance, the resistance was controlled by a single dominant gene. The same investigators demonstrated that in the greenhouse the distinction between resistant and susceptible plants was best determined 20 days after inoculation. All resistant plants became infected and showed symptoms, but symptoms were mild and tended to disappear. In susceptible plants the leaves remained smaller, foliage was intensively mottled, internodes became shorter, and fruit became mottled. In the field, plant stunting and mottling were the criteria of susceptibility. Resistance was not exclusive or necrotic.

*Sources and Range of Resistance in Varieties.*    Tokyo Long Green and Chinese Long are sources of resistance to CMV. A range of resistance would include: susceptible—Model; resistant—Wisconsin SMR 18; highly resistant—Tablegreen.

*Collecting and Preparing Inoculum.*    Virus must be collected from living plants. Pure CMV may be obtained by inoculating pure CMV into nurse plants of summer squash (*Cucurbita pepo*) which are kept in insect-proof cages. Young infected leaves are macerated and strained through double-layer cheesecloth. The virus suspension is then either used immediately, or it can be frozen until needed. If bulk inoculum quantities of questionable purity are desired, infected leaves in the field, previously inoculated with pure CMV, may be collected and processed as described above.

*Inoculation in Greenhouse and Field.*    Inoculum is diluted 1:10 with water and buffered with .05 M phosphate buffer having a 7.5 *p*H to negate the presence of any virus inhibitor. In the greenhouse, cotyledons are dusted with 600-mesh carborundum, and the upper surfaces are wiped with cheesecloth pads soaked with inoculum. Inoculation is done at the first true leaf stage of development. The inoculated plants should be held at a temperature of 26-28 C and under a high light intensity, if possible (Sinclair and Walker, 1956). In the field, inoculation can be made from the first to the third true leaf stage via the greenhouse procedure, or one can use compressed air. If compressed air is utilized, 3/F-mesh carborundum is added to the tank (15 cc/gal) and the mixture is sprayed onto the plants at 60 psi. A virus rating system would be: 1—slight mottle; 2—moderate to extensive mottle, little or no distortion; 3—severe or pronounced mottle plus distortion; 4—death.

## Multiple-Disease Resistant Cucumbers

As diseases are conquered by breeding it is practical to combine separate resistance characteristics into one multiple-disease resistant variety. Walker and Pierson (1955) produced one of the earliest successful efforts by combining scab and cucumber mosaic resistance in the Wisconsin SMR pickle series. Barnes (1961, 1966) has not only produced several pickle and slicer varieties with multiple disease resistance, but he has produced varieties which are also hybrids. The same procedure has been successful for watermelon and cantaloupe.

When widely divergent multigenic characteristics are involved, no one item will have all the desired horticultural and resistance genes. Barnes used the "recombination cross," in which crosses were made in the $F_3$ to $F_5$ between plants having desirable complementary characters, thus circumventing the loss of resistance encountered by backcrossing. The entire gene complex is available. Plants were inoculated with 2 or 3 different pathogens in the same inoculum. Individual breeding lines were inoculated separately to determine the influence of any interference between 2 pathogens used simultaneously or successively. Virus was inoculated last because it may tend to protect the plant from other pathogens.

Examples of multiple-disease resistant cucumber varieties are: pickle—Chipper (downy mildew, powdery mildew, angular leaf spot, anthracnose, and cucumber mosaic); slicer—Poinsett (downy mildew, powdery mildew, angular leaf spot, and anthracnose).

# Goals to be Achieved

Cucurbit plant breeders are continually developing better varieties that are better adapted to a wider range of growing conditions, have multiple disease resistance, and meet the fluctuating demands of economics and consumers. With all cucurbits, work must be directed to the development of smaller, compact vines for mechanical harvest.

Specific goals in cucumber plant breeding are: better resistance to powdery mildew in the greenhouse and in the tropics; resistance to gummy stem blight, fruit rot, *Fusarium* wilt, and bacterial wilt; and investigation on the relationship between watermelon mosaic virus and cucumber.

Specific goals for watermelon plant breeding are: resistance to downy mildew and gummy stem blight; development of triploid and diploid seedless hybrids; clarification of the physiological basis of resistance to *Fusarium* wilt; and establishment of a precise method for measuring resistance to *Fusarium* wilt in both greenhouse and field.

Specific goals for cantaloupe plant breeding are: a higher level of resistance to gummy stem blight, *Alternaria*, and viruses; investigation of the relationship between cantaloupe and powdery mildew; and incorporation of the gynoecious character into cantaloupe for hybrid seed production.

Specific goals for squash and pumpkin plant breeding are: resistance to watermelon, squash, and cucumber mosaic viruses; resistance to downy mildew and powdery mildew; and the incorporation of the gynoecious character into squash for hybrid seed production.

# References

Bailey, R. M., and I. M. Burgess. 1935. Breeding cucumbers resistant to scab. Proc. Amer. Soc. Hort. Sci. 36:645-646.

Barnes, W. C. 1948. The performance of Palmetto, a new downy mildew resistant cucumber variety. Proc. Amer. Soc. Hort. Sci. 51:437-441.

―――. 1961. Multiple disease resistant cucumbers. Proc. Amer. Soc. Hort. Sci. 77:417-423.

―――. 1966. Development of multiple disease resistant hybrid cucumbers. Proc. Amer. Soc. Hort. Sci. 89:390-393.

―――, and W. M. Epps. 1952. Two types of anthracnose resistance in cucumber. Plant Dis. Reporter 36:479-480.

――― ―――. 1954. An unreported type of resistance to cucumber downy mildew. Plant Dis. Reporter 38:620.

―――, C. N. Claxton, and J. M. Jenkins, Jr. 1946. The development of downy mildew resistant cucumbers. Proc. Amer. Soc. Hort. Sci. 47:357-360.

Bohn, G. W., and T. W. Whitaker. 1964. Genetics of resistance to powdery mildew race 2 in muskmelon. Phytopathology 54:587-592.

Braun, A. E. 1942. Resistance of watermelon to the wilt disease. Amer. J. Bot. 27:683-684.

Busch, L. V., and J. C. Walker. 1958. Studies of cucumber anthracnose. Phytopathology 48:302-304.

Chand, J. M., and J. C. Walker. 1964a. Relation of age of leaf and varietal resistance to bacterial multiplication in cucumber inoculated with *Pseudomonas lachrymans*. Phytopathology 54:49-51.

――― ―――. 1964b. Inheritance of resistance to angular leafspot of cucumber. Phytopathology 54:51-54.

Cochran, F. D. 1937. Breeding cucumbers for resistance to powdery mildew. Proc. Amer. Soc. Hort. Sci. 35:541-543.

Crall, J. M. 1963. Physiologic specialization in *Fusarium oxysporum* v. *niveum*. Phytopathology 53:87.

DeCandolle, A. 1882. Origine des planter cultives. Germes Bailliere, Paris. 337 p.

Dutta, S. K., C. V. Hall, and E. C. Hayne. 1960. Observations on physiological races of *Colletotrichum lagenarium*. Bot. Gaz. 121:163-166.

Goode, M. G. 1958. Physiological specialization in *Colletotrichum lagenarium*. Phytopathology 48:79-83.

Harwood, R. R., and D. Markarian. 1966. Genetics of resistance to powdery mildew in the Michigan cantaloupe breeding program. Proc. XVII Int. Hort. Congr. 1:454.

Hibbard, A. D. 1940. Fruit thinning the watermelon. Proc. Amer. Soc. Hort. Sci. 37:825-826.

Hutton, E. M. 1941. A new method of tomato and cucumber seed extraction. J. Coun. Sci. and Ind. Res. Aust. 16:97-103.

Ivanoff, S. C. 1944. Resistance of cantaloupes to downy mildew and the melon aphid. J. Hered. 35:35-39.

Jagger, I. C., and G. W. Scott. 1937. Development of Powdery Mildew Resistant Cantaloupe No. 45. Circ. U.S. Dept. Agr. 441. 5 p.

————, T. W. Whitaker, and D. R. Porter. 1938. A new biological form of powdery mildew on muskmelons in the Imperial Valley of California. Plant Dis. Reporter 22:275-276.

Jenkins, J. M., Jr. 1946. Studies on the inheritance of downy mildew resistance and of other characters in cucumber. J. Hered. 37:267-271.

Jenkins, S. F., N. N. Winstead, and C. L. McCombs. 1964. Pathogenic comparison of three new and four previously described races of *Glomerella angulata* var. *orbiculare*. Plant Dis. Reporter 48:619-623.

Kooistra, E. 1968. Powdery mildew resistance in cucumber. Euphytica 17:236-244.

Littrell, R. H., and W. M. Epps. 1965. Standardization of a procedure for artificial inoculation of cucumbers with *Colletotrichum lagenarium*. Plant Dis. Reporter 49:649-653.

Mann, L. K., and J. Robinson. 1950. Fertilization, seed development, and fruit growth as related to fruit set in the cantaloupe (*Cucumis melo* L.). Amer. J. Bot. 37:685-687.

Mortensen, J. A. 1959. The inheritance of Fusarium resistance in muskmelon. Diss. Abstr. 19:2209.

Orton, W. A. 1911. The development of disease resistant varieties of plants. IV Conf. Int. Genet. C. R. et Rapp., Paris. 247-265.

Pierson, C. F., and J. C. Walker. 1954. Relation of *Cladosporium cucumerinum* to susceptible and resistant cucumber tissue. Phytopathology. 44:459-465.

Porter, D. R., and I. E. Melhus. 1932. The pathogenicity of *Fusarium niveum* (E. F. Sm.) and the development of wilt-resistant strains of *Citrullus vulgaris* (Schrad.). Res. Bull. Iowa Agr. Exp. Sta. No. 149. 183 p.

Porter, R. H. 1928. Further evidence of resistance to cucumber mosaic in the Chinese cucumber. Phytopathology 18:143.

Pryor, D. E., and T. W. Whitaker. 1942. The reaction of cantaloupe strains to powdery mildew. Phytopathology 32:995-1004.

Rhodes, A. M. 1964a. Species hybridization and interspecific gene transfer in the genus *Cucurbita*. Proc. Amer. Soc. Hort. Sci. 74:546-552.

————— 1964b. Inheritance of powdery mildew resistance in the genus *Cucurbita*. Plant Dis. Reporter 48:54-56.

Robinson, R. 1960. Genetic studies of disease resistance and a mineral element deficiency in cucumbers. Proc. Amer. Soc. Hort. Sci. 57:21.

Seaton, H. L., and J. C. Kremer, 1939. The influence of climatological factors on anthesis and anther dehiscence in the cultivated cucurbits. A preliminary report. Proc. Amer. Soc. Hort. Sci. 36:627-631.

Shifriss, O. C., C. H. Myers, and C. Chupp. 1942. Resistance to the mosaic virus in the cucumber. Phytopathology 32:773-784.

Sinclair, J. B., and J. C. Walker. 1956. Assays for resistance to the cucumber mosaic in the pickling cucumber. Phytopathology 46:519-522.

Smith, P. G. 1948. Powdery mildew resistance in cucumber. Phytopathology 39-1027-1028.

Tiedjens, V. A. 1928. Sex ratios in cucumber flowers as affected by different conditions of soil and light. J. Agr. Res. 36:731-736.

Walker, J. C. 1950. Environment and host resistance in relation to cucumber scab. Phytopathology 40:1094-1102.

—————, and C. F. Pierson. 1955. Two new cucumber varieties resistant to scab and mosaic virus. Phytopathology 45:451-453.

Wasuwat, S. C., and J. C. Walker. 1961. Inheritance of resistance in cucumber to cucumber mosaic virus. Phytopathology 51:423-424.

Wellman, F. L. 1939. A technique for studying host resistance and pathogenicity in tomato *Fusarium* wilt. Phytopathology 29:945-956.

Whitaker, T. W. 1959. An interspecific cross in *Cucurbita* (*C. lundelliana* Bailey x *C. moschata* Duch.). Madrono 15:4-13.

—————, and G. N. Davis. 1962. Cucurbits. Interscience Publishers, Inc., New York. 250 p.

Winstead, N. N., M. G. Goode, and W. C. Barham. 1959. Resistance in watermelon to *Colletotrichum lagenarium*, Races 1, 2, and 3. Plant Dis. Reporter 43:570-577.

# 19 Crucifers

J. C. Walker[1]
P. H. Williams[1]

## Introduction

The family Cruciferae, so called because the four petals of the flower form a cross, has many genera and species. We shall be concerned here with a few economically important species of vegetables in which the development of disease-resistant cultivars is well advanced. First of these is *Brassica oleracea* L., which occurs as a winter annual at several locations along and close to the North Sea, the English Channel, and the upper Mediterranean Sea. Cultivars derived from this species include those of cabbage, var. *capitata* L.; cauliflower, var. *botrytis* L.; green sprouting broccoli, var. *italica* Plenck.; Brussels sprouts, var. *gemmifera* DC.; kohlrabi, var. *caulo-rapa* DC.; and collard and kale, var. *acephala* DC. Also in the genus *Brassica* are rutabaga, *B. napobrassica* Mill.; turnip, *B. rapa* L.; and Chinese cabbage, *B. pekinensis* Rupr. Another crucifer referred to here is radish (*Raphanus sativus* L.) of which there are many cultivars.

1. Professor Emeritus and Professor, respectively, Department of Plant Pathology, University of Wisconsin.

## Origin of Cultivars

The origin of *Brassica oleracea* is obscure. The wild form now occurs on lime cliffs in southeastern England, various locations along the North Sea, and on the northern shore of the Mediterranean Sea. DeCandolle (1902) was of the opinion that it is of European origin. Boswell (1949) believed it originated on the north shore of the Mediterranean and was transferred by man to other locations. The wild plant resembles most closely certain cultivars of kale and collard. It is a leafy plant that produces no head, as does cabbage, and no swelling of the lower stem, as does kohlrabi. The stem may reach a meter or more in height. All cultivars of kale and collard are sufficiently frost resistant to grow well in areas with mild winters such as those where the wild plant is now known in Europe.

In 1860, according to Boswell (1949), some selections were made at the Cirencester Agricultural College in southern England from the wild cabbage on the Dover coast. Crude kales, forms of broccoli, and other related cabbage-like forms were developed, demonstrating their common ancestry.

Probably kale and collard were the first to be adapted to human and animal consumption from the wild plant. There are many cultivars now in this subspecies. In the British Isles, a type known as marrow-stem kale is used as stock feed. In the Puget Sound area, a type known as Thousand-headed Kale is grown as winter forage for poultry. These types and many other cultivars have relatively smooth leaves. Other cultivars of kale grown for human food have leaves curled or crinkled to various degrees. They, along with collard, are commonly grown as winter crops for human food in the southeastern United States. While stock kales and collard have stems up to a meter in height in their vegetative stage, other kales grown for human food are sessile in habit of vegetative growth. The latter type was described as early as 350 B.C. by Theophrastus, supporting the theory that kale was the first group of cultivars to be selected and adapted by man.

The soft or loosely headed type of cabbage was probably brought into western Europe by the Celts. Pliny in the first century A.D. gave particular attention to Savoy cabbage, which probably first came into prominence in southeastern France. Hard-headed varieties of cabbage are not mentioned until the time of Charlemagne, who died in 814 A.D. Red cabbage was described in England in 1570.

Sprouting broccoli came into use early in the Christian era or before in southern Europe, and it was probably imported from southeastern

Europe or Asia Minor. The Romans regarded it highly, but it was not mentioned in England until 1720. It is less resistant to cold weather than some other forms of *B. oleracea*. Broccoli may be considered an annual, in contrast to most forms, which are biennial. The stem elongates promptly, and fleshy branches bearing inflorescences appear, which are harvested before blossoms open. If the branches are not harvested promptly, the flowers open and pollination by bees takes place promptly.

According to Magruder (1937), the only consistent difference between winter (not sprouting) broccoli and cauliflower is that broccoli will produce marketable curds during cool weather, whereas cauliflower requires warmer weather. Both are sold as cauliflower on the market. Magruder (1937) believes these are relatively recent in development, but Boswell (1949) states that cauliflower has a cultural history that goes back to the sixth century B.C. Three varieties were described in Spain as having been introduced from Syria in the twelfth century A.D. Cauliflower was described in England in 1586, and it became common on the market by 1600.

Kohlrabi came to notice later than the several subspecies just mentioned. It apparently was derived from a marrow-stem, nonheading type of kale. It was first described in Europe in 1485 and was first grown in Ireland on a field scale in 1724, and in England in 1837. Because the two most popular varieties still used today are White Vienna and Purple Vienna, Magruder (1937) suggests it may have come from southeastern Europe by way of Austria or Czechoslovakia.

Brussels sprouts was, like kohlrabi, relatively late in its appearance as a commercial vegetable. It was described in 1587 when it was prominent in Belgium. It was apparently rapidly adopted in Great Britain, where it has long been the chief subspecies of *Brassica oleracea* grown.

Among the cruciferous root crops, we are concerned here with turnip, rutabaga, and radish. DeCandolle (1902) concludes that the first two were natives of temperate regions of Europe. Both were brought into cultivation very early and were disseminated widely in Europe and Asia. It has been suggested that rutabaga ($n = 19$) was derived from a cross of turnip ($n = 10$) with some form of cabbage ($n = 9$). Both have become prominent in western Europe, eastern Canada, and New Zealand for stock feed. Practically all cultivars are biennial. One Japanese cultivar of turnip, Shogoin, is an annual.

Radish may have been derived from a wild species such as *Raphanus raphanistrum*, with which it crosses readily. It is believed to be of Chinese origin. It appeared in Egypt before the Pyramids and was described by Greeks about 300 B.C. Apparently one of the first Old World species transferred to America, it was seen in Mexico about 1500 and in Haiti in 1560.

Chinese cabbage is thought to be a native of China. The first written record in that country was in the period of the Chin Dynasty, 290-307 A.D.


## Methods of Breeding

Most cultivars of *Brassica oleracea* are biennials or winter annuals. Plants that have matured to the normal vegetative stage require a cool period of 2-3 months with an average temperature of about 5 C, during which flower primordia are formed. This is usually accomplished by planting in early summer so as to bring them to vegetative maturity or nearly so when winter climate sets in. Obviously, this must be done in regions where the cultivar concerned will withstand the winter conditions. Most American cabbage, turnip, and rutabaga seed is now grown in the Skagit Valley of northwestern Washington. Cauliflower requires a milder winter climate, and seed is grown in the cool areas of west central California. Breeders now shorten the seed-to-seed period of cabbage by removing the plants from the field to the greenhouse, where they are potted in suitable soil, and where the temperature is maintained at 5 C ±2 C for about 10 weeks. As the need for selecting for internal as well as external character became greater, the procedure was modified by cutting the head in the field at maturity and moving the stem and root to the greenhouse. When plants show evidence of having broken dormancy, the temperature is gradually raised to about 20 C, as flower raceme stems begin to form from buds above old leaf scars on the stem. In this manner seed can be matured in May in the temperate zone, and seed can be produced in one year, rather than in the customary two-year period. By decapitation with retention of a portion of the curd, flowering was obtained in approximately 65 days.

Unlike the *Brassica oleracea* cultivars, those of radish grown in the United States do not require a cool period to induce flowering. Long days and high temperatures seem to promote flowering in radish. For this reason, radish crops are generally grown during the cooler months of the year in Florida and southern California, and in the spring and fall in the northern United States. Seed is produced under irrigation in the warm, dry, intermountain valleys of Wyoming, Colorado, Idaho, and central and eastern Washington, where it is sown in the spring and harvested in the fall. Although western Washington, Oregon, and California have an ideal climate for the production of radish seed, the widespread prevalence of the weed *Raphanus raphanistrum,* which crosses freely with *R. sativus,* results in contaminated seed lots containing undesirable "off types."

As with a number of species which have a rosetted leaf arrangement, radish can be induced to flower more rapidly with applications of gibberellic acid (A-4, A-7). Although the time from seeding to flower is normally five weeks in radish, this can be shortened by a week or more with applications of 1,000 mg/liter gibberellic acid at five-day intervals after the first true leaf has emerged. Under this treatment the plants will not form the normal enlarged radish root, but will produce an elongated flowering stalk without entering the rosette stage of development. With the aid of gibberellic acid, four generations of radish may be obtained in a year. Although it is impossible to make selections for root type when gibberellic acid has been applied to radish seedlings, the method is useful in advancing breeding material during the early stages of a backcrossing program.

When horticultural type and quality are evaluated, the crop is grown to the peak of root maturity. Plants are pulled from the ground, the roots are washed, and selections made. All but 1-2 inches of the tops are cut off, and the root can be then potted or replanted in the field. The replanted radishes rapidly produce new leaves and roots and begin to flower in a few weeks. As new leaf growth is made, the application of gibberellic acid hastens flowering.

Since all cultivated subspecies of *Brassica oleracea* cross with each other readily, complete isolation of seed fields becomes necessary. Turnip and rutabaga rarely cross with each other, and when this does take place the $F_1$ plants can be detected by their greater height and malformed roots. The $F_1$ plants are practically always sterile and therefore may be disregarded in seed production.

# Floral Development

The flowers of *Brassica oleracea* are borne in stout racemes, whereas in cauliflower and broccoli these racemes are often modified and shortened. A cauliflower or broccoli plant will open its entire crop of 5,000 to 8,000 flowers within a period of 10-14 days, while the blooming season of a cabbage plant will extend over two months, or even longer if seed is not formed.

Bees are the chief insects in cross-pollination. The floral structures of cruciferous root crops differ little from those of *B. oleracea*. Pearson (1932) has documented the development of the flower.

# Incompatibility

Individual plants of *Brassica oleracea* and *Raphanus sativus* produce little or no seed when grown in isolation because they are self-incompatible (Pearson, 1929). Pearson showed that if flower buds were opened two days before anthesis and pollen from an older blossom of the same plant was placed on the pistil, a normal or near-normal set of seed would follow. Pollination in nature is carried out chiefly by bees. Incompatibility assures crossing, and as a result cultivars are heterozygous for many characters. If a breeding line is selfed artificially for a number of generations, it becomes homozygous for many characters and thus more uniform. Selecting for incompatibility makes it possible to produce sib-incompatible (SI) lines. They are essential for production of hybrid varieties. In seed production, rows of an SI line are interspersed with rows of a previously tested normal line or with another SI. The seed produced on SI plants is, therefore, "hybrid"; that is, $F_1$ seed from crossing of the two parental lines.

The incompatibility system in crucifers is sporophytic rather than gametophytic, as originally supposed (Bateman, 1955). A plant heterozygous for compatibility produces pollen of two genotypes. If such genotypes are the same in the stigma of such a plant, incompatibility results. If they differ, the cross is successful. An inbred SI line is, therefore, homozygous for the *S* allele concerned, but the chances are good that the *S* alleles in another SI line or in a non-SI line used as the pollen parent are different.

The genetics of SI as compared with non-SI is still not fully understood. The SI character in a given line that is effective in one environment may not function well in another. Thus, lines found to be SI in one locality, and thus suitable for production of hybrid seed, may be more or less sib-compatible in another environment. The proportion of $F_1$ seed is reduced accordingly. Since sib-incompatibility is usually not expressed completely under ideal conditions, a small percentage of "selfs" resembling the SI parent may sometimes be found in any hybrid seed lot. Current reports on compatibility may be found in papers by Wallace and Nasrallah (1968).

Compatibility and incompatibility in other types of *Brassica oleracea* and in *Raphanus* appear to follow the same general pattern as in cabbage. In Brussels sprouts, Johnson (1966) reported in connection with the hybrid variety Avencross that partially sib-compatible inbred lines were used. The percentage of sibs in the field was acceptably low because such plants could be detected and rejected at time of transplanting because of their smaller size. In kale (Thompson and Howard, 1959; Johnston, 1965), where incompatibility is also sporophytic and controlled by a multiple series of *S* alleles, the production of single-cross seed is not practical, and the use of double-cross seed is suggested. In green sprouting broccoli Martin (1962) found a range of complete self-incompatibility to extreme self-compatibility, indicating that within the former group sib-incompatible, cross-fertile lines could be developed. In radish, Sampson (1964) found that incompatibility was controlled by *S* alleles at one locus, and that self-compatibility was absent. Watts (1963) in England reported that summer cauliflowers are largely self-compatible, while both autumn and winter varieties are intermediate between self-compatibility and self-incompatibility. According to Poole (1937), self-incompatibility is common in turnip and rutabaga. Yarnell (1956) has given a thorough review of the researches on genetics of crucifers up to 1956.

The system involving genes that maintain sterility in certain types of cytoplasm, successfully used in the development of many hybrid crops (for example, carrots and onions), has not been found in *B. oleracea* (Nieuwhof, 1969).

# Development of Resistant Cultivars

## Cabbage Yellows

The first attempt to control a cabbage disease by selection of resistant lines was the work of Jones and Gilman (1915) at the University of Wisconsin with the disease known as yellows [ *Fusarium oxysporum* f. Sp. *conglutinans* race 1 (Wr.) Snyder & Hansen].

During the first decade of the twentieth century, this disease appeared in severe form in the cabbage-growing regions of Ohio, Indiana, Illinois, and Wisconsin. The causal organism is a natural inhabitant of the soil, where it persists indefinitely, at least for many years. The persistence is apparently made possible by the ability of the fungus to invade fibrous roots of nonhosts, where it forms chlamydospores which serve to carry it over periods of unfavorable environment. These spores germinate and produce more chlamydospores when conditions are more favorable, and it is possible that exudates from the roots of host plants stimulate germination.

Jones and Gilman worked in southeastern Wisconsin. In severely affected fields they found occasional survivors that appeared to withstand the disease. Individuals were selected within this group for type characteristics. They were planted in a group in an area free from other seed plants of *B. oleracea.* It is well to emphasize again that the various subspecies of *B. oleracea* (*acephala, botrytis,* and so forth) hybridize readily with each other, and isolation from any member of the species is necessary. At maturity seeds from each plant were harvested separately. In the following year the progenies from the mass cross were planted with suitable controls on land with a severe yellows history. The increase in percentage of resistant plants was rapid. After three generations of selection most lines contained up to 90% resistant individuals on land where susceptible varieties were nearly a complete loss.

One important observation by these workers was that resistant lines were never free from a varying percentage of susceptible plants. In the meantime, it had been shown that yellows was a warm weather disease. Under controlled soil temperatures, the disease was nearly prevented in plants grown on infested soil kept at a constant temperature of 16 C, while it increased in severity up to an optimum at about 28 C. This was in accord with observations made in southeastern Wisconsin, where the disease was most severe in midsummer. It was shown that it was less severe in cool summers than in unusually warm ones. Also, it was observed that the disease was least destructive in the cooler parts of the Great Lakes regions (northeastern Wisconsin and northwestern New York).

When young seedlings of resistant lines were exposed to the controlled optimum soil temperature in the greenhouse, there was little difference in rate and amount of disease development from that with susceptible lines. This situation discouraged the testing for disease resistance elsewhere than in the normal field trials. For several years the same methods of field selection and testing described above were followed. The first yellows-resistant variety, released in 1916 under the name Wisconsin Hollander, is still in use in southeastern Wisconsin. In 1916, which was a warmer than average summer, the resistant variety averaged 24.3% diseased plants in 20 different fields in southeastern Wisconsin, while commercial (susceptible) Hollander averaged 89.0%.

Two slightly earlier varieties of drumhead type, grown especially for sauerkraut production, were released under the names Wisconsin All Seasons and Wisconsin Brunswick in 1920. Since that time a large number of resistant varieties have come into use.

In 1922 three resistant plants of All Head Early, the parent of All Head Select, were grown to seed in an isolated spot at Madison, Wisconsin. Two of the plants died before blossoms appeared. The solitary one produced a fairly heavy set of seed. It appeared to be the first fully self-compatible plant of cabbage recorded. Such plants have not been observed since. When the progeny of this plant were grown on heavily infested soil near Racine, Wisconsin, along with many other selections, they were very uniform in type and outstanding in that no diseased plants in a total of 50 were observed. This was in contrast with other selected lines, which varied in the percentage of resistant plants, and with the susceptible control, in which more than 90% of the plants were diseased. Selections made from the All Head Early line were continued during the next two generations (greenhouse seed propagation was carried out). In the second generation out of 260 plants, three were diseased. Of these one died, one was slightly affected, and one recovered. In the third generation, except for one progeny which had obviously resulted from pollen contamination, all plants were resistant.

Since the line derived from All Head Early proved to be practically 100% resistant in the field, it was tested further in the greenhouse. Here it responded differently from Wisconsin Hollander in that very young plants set in heavily infested soil showed no sign of disease at a constant soil temperature of 24 C or slightly higher. At the same temperature, young Wisconsin Hollander plants all died. This showed that two types of resistance to yellows occur in cabbage. That recovered from All Head Early is known as Type A and was determined later to be controlled by a single dominant gene. Resistance in Wisconsin Hollander, known now as Type B, is polygenic and recessive, or at most very slightly dominant. Seedlings of this variety succumb completely to yellows when grown at 24 C constant soil temperature.

It soon became obvious that the old method of selection for yellows resistance in the field did not permit distinction between Type A and Type B resistance. Since Type A is the more desirable, the testing of progenies in the seedling stage under controlled environment is the safest method. This is now the manner in which the yellows-resistant lines are developed. Since the introduction of Wisconsin Hollander, Wisconsin All Seasons, and Wisconsin Brunswick, all resistant varieties have had the gene for Type A resistance.

The use of infested soil has been supplanted by growing seedlings in sand supplemented by a solution of mineral nutrients. Hoagland's solution is satisfactory. Seed is sown fairly thickly in quartz sand. The fungus is grown in Czapek's solution in Ehrlenmeyer flasks on a shaker. When the cotyledons have fully expanded, seedlings are removed and the roots are dipped in the inoculum, which consists of a 10-day-old culture, the mycelium of which is filtered off and homogenized in a blender. Plants are then set in quartz sand in shallow stainless steel trays about 2.5 inches deep. They are set about 3/4 inch apart in rows about 2 inches apart. The trays are set in soil temperature tanks at 24 C. A suitable susceptible line is included as a control. Readings can be made within 30 days. Illustrations may be found in Walker's *Plant Pathology,* Third Edition, pp. 793-796.

The Type A resistant gene has been shown to occur in wild cabbage from southern England, as well as in Brussels sprouts and kohlrabi. Resistant varieties of the last two have not been developed.

## Radish Wilt

The cabbage yellows organism is now referred to as *Fusarium oxysporum* f. sp. *conglutinans* race 1, since three other races have been described. Race 2 is the common causal organism of radish wilt. This disease was first described in California and now occurs in a number of regions. It has been studied most extensively by Pound and Fowler (1953) and by Peterson and Pound (1960). The two races are similar in spore morphology and in characteristics on artificial media and may attack a number of species of *Brassica* and *Raphanus,* but race 2 is not pathogenic on cabbage, kale, and cauliflower. Race 2 and race 1 are most active on media at 28 C, but race 2 develops much more rapidly on radish at 16 C and 20 C than race 1 does. The two forms produce identical symptoms on radish. On young seedlings exposed to optimum temperature the disease is characterized by rapid yellowing and wilting of cotyledons and dying of the roots. Under less favorable conditions for the disease older leaves are yellowed, especially along veins, and they often show a unilateral development. Cross sections or longitudinal sections of root, hypocotyl, or stem reveal vascular discoloration. The Red Prince variety, which is highly resistant, was announced by Pound in 1959. It was selected from Early Scarlet Globe. Scarlet Knight is another *Fusarium*-resistant red globe radish. White Spike is an intermediate length white icicle type with high resistance to yellows. Resistance in White Spike was derived from a white mutant of Red Prince. Peterson and Pound showed that the fungus enters usually through the root tip in both susceptible and resistant varieties, but that mycelial concentration in the vascular system of the latter is sparse. Resistance is considered to be polygenic.

The methods used for selection of resistant radish are essentially the same as those used for cabbage yellows. Progenies are screened for resistance in sand culture pans in soil temperature tanks. The optimum temperature for disease development is 24 C. Seedlings are started in clean sand. After 7 days they are removed, washed, and the roots dipped in a mycelial-spore suspension of the fungus grown on Czapek's or potato dextrose medium. Pans are kept at 24 C during the test. Seedlings are rated after about 14 days, and survivors are removed and potted in 4-inch pots for flowering.

## Clubroot

Clubroot is perhaps the best known disease of crucifers. The first intensive study of it was by Woronin at St. Petersburg, Russia, where the malady was extremely destructive on cabbage. The fungus, *Plasmodiophora brassicae* Woronin, is a soil invader, many of the resting spores remaining several years before they germinate and infect the root hairs of the suscept. Many species and genera of the crucifers are susceptible. As early as 1853, Anderson in England noted that certain varieties of turnip were more resistant than others to clubroot. The first intensive work directed toward development of desirable resistant varieties was with turnip and rutabaga in Scandinavia and the British Isles. Among resistant varieties released many years ago are May, Bruce, and Immuna turnip, and Herning and Studsgaard rutabaga. A resistant variety of kale was announced by McDonald (1935). Two resistant varieties of cabbage, Bohmerwald and Bindsachsener, were reported in Germany by Gante (1951). Badger Shipper cabbage was the first American introduction. While highly resistant in the area of southeastern Wisconsin where it was developed, it soon became very susceptible on the same fields where a newly observed pathogenic race 7 built up rapidly, aided by monoculture of cabbage. Strandberg and Williams (1967) reported a line of Chinese cabbage in which resistance to races 6 and 7 is controlled by a single dominant gene.

Almost inevitably, varieties of *B. oleracea* found to be highly resistant in the locality of their origin were eventually found to be susceptible in some other area, or in the locality of their origin. *P. brassicae* has a large number of pathogenic races. Ayers, in Prince Edward Isle, Canada, described six races, of which race 6 was sent to him from Wisconsin. In Wisconsin, Seaman isolated race 7, to which resistant Badger Shipper is susceptible.

Williams (1966) offered a system for the determination of races of *Plasmodiophora brassicae* that infect cabbage and rutabaga. In addition to field assays on heavily infected soils, laboratory methods for assay of progenies as outlined below could well be followed by one carrying on breeding for clubroot resistance.

A mixture of one part of finely divided sphagnum peat moss and one part of black organic peat (muck) soil is passed through a 1/8-inch mesh screen, mixed thoroughly, and steamed at 90 C for 2 hr on each of two consecutive days. Spores are secured from clubs of known origin stored at –10 C. One hundred grams of frozen clubroots in 400 ml of distilled water are macerated for 3 min in a high-speed blender. The mixture is passed through 8 layers of cheesecloth, and the filtrate is centrifuged at 2,000 $g$ for 7 min. The pellet is suspended in 100 ml water. Centrifugation and washing are repeated three times, after which the pellet containing spores is light gray. After the final pellet is suspended in 100 ml distilled water, the spores are counted with a bright line haemocytometer and stored at 2 C until used.

The spores are mixed with soil described above to obtain a concentration of $10^8$ spores/cc of soil. Seven-day-old seedlings are transplanted from vermiculite into 2.5-inch pots, 5 plants per pot. Pots are placed in aluminum pans half filled with steamed vermiculite. The vermiculite is kept moist by regular watering with 1 N Hoagland's solution. The pans are kept under light of about 1,000 foot candles at about 22 C. After 35 days plants are removed, and soil is carefully shaken from the roots. The latter are then washed and rated for disease classification into one or another of the following categories. Roots with no clubs are in Class 0; those with a few small clubs on the secondary roots are in Class 1; those with considerable clubbing on the secondary roots are in Class 2; those with severe clubs on primary and secondary roots are in Class 3. To secure a disease index for each lot, those in Class 0 are multiplied by 0, those in Class 1 by 10, those in Class 2 by 60, and those in Class 3 by 100. The sum of the products is divided by the number of plants to secure the disease index. Plants of cabbage (Jersey Queen and Badger Shipper varieties) and rutabaga (Laurentian and Wilhelmsburger varieties) should be included as controls. Plants in Class 0 are transplanted and continued for breeding work.

## Rhizoctonia Bottom Rot and Powdery Mildew of Cabbage

Rhizoctonia bottom rot, incited by the common soil fungus *Rhizoctonia solani* Kühn, is usually a minor disease of cabbage. Where cabbage is close cropped, however, it may reach alarming proportions. The same fungus incites wirestem of young plants, which appears as black to brown lesions on the hypocotyl. Bottom rot appears on plants from the time of early heading to when heads approach maturity. The fungus migrates up the main stem to the outer leaves, where the midribs decay and the leaves droop and eventually drop. If conditions are very favorable when the head is beginning to form, the entire head may succumb to rapid soft rot. If the head is near maturity when the fungus attacks, dark lesions of various shapes and sizes appear on the outer and one or two next inner leaves, and soft rot is not then evident. In work at the University of Wisconsin directed primarily toward development of lines resistant to internal tipburn, where continuous cropping of the trial plots was desirable, Williams and Walker (1966) noted that in some selfed lines all plants were severely diseased with bottom rot, while in other lines the disease was absent. It was shown by further work that resistance was controlled by a single dominant gene.

In the same plots powdery mildew became prevalent on outer leaves and to some extent on outer head leaves. This is incited by a certain race or races of *Erysiphe polygoni* DC. It is favored by cool weather when heavy dew is likely to occur. As in the case of Rhizoctonia bottom rot, Walker and Williams (1965) observed that some selfed lines were free of mildew and others were uniformly diseased. $F_1$ progenies were either free from mildew or showed slightly susceptible plants. $F_2$ progenies segregated into three resistant or slightly susceptible plants to one severely affected plant, indicating a monogenic dominant character.

The variety Globelle is homozygous for resistance to bottom rot and powdery mildew at Madison, Wisconsin.

*Rhizoctonia solani* and *Erysiphe polygoni* are each known to consist of numerous pathogenic races. How the two resistant genes express themselves in localities where the race population may be different is not known. The bottom rot resistance of Globelle is not noticeably effective in the seedling disease known as wirestem. Only natural field tests at Madison have been used.

## Downy Mildew

Downy mildew incited by *Peronospora parasitica* (Pers.) ex Fr. is widespread throughout cabbage-growing areas of the world, where it causes varying degrees of damage to the crop. It is perhaps most severe in the southeastern United States on young plants in the seedbeds. High humidity, warm days, and cool nights followed by heavy dews favor the spread and development of the pathogen. On mature plants, irregular chlorotic lesions may form on the leaves, whereas dark, irregular lesions ranging in size from a few mm to several cm disfigure the head. Occasionally the pathogen may grow systemically throughout the head, causing an internal darkening of tissues surrounding the veins. Downy mildew causes considerable damage to other subspecies of *B. oleracea.* On cauliflower and sprouting broccoli the curd or sprouts may be darkened, whereas in Brussels sprouts a rot of the sprouts may develop.

A number of biological races of *P. parasitica* occur on various genera of the Cruciferae. Natti, Dickson, and Atkin (1967) described two on *B. oleracea,* and they discussed genes governing resistance to them. Inoculum is prepared from freshly produced conidia. The fungus can be maintained in the leaves of an older susceptible cabbage. Conidia can be produced by placing the infected plant or infected leaves taken from the field in a mist chamber for 24 hr at 20 C. Conidia are collected by dipping the sporulating portions of the leaves in distilled water. These conidia are then sprayed with an atomizer directly onto the cotyledons or leaves of young seedlings, where they germinate directly when held in the mist chamber at 100% relative humidity at 20 C.

The fungus spreads throughout susceptible seedlings in 4-6 days, and abundant conidia and conidiophores occur on the lower sides of the cotyledons. Resistant plants will not show any fungus growth. Upon removal from the mist chamber, the infected seedlings will quickly collapse and die, whereas resistant individuals grow on vigorously. Susceptibility of seedlings is not necessarily equated to mature plant susceptibility. Natti (1958) described a "moderate" form of resistance in broccoli which is not expressed in the seedling stage. In this instance, inoculation would have to be withheld until the plants reached adequate maturity for resistance to be expressed.

## White Rust of Radish

White rust, incited by the obligate parasite *Albugo candida* (Pers.) Kuntze, is restricted to the families Cruciferae and Capparidaceae. Although white rust is most commonly seen on cruciferous weeds such as shepherd's purse in the United States, it is mainly an economic problem in the production of radish seed. The fungus enters the racemes, causing large galls and inducing a reversion to vegetative structures in the flowering parts. Occasionally on leaf mustard, *Brassica juncea* (L.) Coss., white rust can destroy the foliage. In Europe the fungus can be a problem on the various subspecies of *B. oleracea.*

*A. candida* exists in a number of specialized pathogenic races. Pound and Williams (1963) have described six races, each of which is restricted in nature to a genus of the Cruciferae. Thus, in collecting conidia careful attention should be given to the source of inoculum.

In contrast to the conidia of *P. parasitica,* which germinate and penetrate the host directly, *A. candida* germinates to produce motile zoospores which become the infective units. Zoospores require a film of surface moisture on the leaf in which to swim, encyst, germinate, and penetrate the host. Thus, inoculations of this fungus are also made in a mist chamber. Conidia of *A. candida* may be collected from the erumpent white rust pustules with the aid of a cyclone spore collector. They can be stored dry up to 6 months at −20 C.

In order to produce zoospores for inoculation, a few mg of conidia are shaken with 100 ml of distilled water and placed at 12-16 C for 4 hr. The water then is examined under the microscope for motile zoospores. The suspension containing conidia and zoospores is gently sprayed with an atomizer onto the cotyledons of week-old seedlings in a mist chamber at 20 C. Older plants may be inoculated in the same way. Plants are held in the mist chamber for 12-24 hours, then returned to the greenhouse bench. After 7 days abundant white pustules will develop mainly on the lower surface of the susceptible cotyledons, and the plants will become stunted. Resistant plants may show minute necrotic areas on the cotyledons, where the invading fungus induces a hypersensitive host response.

In *Raphanus sativus* Williams and Pound (1963) described monogenic dominant resistance derived from the cultivars Round Black Spanish and China Rose Winter. All of the red globe and white icicle types of radish commercially available were susceptible. A program to incorporate white rust resistance into these types is under way.

## Cabbage Mosaic

Cabbage mosaic is usually incited by the combined effect of turnip mosaic virus (cabbage virus A) and cauliflower mosaic virus (cabbage virus B). The disease symptoms associated with the former are most pronounced at relatively high temperatures, and those of the latter at relatively cool temperatures. Both viruses are transmitted mechanically and by aphids. The relative resistance in cabbage varieties is reported by Pound and Walker (1951). Resistance to both viruses is incompletely dominant and inherited independently. Resistant plants showing little or no symptoms may still harbor the viruses. Working with a yellows-resistant variety of Wisconsin All Seasons which had first been selected for monogenic resistance to yellows, several varieties and inbred lines carrying high resistance were developed.

The screening method used is as follows: each virus component of cabbage mosaic is maintained independently in a favorable host plant at its optimum temperature. Cabbage virus A is maintained in *Nicotiana glutinosa* L. or in a very susceptible cabbage variety at about 29 C. Cabbage virus B is maintained in Chinese cabbage, *Brassica pekinensis* Rupr., or in a very susceptible cabbage variety at about 20 C. The juice of the infected plants is extracted by grinding for several minutes in a blender and is then filtered through several layers of cheesecloth. The two extracts are combined, and a small amount of carborundum powder is added.

Test plants are grown in flats in greenhouse or cold frame. They are inoculated in the third-leaf stage by rubbing gently the second and third leaves with a glass spatula which is first dipped in the virus extract. The plants are grown to transplanting stage at about 25 C, when they are rogued according to relative symptoms. The survivors are transplanted to the field.

When planting in flats is not feasible, plants may be inoculated soon after they recover from transplanting in the field. The development of 10 inbreds with high resistance to mosaic and homozygous for Type A resistance to yellows is described by Pound, Williams, and Walker (1965).

## Internal Tipburn of Cabbage

In the late 1950s widespread internal tipburn caused heavy losses, especially in cabbage grown for sauerkraut. It consists of from brown to black necrosis beginning at the margins of leaves near the center of the head. It was shown to be a physiological rather than a parasitic disease, and it appears principally in cabbage grown for kraut, which is allowed to reach full maturity before harvest. It is most common in crops growing vigorously on well fertilized soil and under favorable climatic conditions. It appears in tissue that is normally low in calcium, and in this respect it resembles blossom end rot of tomato. Various approaches to control through soil amendments and calcium sprays were not successful.

Selection of plants within the yellows-resistant kraut varieties (Globe and Resistant Glory) was begun in 1960 and led to two highly resistant varieties: TBR Globe and Globelle (Walker, Williams, and Pound, 1965). Resistance is polygenic and recessive. Since all selections must be made in the field by examining heads internally, it is desirable to use a plot which has a record of producing cabbage unusually high in tipburn. Watering during dry spells is favorable to disease development.

# What's Ahead in Disease Resistance in Crucifers?

A sufficient number of resistance genotypes in cabbage are available to permit the development of a large number of multi-resistant hybrid combinations. The first of these to be produced are the hybrids Hybelle and Sanibel, released by Williams, Walker, and Pound (1968). Mating sib-incompatible inbreds resistant to yellows, internal tipburn, and mosaic with a sib-compatible variety Globelle resistant to yellows, mosaic, internal tipburn, powdery mildew, and Rhizoctonia head rot, produces $F_1$ hybrids with resistance to yellows, mosaic, internal tipburn, powdery mildew, and Rhizoctonia head rot.

Resistance to certain races of *Plasmodiophora brassicae* have been incorporated into swede, turnip, and cabbage, but the wide variation in pathogenicity of the organism places a definite limit on so called resistant varieties.

Bain (1955) has shown that there are genes for resistance to black rot [ *Xanthomonas campestris* (Pam.) Dows.] in certain Japanese cabbage varieties. These offer promise and are being worked upon intensively by Williams.

Undoubtedly other important genes are to be found. A dire need is that of resistance in subspecies of *B. oleracea,* and in turnip and rutabaga, to blackleg [ *Phoma lingam* (Fr) Desm.] [ *Leptosphaeria maculans* (Desm.) Ces. & deNot.]. Wallace (1943) has reported that the N.Z. resistant swede, which is resistant to clubroot, is also somewhat resistant to blackleg.

Another lead to be explored is resistance to the anthracnose leafspot (*Colletotrichum higginsianum* Sacc.) of crucifers, which is sometimes serious in the southeastern United States on turnip, radish, mustard, and Chinese cabbage. Scheffer (1950) reported that Southern Giant Curled mustard was highly resistant. No turnip varieties tried were resistant, but American Purple Top rutabaga was highly resistant. One P.I. accession, 161429, was very highly resistant. Another accession, 161425, had some highly resistant plants.

There needs to be constant exposure of foreign accessions of *Brassica* and *Raphanus,* as well as established cultivars, in the search for resistant genes. While interspecific crosses between cabbage and other *Brassica* species are not common, certain success has been achieved. Fertile hybrids of rutabaga and turnip, while uncommon, have been produced. It remains to be determined what other sources of resistance can be found. At the moment black rot and blackleg resistances in

*Brassica oleracea* and in other cultivated Brassicas are direly needed. Clubroot resistance in all cultivars is highly desirable. Downy mildew resistance should be incorporated into all cultivars as needed, especially for crops grown in regions of high humidity, as illustrated by those along the Atlantic and Pacific coasts of the United States.

# References

Bain, D. C. 1955. Resistance of cabbage to blackrot. Disappearance of blackrot symptoms in cabbage. Phytopathology 45:35-37; 55-56.

Bateman, A. J. 1955. Self-incompatibility systems in angiosperms. III. Cruciferae. Heredity 9:53-68.

Boswell, V. R. 1949. Our vegetable travellers. Nat. Geog. 96:145-217.

DeCandolle, A. 1902. Origin of cultivated plants. Appleton & Co., New York 468 p.

Dickson, M. H. 1970. A temperature sensitive male sterile gene in broccoli, *Brassica oleracea* L. var. *italica*. J. Amer. Soc. Hort. Sci. 95:13-14.

Gante, T. 1951. Hernieresistenz bei Weisskohl. Z. Pflanzenzücht. 30:188-197.

Johnson, A. G. 1966. Inbreeding and the production of commercial $F_1$ hybrid seed of Brussels sprouts. Euphytica 15:68-79.

Johnston, T. D. 1965. Inbreeding and hybrid production in marrow-stem kale (*Brassica oleracea* L. var. *acephala* DC.). 3. The development and production of hybrids for commercial use. Euphytica 14:120-124.

Jones, L. R., and J. C. Gilman. 1915. The control of cabbage yellows through disease resistance. Wisc. Agr. Exp. Sta. Res. Bull. 38.

Magruder, R. 1937. Improvement in the leafy cruciferous vegetables. U.S. Dept. Agr. Yrb. 1937:283-299.

Martin, F. W. 1962. Factors affecting seed set in cross-pollinations of green sprouting broccoli (*Brassica oleracea* var. *italica*). Euphytica 11:81-86.

McDonald, J. A. 1935. Plant pathology. The resistance of marrowstem kale to finger and toe. Scot. J. Agr. 18:164-165.

Natti, J. J. 1958. Resistance of broccoli and other crucifers to downy mildew. Plant Dis. Reporter 42:656-662.

———, M. H. Dickson, and J. D. Atkin. 1967. Resistance of *Brassica oleracea* varieties to downy mildew. Phytopathology 57:144-147.

Nieuwhof, M. 1969. Cole crops, botany, cultivation, and utilization. CRC Press, Cleveland. 353 p.

Pearson, O. H. 1929. Observations on the type of sterility in *Brassica oleracea* var. *capitata*. Proc. Amer. Soc. Hort. Sci. 26:34-38.

———. 1932. Breeding plants of the cabbage group. Calif. Agr. Exp. Sta. Bull. 532.

Peterson, R. L., and G. S. Pound. 1960. Studies on resistance of radish to *Fusarium oxysporum* f. *conglutinans*. Phytopathology 50:807-816.

Poole, C. F. 1937. Improving the root vegetables. U.S. Dept. Agr. Yrb. 1937:300-325.

Pound, G. S. 1959. Red Prince is new radish. Wisc. Agr. Exp. Sta. Bull. 538, Part II:93.

————, and D. L. Fowler. 1953. *Fusarium* wilt of radish in Wisconsin. Phytopathology 43:277-280.

————, and J. C. Walker. 1951. Mosaic resistance in cabbage. Phytopathology 41:1083-1090.

————, and P. H. Williams. 1963. Biological races of *Albugo candida*. Phytopathology 53:1146-1149.

———— ————, and J. C. Walker. 1965. Mosaic and yellows resistant inbred cabbage varieties. Wisc. Agr. Exp. Sta. Res. Bull. 259.

Sampson, D. R. 1964. A one-locus self-incompatibility system on *Raphanus raphanistrum*. Can. J. Genet. and Cytol. 6:435-445.

Scheffer, R. P. 1950. Anthracnose leafspot of crucifers. N.C. Agr. Exp. Sta. Tech. Bull. 92.

Strandberg, J. O., and P. H. Williams. 1967. Inheritance of clubroot resistance in Chinese cabbage. Phytopathology 57:330.

Thompson, K. F., and H. W. Howard. 1959. Self-incompatibility in marrowstem kale (*Brassica oleraceae* var. *acephala*). II. Methods for the recognition in inbred lines of plants homozygous for S alleles. J. Genet. 56:325-340.

Walker, J. C. 1969. Plant pathology, 3rd ed. McGraw-Hill, New York. 819 p.

————, and R. H. Larson. 1960. Development of the first clubroot resistant cabbage variety. Wisc. Agr. Exp. Sta. Bull. 547:12-16.

————, and P. H. Williams. 1965. The inheritance of powdery mildew resistance in cabbage. Plant Dis. Reporter 49:198-201.

———— ————, and G. S. Pound. 1965. Internal tipburn of cabbage. Its control through breeding. Wisc. Agr. Exp. Sta. Res. Bull. 258.

Wallace, D. H., and M. E. Nasrallah. 1968. Pollination and serological procedures for isolating incompatibility genotypes in the crucifers. N.Y. (Cornell) Agr. Exp. Sta. Memoir 406.

Wallace, J. O. 1943. N. Z. resistant swede. N. Z. J. Agr. 67:341, 343.

Watts, L. E. 1963. Investigation into the breeding system of cauliflower (*Brassica oleracea* var. *botrytis* L.). I. Studies of self-incompatibility. Euphytica 12:323-340.

————. 1965. Investigation into the breeding system of cauliflower (*Brassica oleracea* var. *botrytis* L.). II. Adaptation of the system to inbreeding. Euphytica 14:67-77.

Williams, P. H. 1966. A system for the determination of races of *Plasmodiophora* that infect cabbage and rutabaga. Phytopathology 56:624-626.

————, and G. S. Pound. 1963. Nature and inheritance of resistance to *Albugo candida* in radish. Phytopathology 53:1150-1154.

————, and J. C. Walker. 1966. Inheritance of *Rhizoctonia* bottom rot resistance in cabbage. Phytopathology 56:367-368.

———— ————, and G. S. Pound. 1968. Hybelle and Sanibel, multiple disease-resistant $F_1$ hybrid cabbages. Phytopathology 58:791-796.

Yarnell, S. H. 1956. Cytogenetics of the vegetable crops. II. Crucifers. Bot. Rev. 22:81-166.

# 20 Peas

## D. J. Hagedorn[1]

## *Introduction*

Peas, *Pisum sativum* L., are one of the most important vegetable crops. They are used commercially for many purposes, including canning, freezing (peas and edible pods), soup making, and as a fresh-market vegetable and a cover crop. In addition, many home gardens have one or two plantings of peas for table use as a fresh vegetable or for home canning or freezing.

In Europe and North America the most extensive pea acreage is planted for canning purposes. In the United States there are three important growing areas for the production of processing (canning and freezing) peas: the East, including such states as Delaware, Maryland, and New York; the Midwest or Central area, including Wisconsin, Minnesota, Illinois, Michigan, and others; and the West, which includes primarily the states of Washington, Oregon, Idaho, and California. In 1970 the total harvested processing pea acreage in the U.S. was 383,850. This acreage produced 476,250 tons of green shelled peas worth $51,889,000 to the growers. The pea production area in the East har-

1. Professor of Plant Pathology, University of Wisconsin.

vested 13,450 acres for canning and 21,360 acres for freezing. Harvested acreage in the Central area was 194,840 for canning and 21,950 for freezing. (Wisconsin led all states with 118,000 harvested acres.) In the West 43,950 acres were harvested for canning and 88,300 for freezing. The 1970 average yield per acre of shelled peas for processing in the U.S. was 1.24 tons; for the East 1.56 tons; for the Central area 1.15 tons; and for the West 1.31 tons.

The third major use for peas is the processing of dry peas for soup making by commercial companies or by the housewife. They are used whole or as split peas with the seed coat removed. Large acreages of peas are grown for this purpose, especially in Northern Idaho and Eastern Washington. The total U.S. acreage devoted to this use in 1970 was about 290,000 acres.

The harvest season for processing peas (those used for canning and freezing) is extended by planting early- and late-maturing cultivars in addition to the regular mid-season pea strains. For many years the earliest cultivar used for this purpose was Alaska, which is a round- and smooth-seeded pea, with indeterminate growth habit and light green foliage. Most such peas produce one pod on each flowering node. During the 1940s the Alsweet pea strains came into prominence as an early canning pea type. They are also light green and indeterminate, but are dimple-seeded and possess a higher yield potential. Although some dark green-seeded strains of Alaska have been used for freezing, Alaska and Alsweet strains are, for the most part, considered canning peas. Early freezing pea cultivars are often wrinkle-seeded strains referred to as "sweets." The majority of main season pea cultivars (whether their immature peas are "conventional" green and used for canning, or dark green and used for freezing) are also wrinkle-seeded, possess more sugar in the cotyledons, and have a determinate plant habit whereby the pods are borne in a concentrated manner on 3-5 upper nodes. These determinate strains often produce two pods per flowering node.

Most processing and garden peas have white flowers, while those grown for other purposes may have colored (often purple) flowers. There is an oblique calyx tube of five sepals which are undiverged at their bases with 4-5 unequal marginal lobes. The corolla has five petals: the large broad upper petal is known as the "standard," and it encloses the others (two lateral "wings" and two lower petals, the "keel"). The keel surrounds the single-carpelled pistil and the 10 diadelphous stamens. There is a one-celled superior ovary containing two rows of ovules along parallel placentae. The style is almost at a right angle to the ovary, and the stigma is bearded on its lower inner surface. The flowers are self-fertile.

Peas have 7 haploid and 14 diploid chromosomes. Several chromosomes have been reported to have "secondary constrictions." Seven pairs may be observed in the pollen mother cells, and meiosis proceeds normally through both divisions. The gene linkage groups as determined from a study of commercial cultivars are regarded as representing the normal chromosomal arrangement.

Disease resistance has played a major role in the successful pea production enjoyed by pea growers in many parts of the world, especially in Europe, North America, and Australia. The development of wilt (*Fusarium oxysporum* f. *pisi* (Linford) Snyd. & Hans., race 1) resistant cultivars, pioneered by the Wisconsin Agricultural Experiment Station, has kept peas growing in areas which would have been out of production many years ago because of widespread and severe infestation by the causal fungus.

# Fungus Diseases

## Aphanomyces Root Rot

The most serious pea disease in the United States is common root rot, or *Aphanomyces* root rot, caused by the phycomycetous fungus, *Aphanomyces euteiches* Drechs. It is especially important in the Midwest, where processing peas are grown in heavy soils with abundant moisture and warm temperatures during the latter part of the growing season. There has been keen interest in resistance to this disease since it was first described in 1925. Early studies on resistance concluded that resistance observed in the field did not hold up in the greenhouse when peas were grown in artificially infested sand. Since that time many researchers have investiaged the possibility that resistant strains of peas existed in one of the large collections of *Pisum* germ plasm available for study. Substantial research programs were undertaken at the New York, Michigan, Minnesota, and Wisconsin Agricultural Experiment Stations. These studies were often made in naturally infested soil in the field, and sometimes in artificially infested soil or sand in the greenhouse. Sometimes pea strains that looked promising in the field were retested in the greenhouse, only to reveal that field-tolerant lines showed a mediocre or poor level of resistance in greenhouse tests. Under some circumstances the opposite was true.

Even though there are no pea cultivars resistant to *Aphanomyces* root rot in commercial use, it should be noted that tolerance to *A.*

*euteiches* has been noted in several pea plant introductions (P.I.) including 166159, 173059, and 180868. Others which appear to have a good degree of tolerance to the disease are P.I. 167250, 169604, and 180693.

## Ascochyta Diseases

There are three diseases of processing pea caused by species of *Ascochyta: Ascochyta pisi* Lib. incites *Ascochya* leaf and pod spot; *A. pinodella* Jones causes *Ascochyta* foot rot; and *Mycosphaerella pinodes* (Berk. & Blox.) Stone incites *Ascochyta* blight. Because these diseases are seed-borne and are widespread in the pea-growing areas of the world, much research on disease resistance has been undertaken.

In northeastern Europe it was found that several pure lines from "Victoria Heine" were far more resistant to *A. pisi* than the parent cultivar. Wark in Australia (1950) tested 313 pea lines for reaction to *A. pisi* and found none of them as resistant as Austrian Winter. When this pea was crossed with susceptible cultivars, segregation in the F$_2$ and later generations indicated that resistance was due to a combination of three dominant factors. Some progenies showed very good resistance but possessed some of the undesirable plant characteristics from the Austrian Winter parent. Somewhat different results were obtained by Lyall and Wallen (1958) in Canada, where they crossed the resistant pea line Ottawa A-100 with susceptible Thomas Laxton. They concluded that resistance to *A. pisi* was conferred by duplicate dominant genes, either of which gave resistance. Later other workers showed that the resistance of the A-100 pea to this pathogen was due to the physical barrier of the cuticle. In contrast, the resistance of O.A.C. 181 was physiological.

German researchers found that plants from colored seed were more resistant to *A. pinodella* than plants from colorless seed. Most promising pea cultivars from colored seed were Graue Buntblühende, Schweizer Riesen, and Lucienhofer winter pea. The most satisfactory colorless-seeded peas were Zeiners Kurz, Gut, and Onsa Schal. Weimer (1947) described methods of value in breeding Austrian Winter peas for disease resistance in southern United States, and he reported the results of ten years' tests with 160 pea strains against infection by *A. pinodella* and *M. pinodes*. Austrian Winter showed the greatest resistance of the peas studied. A strain of *Lathyrus hirsutus* was immune, and *L. sativus* was resistant. In Wisconsin, 100 canning pea strains were tested for reaction to all three *Ascochyta* pathogens, but there was not a high degree of resistance in any of the peas studied.

## Downy Mildew

Pea downy mildew, incited by *Peronospora pisi* Syd. emend Campbell, is sometimes troublesome to pea crops growing under moist, warm conditions. White and Raphael (1944) reported that the pea cultivars W. F. Massey, Radium, Perfection, and Perfection S.E.S. 1 showed no downy mildew under conditions conducive to disease development in Tasmania in 1943 and 1944. Several resistant processing peas were listed in the 1962, 1963, and 1967 annual reports of the Pea Growing Research Organization, Peterborough, England. These included Greenland 77, Freeze-Elite, Friends Glory, 402/59, Minigolt, Pauli, and Puget.

## Powdery Mildew

This troublesome pea disease is caused by *Erysiphe polygoni* DC. and becomes of real concern to pea growers under prolonged warm, dry daytime conditions and when nights are cool enough to cause dew formation. Harland in 1948 reported that about 10% of the plants in a collection of Peruvian peas grown in that locale were resistant to powdery mildew. Results of crosses made to determine the mode of inheritance of this resistance indicated that it was controlled by a single recessive gene designated *er*. Pierce (1948) reported resistance to pea powdery mildew in a 'Stratagem' selection. As a result of crosses made between the resistant Stratagem and Glacier and Shasta, he too concluded that resistance and susceptibility were controlled by a single gene pair, with susceptibility being dominant. Cousin (1965*b*) assessed the reaction of about 400 pea cultivars to naturally occuring powdery mildew in France. Certain lines from Mexican and Peruvian peas were practically immune, and 'Resistant Stratagem' was resistant. His genetic studies showed that the resistance of 'Mexique' and 'Resistant Stratagem' was monofactorial and recessive. The $F_1$ from a cross between these cultivars was resistant, indicating that resistance was dependent upon the same gene in both cultivars.

Schroeder and Provvidenti (1965) reported the breakdown of the *er er* resistance to powdery mildew in peas by an isolate of the pathogen obtained from an *Er Er* genotype in the field. This new, more virulent isolate of *E. polygoni* caused mildew-resistant pea lines from several sources to become severely diseased in repeated greenhouse tests. Host range of the new isolate included *Vicia faba, V. sativa, V. villosa, V. monantha, Lathyrus odoratus, L. cicera, Lupinus albus, L. luteus, Trifolium hybridum,* and a *Robinia* sp.

In 1969, Heringa, Van Norel, and Tazelaar found that pea lines from several countries showed adequate resistance to powdery mildew in The Netherlands. This resistance could be conferred by a single recessive gene, $er_1$. The resistance in progenies from Peruvian material was confined to the leaves and was due to a second rescessive gene, $er_2$. They recommended the use of Stratagem or line SVP942 as sources of resistance to powdery mildew under Dutch conditions.

In the United States, where widely used cultivars of processing peas are largely the result of commercial pea breeding programs, there are practically no such peas with powdery mildew resistance. This is true in spite of the fact that the New York and Wisconsin Agricultural Experiment Stations have both made several releases of resistant germ plasm of good type. It is likely that the sporadic occurrence of ecnomically important powdery mildew epiphytotics has tended to keep interest in resistance at a level too low to make best use of this germ plasm.

## Fusarium Root Rot

This disease is caused by *Fusarium solani* f. *pisi* (F. R. Jones) Snyd. & Hans. It is occasionally of economic importance, primarily under conditions of prolonged unusually warm temperatures. Hagedorn in 1960 devised a technique for testing pea seedlings for their reaction to the pathogen under controlled conditions. He studied 280 pea cultivars from 17 seed companies. No immune or highly resistant pea cultivars were found, but a few showed varying degrees of tolerance, with 'Wando' being the most tolerant. King et al. (1960) obtained similar results by field-testing 150 commercial cultivars, but 8 out of 391 P.I.s contained plants which were "more resistant than currently used canning varities." These P.I.s were 164417, 164837, 164971, 165577, 165965, 169606, 171816, and 173057. Promising results were obtained when the selected resistant plants were crossed with one another within a selection line, and with resistant plants from other selection lines. Knavel (1961) found that the resistance of P.I. 140165 was dominant when this pea was crossed with the susceptible Early Perfection. $F_3$ segregates from wrinkled seeds and seeds with transparent coats, characteristic of Early Perfection, were similar in resistance to plants from color-coated and smooth seeds. Short internode Perfection-type progeny of the $F_3$ populations were slightly more resistant than the long internode 140165 type. Kraft and Roberts (1969) evaluated 520 P.I.s for reaction to the *Fusarium-Pythium* root rot complex in the greenhouse. Fifty-eight were more resistant than the Dark Skin Perfection control, and they were tested again in a growth chamber. The following P.I.s were somewhat resistant to both pathogens: 140165, 140295, 223285, 234263, and 257593.

Although there are no highly resistant processing pea cultivars in general production, several canning peas with *F. solani* f. *pisi* tolerance are being used. These include 3019 and 3040, which were developed by Dr. E. J. Renard, of the Canners Seed Corp., Lewisville, Idaho, who repeatedly utilized a highly infested plot of land for his pea-breeding nursery.

## Fusarium Wilt

The most important disease ever to attack pea was Fusarium wilt caused by *Fusarium oxysporum* f. *pisi* (Lindf.) race 1 Snyd. & Hans. Although it generally has been controlled through the development of resistant cultivars, there are still restricted, local losses caused by the disease where susceptible cultivars are grown. When the disease was first described certain pea cultivars, including Horal, Green Admiral, and Yellow Admiral, were considered resistant. A program of selecting healthy plants from diseased fields of Alaska was begun, and these plants proved to be resistant in subsequent tests.

Wade (1929) reported that resistance was due to a single dominant gene designated *Fw*. He used resistant Horal, Green Admiral, Alaska, Improved Surprise, and Fasciated Sweet in crosses with susceptible Alaska, Surprise, Horsford, Perfection, and Acme. No linkage was found between the gene for sugary cotyledon and resistance, but a loose 31% linkage was noted between the *Le* gene for tallness and *Fw*. In 1931 Walker reported on wilt resistance in a large number of garden, canning, and field peas. There was a great variation in the reaction of 243 Alaska samples in the field: only one was wholly resistant. All of the 199 samples of Perfection were susceptible. In other garden and canning cultivars resistance varied from 100% resistant to 100% susceptible. He listed 10 resistant cultivars used by Wisconsin canners and later pointed out that new wilt-resistant canning peas included Alcross, Wisconsin Perfection, Wisconsin Early Sweet, Penin, Climax, Pride, Ace, Canner King, and Merit. Wade, Zaumeyer and Harter (1938) published on the reaction of 1,024 pea strains from all over the world to Fusarium wilt in the field near Fairfield, Washington, in 1931-1933. They concluded that resistance was scattered widely but not commonly among the peas of the world, occurring in presumably primitive types and in older cultivars. Most important American pea cultivars were susceptible. The dwarf, early market-garden cultivars were all completely susceptible.

Resistance to the pea wilt disease has been the object of research efforts by scientists in many parts of the world. For instance, Hubbeling (1956) in The Netherlands listed the pea cultivars resistant to Fusarium wilt in that country; and Yen and Cruickshank in 1959 described the development of new wilt-resistant peas for New Zealand using the backcross method of breeding.

The outstanding success of plant pathologists and pea breeders in their efforts to develop Fusarium wilt resistant peas of many types is a classic in the area of disease control through cultivar resistance.

## Near-Wilt

About the time that Fusarium wilt was being successfully controlled, a troublesome new vascular wilt disease was discovered in the processing-pea growing areas of the Midwest. Because of symptom similarity to wilt, it was called "near-wilt" (*F. oxysporum* f. *pisi* race 2). The wilt-resistant cultivars were susceptible to this new disease. An intensive research program to develop processing peas resistant to both wilt diseases was undertaken at the University of Wisconsin. Hare et al. (1949) selected a promising pea line from a cross between two Admiral strains. This was crossed with Pride, and the progeny gave rise to Delwiche Commando, the first pea cultivar resistant to both wilt and near-wilt. Resistance to near-wilt was governed by a single dominant gene designated *Fnw*.

Wisconsin researchers also developed a greenhouse technique for determining the disease reaction of parental lines and segregating populations in the seedling stage. This technique was used by Wells, Walker, and Hare (1949) to show that the *Fw* and *Fnw* genes are on the same chromosome and are associated with a loose linkage of $40.2 \pm 3.4$ per cent. The gene for near-wilt resistance appeared to be inherited independently of the *Le* gene for tallness.

Hagedorn (1953) described the development of the New Era canning pea which he had released earlier as being resistant to wilt, near-wilt, and bean virus 2 pea mosaic. A few years later he released two more canning pea cultivars, New Season and New Wales, with similar disease resistance but different plant characteristics. He subsequently released other wilt and near-wilt resistant peas, including C165, Wis. 729 and Wis. 741, the latter two in cooperation with E. T. Gritton. Wisconsin's New Era was the resistant parent and Greenfeast the susceptible, but locally important parent of Green Era, a new near-wilt resistant green pea developed for use in South Australia. Hubbeling (1966) found that the resistance in near-wilt resistant peas developed in America could be broken down by soil conditions of low $p$H plus calcium deficiency. Koroza, Ivora, Emigrant, and Maro were resistant even under those extreme conditions. He had earlier demonstrated the presence of near-wilt in The Netherlands and reported that Rovar, New Season, Kelva, and Koroza were resistant.

### Pythium Damping-Off

The first pathogen to attack the pea "plant" after the seed is sown is *Pythium ultimum* Trow. or other *Pythium* spp. Peas are very susceptible to these damping-off fungi, especially under cool, wet conditions, and they must have the benefit of a proper seed protectant in order to emerge normally. Resistance to these damping-off pathogens would be a very desirable character because of the savings in cost of treating the seed, and because untreated seed may well respond more significantly to inoculation with nodule-forming bacteria. McDonald and Marshall in 1961 reported on their studies with 105 pea cultivars and damping-off in Manitoba. Most colored-flowered cultivars were resistant. Ewing (1959) concluded that resistance was due to the presence in the testa of the basic gene *A* for anthocyanin. The gene worked solely through the testa. Seeds enclosed in *A* testa were resistant regardless of whether cotyledons were round *R* or wrinkled *r*, whether they were yellow *I* or green *i*. Resistance was intimately associated with colorless, phenolic compounds belonging to the leucoanthocyanin group. So far this resistance has not been successfully transferred to commonly used processing pea varieties.

### Septoria Blotch

Zaumeyer in 1942 reported the results of testing the reaction of 134 pea strains to *Septoria pisi* West. in Colorado during the growing seasons of 1936-1939. Only two pea cultivars, Perfection and a pea imported from Puerto Rico, exhibited a high level of resistance. In general, the Perfection canning-pea strains and later-maturing cultivars were more tolerant than market garden or early maturing peas.

# Virus Diseases

### Common Pea Mosaic

This pea disease is one of the first virus diseases reported on this crop plant. It has occasionally been of economic importance, but generally it is not regarded as a serious disease of pea, perhaps because resistance has been known for a long time.

About the time the disease was described, Wisconsin researchers tested 34 pea cultivars in the greenhouse for their reaction to three strains of the pea common mosaic virus (PCMV). Twenty-eight of them were resistant, including Alaska, Prince of Wales, Surprise, Thomas Laxton, Alderman, and Canada Field Pea. All three strains of the virus reacted similarly. Idaho researchers found 44 susceptible and 21 resistant garden peas, the later including such cultivars as Horal, Canners Gem, Little Marvel, Morse Market, Onward, Perfection, Surprise, Thomas Laxton, Wisconsin Early Sweet, and Tom Thumb. Most field pea strains were susceptible.

Cousin (1965a) tested 405 pea cultivars in France and found 118 of them resistant, including American cultivars Canner Pefection, Darkskin Perfection, Early Perfection, Midfreezer, New Era, New Season, Perfected Wales, Small Sieve Freezer, Wisconsin Merit, and Wisconsin Perfection. Data from various crosses involving resistant Relance and Juwel and susceptible Mignon and Diamant indicated that resistance was controlled by a single recessive gene. He noted the similarity between his results and those obtained with bean yellow mosaic virus (BYMV) on pea.

When Chamberlain (1939) in New Zealand studied 34 garden pea and 22 field pea cultivars, he found that 12 of the former and two of the latter were immune to PCMV. Among the immune garden peas were such cultivars as Hundredfold and Little Marvel. It has been concluded that Perfection peas in general were resistant to PCMV, and thus this pea has often been one of the differential hosts used by legume virus researchers in virus identification.

## Bean Virus 2 Pea Mosaic

Another mottling or mosaic disease of pea is caused by various strains of bean virus 2 (BV2), often referred to as the bean yellow mosaic virus (BYMV). Because of several similarities, some legume virus researchers feel that the pea common mosaic virus is merely a strain of BYMV.

Hagedorn and Walker (1954) studied 18 pea cultivars and found 12 of them resistant, including Bonneville, Canner King, Delwiche Commando, Horal, Improved Penin, Loyalty, Merit, Pride, Resistant Famous, Wasatch, Wisconsin Early Sweet, and Wisconsin Perfection. Alaska, Alderman, Glacier, Perfected Wales, Profusion, and Thomas Laxton were susceptible.

The resistance of Perfection-type peas to BYMV has received close scrutiny by several researchers. Hagedorn (1951) conducted such a study using four isolates of BYMV from Wisconsin and 36 Perfection type peas supplied by seven important seed companies. Perfected Wales was the susceptible, and Wisconsin Perfection the resistant control. He found 28 cultivars resistant and 8 susceptible, including Early Perfectah and rarely Davis Perfection. Of the 8 susceptible cultivars, 4 included the word "early" in the varietal name, suggesting that some of the pea geneticists, in breeding for earliness in Perfection peas, used a parent which was susceptible to BYMV. In a later study of this kind Ford (1963) tested 62 peas, including Perfection types, plant introductions, and breeding lines, using two isolates of BYMV, one from Oregon and one from Washington. About 30% of the plants of 16 susceptible cultivars were infected by one isolate, and 80% were infected by the other isolate. The most extensive study of this kind was made by Schroeder and Provvidenti (1966), who studied the reaction of 248 commercial pea cultivars and breeding lines of several types to BYMV. Specific conditions for greenhouse testing were described, with emphasis on 80-85 F post-inoculation temperature. This temperature was found important for disease symptom development in heterozygous pea populations especially. True susceptible $Mo\ Mo$ genotypes were often slow to develop symptoms at lower temperatures, resulting in a confusion of the true genetic evaluation of a given pea seed stock. Many pea cultivars were susceptible, but most Perfection types were resistant.

Yen and Fry (1956) reported that immunity to pea mosaic virus in Australian pea cultivars, William Massey and Onward, was controlled by a single recessive gene, $mo$. Both isolates of the virus used were infectious to bean, so it is likely that it was BYMV instead of PCMV. At the same time Johnson and Hagedorn (1958) were studying the inheritance of resistance to BYMV in American peas, including New Era, Pride, Wisconsin Early Sweet, and Wisconsin Perfection. They also found that resistance was controlled by a single recessive gene. Both teams of researchers observed delayed symptom expression in the susceptible heterozygotes.

Schroeder et al. (1966) reported evidence that symptom expression in $F_2$ heterozygotes from similar crosses was governed by temperature. Phenotypic resistance in $F_2$ progenies was dominant at air temperatures of 18 C or below, and recessive at air temperatures of 27 C or above. Between these two temperatures, symptom expression was delayed in the heterozygotes as reported above. This temperature effect was not observed in homozygous resistant or homozygous susceptible plants. Symptoms in heterozygotes held at 27 C disappeared when plants were removed to air temperatures of 18 C or below, and they reappeared on the new growth that developed when plants were again placed at 27 C.

Heterozygotes could be separated from homozygotes by manipulating temperatures. Results were identical with two strains of BYMV and were similar to those obtained with PCMV.

Barton et al. (1964) studied analogous reactions of clonal cuttings from segregating progenies of susceptible x resistant crosses in garden pea to BYMV and PCMV. Results indicated that resistance to both viruses was conditioned by the same genotype, *mo mo*. This resistance was considered tantamount to immunity from infection by PCMV, but BYMV could be recovered from resistant plants under certain conditions. High temperature influenced symptom expression among heterozygotes, as noted above.

## Early Browning

This virus disease of peas is potentially important because of two unique characteristics regarding transmissibility; that is, the inciting virus, pea early browning virus (PEBV), is seed-borne and soil-borne. So far, distribution has been limited mostly to The Netherlands and Great Britain. When the disease was first described in The Netherlands, all 21 pea cultivars tested were considered susceptible, including round blue peas, marrowfats, gray peas, and horticultural wrinkled green peas. The latter type, including Excelsior, Juwel, and Wyola, showed less severe reaction to the virus.

A very comprehensive study of the reaction of many pea cultivars to PEBV was made by Hubbeling and Kooistra (1963). They found that most of them were highly susceptible, many were moderately susceptible, 12 cultivars were classified as slightly susceptible, and 35 were classified as resistant. The resistant strains included such processing cultivars as Early Perfection, Perfection 25, Pride, Prince of Wales, Profusion, and Wyola. Cultivars that developed no symptoms were Choscone, Druses, Fabula, Magnum Bonum, Nain Hatif Ameliore, and Superlative.

A study of the reaction of pea cultivars to the British strain of PEBV has shown that the severity of symptoms caused by PEBV from Britain was not always correlated with the severity of reaction to PEBV from The Netherlands. When 31 pea cultivars that were resistant or tolerant in The Netherlands were tested against the British form of the virus, almost all of these varieties developed local or systemic symptoms or both. This indicated that there was no correlation between field resistance in The Netherlands and resistance to infection by manual inoculation with PEBV from Britain.

## Enation Mosaic

This striking virus disease of pea, caused by the pea enation mosaic virus (PEMV), is one of the most important pea diseases in certain pea growing areas. In the U.S. it has been especially troublesome to Eastern and Western pea production.

In Wisconsin all 34 of the pea cultivars tested for reaction to PEMV under greenhouse conditions were susceptible, Perfection included. Plant pathologists at Geneva, New York, studied the reaction of 171 pea P.I.s and 47 commercial cultivars from six seed companies to PEMV in the field during the 1950 growing season. Enation pea mosaic developed in epidemic form, and only one pea line, P.I. 140295, showed any healthy plants at the season's end. Alaska and Bridger showed only mild enation symptoms. Bonneville and Horal developed enations, but not other symptoms. Schroeder and Barton (1958) studied the inheritance of the resistance (but not immunity) to PEMV obtained by the selection of G168 from P.I. 140295. Resistance was conditioned by a single dominant gene, *En.* It was not closely linked to *Le* (tallness) or *R* (round seed).

## Leafroll or Top Yellows

This aphid-transmitted virus disease of pea was first called "leafroll" in Germany, but it has been referred to repeatedly as "top yellows" in The Netherlands.

Hubbeling (1956) published the results of an extensive study to determine the reaction of 273 pea cultivars to the leafroll disease. He found that 95 of them were resistant, including Abundance, Alderman, Big Ben, Small Late Canner, Excelsior, Climax, Confidence, Early Duplex, Elf, Everbearing, Foremost, Gradus, Heralda, Hundredfold, Jubilee, Juwel, Koroza, Laxton's Progress, Lincoln (Greenfeast), Loyalty, Wisconsin Merit, Onward, Perfection, Rapida, Salzmunder Edelperle, Senator, Splendor, Stijfstro C. B., Improved Stratagem, Superlaska, Surpass, Telephone, Wando, Wyola, and Zelka.

In New Zealand, Greenfeast, Massey, and Victory Freezer were resistant, while Onward and Darkskin Perfection were susceptible to the leafroll virus.

The inheritance of resistance in peas to the pea leafroll virus was studied in The Netherlands by Drijfhout (1968), who made crosses between Cobri (resistant) and Gloire de Quimper (susceptible), and between Wyola (resistant) and Conserva (resistant). The result of testing over 300 plants led to the conclusion that resistance was inherited as a single recessive character. The gene for susceptibility to the pea leafroll virus was suggested to be *Lr*.

## Streak

This widespread virus disease can be caused by several pea streak viruses, including pea streak virus 1, Wisconsin pea streak virus (WPSV), Idaho pea streak virus (IPSV), alfalfa mosaic virus (AMV), and combinations of other pea viruses, especially BYMV and RCVMV, or red clover vein-mosaic virus.

Commonly available canning and garden cultivars, when tested over a period of two years, were all susceptible to pea streak virus 1. Hagedorn and Walker (1949), in describing the Wisconsin pea streak disease, reported on the reaction of 18 pea cultivars, all of which were considered susceptible, and the virus was easily recovered from each of them. Hagedorn (1968) reported on the reaction of 397 P.I.s of *Pisum sativum* to the Wisconsin pea streak virus. Only three, P.I. 116944, 140297, and 195405, were considered slightly diseased. Zaumeyer and Patino (1959) reported on the reaction of 119 pea strains to inoculation with the Idaho strain of the pea streak virus. Five cultivars, including Yellow Admiral, Eureka, Nome, Pedigree Extra Early, and Willet Wonder, showed mild symptoms.

Kim and Hagedorn (1959) studied the reaction of 16 commonly grown canning and freezing pea cultivars to four streak-inciting viruses collected from various pea growing areas in the United States. None were resistant to any of the four streak viruses tested.

An extensive study of AMV reaction in 397 P.I.s was reported by Hagedorn in 1968. Twenty-four of these peas were only slightly infected by AMV, the most tolerant line being P.I. 180695. A similar study was reported by Ford and Baggett (1965a), who found 31 P.I. lines resistant to AMV, but none of them were among those listed as slightly infected by Hagedorn (1968).

## Stunt

In 1949 a pea stunt disease was described which was troublesome in certain Wisconsin canning pea fields and affected all pea cultivars tested. The inciting virus was later shown to be the red clover vein-mosaic virus (RCVMV).

In 1954 the author tested 215 strains of peas, including commercial cultivars, collection strains, and P.I. accessions, for reaction to RCVMV. Only three pea strains failed to develop pea stunt symptoms, and in later trials they were also shown to be susceptible. Hagedorn (1968) reported the results of testing 397 P.I. accessions of *Pisum sativum* for reaction to the RCVMV. A large number of the pea lines tested were very susceptible, but six were only slightly infected: 10% or less of the inoculated plants became diseased. These included P.I.

116056, 134271, 164837, 194339, 195026, and 195628. Ford and Baggett (1965b) also tested *Pisum* P.I.s against the P-42 virus, which was considered to be related to RCVMV; P.I. 193845, 203066, 212029 and 261677 were resistant.

### Seed-Borne Mosaic

This disease, of potential importance to the pea-growing industry because it is readily seed-borne, is caused by the pea seed-borne mosaic virus (PSbMV). It is sometimes referred to as the "fizzle-top disease." When the disease was first described all pea cultivars tested were considered to be susceptible.

More recently, however, Stevenson, Hagedorn, and Gritton (1970) reported finding two pea P.I.s (among 551 tested) which contained resistant plants. (All 143 processing pea cultivars from nine seed companies were susceptible.) Crosses between the resistant P.I.s 193586 and 193836 and the susceptible commercial cultivars A45, Sprite, Alaska, Alsweet, New Era, and New Season indicated that resistance was conditioned by the presence of a single recessive gene.

# Nonparasitic Diseases

Perhaps the most common nonparasitic disease of pea is freezing injury. This is because "unseasonable" frosts are not uncommon in many of the northern pea-growing areas. It has long been known that Austrian Winter pea types have some tolerance to freezing injury. (This is probably one reason for their popularity as winter cover crops in the South.) This tolerance is confined to leaves and stems. Pea breeders in Europe have developed some pea lines with promising tolerance to frosts, but the only processing pea with this characteristic known to the writer is Wando, a freezing pea developed by Dr. B. L. Wade, U.S.D.A. Plant Geneticist.

Another common nonparasitic disease of pea is water congestion. This disease occurs when peas are grown for several days under conditions of high soil moisture, high humidity, and high temperature. Hagedorn and Rand (1971) recently pointed out that, in general, main-season pea cultivars of the Perfection type were slightly more tolerant to this disease than early peas.

The marsh spot disease, which sometimes develops when manganese is deficient or lacking, is also of concern as a nonparasitic disease.

Early pea cultivars, including Earliest of All, Early White Seedling, First and Best, Union Jack, and William the Conqueror, are apparently more tolerant than late-maturing peas.

# What's Ahead in Pea Disease Resistance

There are reasons to be optimistic about future development of improved pea cultivars, including those with disease resistance. Hagedorn (1967) has summarized the pea-breeding efforts of most of the researchers in the U.S. who are concerned with these activities. Considering the substantial numbers of commercial and governmental programs of this type in America, Europe, and other parts of the world, it is logical to assume that future pea cultivars will be progressively more disease resistant. If the present situation continues, the governmental programs will be largely responsible for discovering new genes for disease resistance and working out techniques for using them. Then commercial pea breeders will develop the new, widely used cultivars. Powdery-mildew resistant peas are on the way, as are peas with enation mosaic resistance and root-rot tolerance. Persistent efforts will materialize these new and welcome developments.

# References

Barton, D. W., W. T. Schroeder, R. Provvidenti, and W. Mishanec. 1964. Clones from segregating progenies of garden pea demonstrate that resistance to BV$_2$ and PV$_2$ is conditioned by the same genotype. Plant Dis. Reporter 48:353-355.

Chamberlain, E. E. 1939. Varieties of garden field peas immune to pea-mosaic. New Zealand J. Sci. Technol. 21:178A-183A.

Cousin, R. 1965a. Etude de la sensibilite des varietes de pois au virus de la mosaique commune du pois. Etude genetique de la resistance. Ann. Amelior. Plantes 15:23-36.

———. 1965b. Etude de la resistance a l'oidium chez la pois. Ann. Amelior. Plantes 15:93-97.

Drijfhout, E. 1968. Testing for pea leafroll virus and inheritance of resistance in peas. Euphytica 17:224-235.

Ewing, E. E. 1959. Factors for resistance to the pre-emergence damping-off in pea (Pisum sativum) incited by Pythium ultimum Trow. Diss. Abstr. 20:1518.

Ford, R. E. 1963. Susceptibility of Perfection-type peas to bean yellow mosaic virus. Plant Dis. Reporter 47:384-388.

————, and J.R. Baggett. 1965*a*. Reactions of plant introduction lines of *Pisum sativum* to alfalfa mosaic, clover yellow mosaic and pea streak virus, and to powdery mildew. Plant Dis. Reporter 49:787-789.

———— ————. 1965*b*. Relative severity of legume viruses in peas measured by plant growth reduction. Plant Dis. Reporter 49:627-629.

Hagedorn, D. J. 1951. The reaction of Perfection-type peas to Wisconsin bean virus 2 isolates from pea. Phytopathology 41:494-498.

————. 1953. The New Era canning pea. Wisc. Agr. Exp. Sta. Bull. 504. 8 p.

————. 1960. Testing commercial pea varieties for reaction to Fusarium root rot, *Fusarium solani* f. *pisi*. Phytopathology 50:637.

————. 1967. Breeding disease resistant *Pisum sativum* in the United States. Proc. 6th Int. Congr. Plant Protection, 119-120.

————. 1968. Disease reaction of *Pisum sativum* plant introductions to three legume viruses. Plant Dis. Reporter 52:160-162.

————, and R. E. Rand. 1971. Water congestion of pea, *Pisum sativum*. Plant Dis. Reporter 55:249-253.

————, and J. C. Walker. 1949. Wisconsin pea streak. Phytopathology 39:837-847.

———— ————. 1954. Virus diseases of canning peas in Wisconsin. Wisc. Agr. Exp. Sta. Res. Bull. 185. 31 p.

Hare, W. W., J. C. Walker, and E. J. Delwiche. 1949. Inheritance of a gene for near-wilt resistance in the garden pea. J. Agr. Res. 78:239-250.

Harland, S. C. 1948. Inheritance of immunity to mildew in Peruvian forms of *Pisum sativum*. Heredity 2:263-269.

Heringa, R. J., A. Van Norel, and M. F. Tazelaar. 1969. Resistance to powdery mildew (*Erysiphe polygoni* DC.) in peas (*Pisum sativum* L.). Euphytica 18:163-169.

Hubbeling, N. 1956. Resistance to top yellows and Fusarium wilt in peas. Euphytica 5:71-86.

————. 1966. Resistance to American vascular disease and 'near-wilt' on pea. Nether. Pl. Path. 72:204-211.

————, and E. Kooistra. 1963. Resistance to early browning in peas. Euphytica 12: 258-60.

Johnson, K. W., and D. J. Hagedorn. 1958. The inheritance of resistance to bean virus 2 in *Pisum sativum*. Phytopathology 48:451-453.

Kim, W. S., and D. J. Hagedorn. 1959. Streak-inciting viruses of canning pea. Phytopathology 49:656-664.

King, T. H., H. G. Johnson, H. Bissonnette, and W. H. Haglund. 1960. Development of lines of *Pisum sativum* resistant to *Fusarium* root rot and wilt. Proc. Amer. Soc. Hort. Sci. 75:510-516.

Knavel, D. E. 1961. The inheritance of Fusarium root rot, *Fusarium solani pisi,* resistance in *Pisum sativum*. Diss. Abstr. 21:2428-2429.

Kraft, J. M., and D. D. Roberts. 1969. Evaluation of pea introductions for resistance to Fusarium and Pythium root rot. Phytopathology 59:1036.

Lockwood, J. L., and J. C. Ballard. 1960. Evaluation of pea introductions for resistance to Aphanomyces and Fusarium root rots. Mich. Agr. Exp. Sta. Quart. Bull. 42:704-713.

Lyall, L. H., and V. R. Wallen. 1958. The inheritance of resistance to *Ascochyta pisi* Lib. in peas. Can. J. Plant Sci. 38:215-218.

McDonald, W. C., and H. H. Marshall. 1961. Resistance to pre-emergence damping-off in garden peas. Can. Plant Dis. Survey 41:275-279.

Pierce, W. H. 1948. Resistance to powdery mildew in peas. Phytopathology 38:21.

Schroeder, W. T., and D. W. Barton. 1958. The nature and inheritance of resistance to the pea enation mosaic virus in garden pea, *Pisum sativum* L. Phytopathology 48:628-632.

————, and R. Provvidenti. 1965. Breakdown of the *er er* resistance to powdery mildew in *Pisum sativum*. Phytopathology 55:1075.

————, ————. 1966. Further evidence that common pea mosaic virus (PV$_2$) is a strain of bean yellow mosaic virus (BV$_2$). Plant Dis. Reporter 50:337-340.

————, ————, D. W. Barton, and W. Mishanec. 1966. Temperature differentiation of genotypes for BV$^2$ resistance in *Pisum sativum*. Phytopathology 56:113-117.

Stevenson, W. R., D. J. Hagedorn, and E. T. Gritton. 1970. Resistance to the pea seed-borne mosaic virus. Phytopathology 60:1315-1316.

Wade, B. L. 1929. The inheritance of Fusarium wilt resistance in canning peas. Wisc. Agr. Exp. Sta. Res. Bull. 97. 32 p.

————, W. J. Zaumeyer, and L. L. Harter. 1938. Varietal studies in relation to Fusarium wilt of peas. U.S.D.A. Cir. 473.

Walker, J. C. 1931. Resistance to Fusarium wilt in garden, canning and field peas. Wisc. Agr. Exp. Sta. Res. Bull. 107. 15 p.

Wark, D. C. 1950. The inheritance of resistance to *Ascochyta pisi* Lib. in *Pisum sativum* L. Australian J. Agr. Res. 4:382-390.

Weimer, J. L. 1947. Resistance of *Lathyrus* spp. and *Pisum* spp. to *Ascochyta pinodella* and *Mycosphaerella pinodes*. J. Agr. Res. 75:181-190.

Wells, D. G., J. C. Walker, and W. W. Hare. 1949. A study of linkage between factors for resistance to wilt and near-wilt in garden peas. Phytopathology 39:907-912.

White, N. H., and T. D. Raphael. 1944. The reaction of green pea varieties to downy mildew and two viruses. Tasm. J. Agr. 15:92-93.

Yen, D. E., and I. A. M. Cruickshank. 1959. The breeding of peas resistant to Fusarium wilt in New Zealand. N.Z. J. Sci. Tech. A. 38:702-705.

————, and P. R. Fry. 1956. The inheritance of immunity to pea mosaic virus. Australian J. Agr. Res. 7:272-280.

Zaumeyer, W. J. 1942. Reaction of pea varieties to *Septoria pisi*. Phytopathology 32:64-70.

————, and G. Patino. 1959. A recently discovered virus-induced streak of peas. Plant Dis. Reporter 43:698-704.

# 21 Tomatoes
## R. E. Webb, T. H. Barksdale, and A. K. Stoner[1]

The cultivated tomato, *Lycopersicon esculentum* Mill., originated as wild forms in the Peru-Ecuador-Bolivia area of the Andes mountains of South America. Numerous wild and cultivated forms of the tomato can be found in this mountainous area today. From South America the tomato was apparently carried into Central America by prehistoric Indians and to other areas of the world by European travelers. It is reported that Europeans first grew tomatoes about 1550. The first report of tomatoes being grown in the United States was in 1781 on the plantation of Thomas Jefferson. Between 1800 and 1850, tomatoes began gaining acceptance as a food in the U.S.

In 1968 over 325,000 acres of tomatoes were grown in the U.S. for processing, and approximately 150,000 acres (including 1,500 in greenhouses) were grown for fresh market. The farm value of the 1968 processing crop in the U.S. was in excess of $220 million, and the fresh-market crop was valued in excess of $230 million.

Early efforts to improve tomatoes consisted of selecting for higher yields and large smooth-fruited types adapted for growth under local

1. Plant Pathologists and Horticulturist, Crops Research Division, Agricultural Research Service, U.S. Department of Agriculture, Beltsville, Maryland.

environmental conditions. Prior to 1910, practically all of the introductions of new tomato varieties were made by private individuals. Since 1910 many public agencies have been involved in tomato varietal development programs with emphasis on resistance to disease. Selection for many other characters (that is, yielding ability, adaptation to local environment, crack resistance, quality, and so forth) has been simultaneous with the selection for disease resistance. Over 1,100 tomato varieties have been developed in the U.S. and Canada since 1900. More than 600 of these have been released since 1935.

## Breeding Techniques

The normal tomato blossom is a complete flower. It has functioning male organs (anthers) and a female organ (pistil). Individual flowers are borne on pedicels (flower stalks), and the flower cluster develops on a peduncle (main flower stalk) that grows out from the stem between the leaf attachments and usually on the side opposite the nearest leaf. Usually, 4-8 flowers develop per cluster.

Because self-pollination usually occurs soon after the first pollen is released from the anthers, it is necessary to remove the anthers from flowers that are to be cross-pollinated before they release any pollen. This emasculation of the flower should be done as soon as the green sepals separate enough to expose the upper part of the anther ring, which should then be changing from green to a deep yellow color.

Theoretically, tomato flowers should be pollinated as soon as possible after they have fully opened. However, satisfactory fertilization occurs when flowers are pollinated immediately after emasculation, even though the flowers would not be fully opened until sometime the following day. Flowers emasculated and pollinated at the same time, a day before they open, saves returning and pollinating each flower a day or two later.

Pollen should be obtained from a freshly matured, open flower. A more abundant succession of flowers for pollen production will be assured if fruit is prevented from setting on plants used as pollen parents. Freshly opened flowers may be kept for several days if they are refrigerated in a moist condition. Pollen will retain its viability for 5-7 days in favorable storage.

It is best to grow the parent plants in a greenhouse where temperature and humidity can be controlled, and where the plants can be more easily handled than out-of-doors. Also, it is easier to make the crosses in a greenhouse and observe the developing fruit. Chance crossing can be prevented by covering the flowers with glassine bags, gelatin capsules, or some other material.

The number of seeds obtained per cross varies greatly, depending on the environmental conditions when the cross was made and the fruit type of the plant being used as the female parent. With favorable environmental conditions and a good seed-yielding parent, 200 or more seeds may be obtained from a single pollination made in a greenhouse. Pollinations made in the field are likely to result in more seeds per cross.

# Breeding Systems and Sources of Genetic Variability

Breeding systems most often used by tomato breeders are pure-line or individual plant selection, and artificial hybridization followed by selection in the $F_2$ generation or a backcross generation. Various modifications of these basic systems are also often used (Allard, 1964).

Pure-line or individual plant selection is limited by the fact that one cannot create genetic variability with this system but can act only on that already present. This system is useful when working with lines that have many desirable traits but are still segregating for some characteristics.

To create genetic variability the usual procedure is to hybridize artificially lines that possess the desired characters. Following hybridization, selection usually begins in the $F_2$ generation. Selection is likely to continue for another 3 or 4 generations. Screening for disease resistance and the other desired characteristics is carried out in each generation.

If practically homozygous plants are crossed, the $F_1$ population should consist of 15-25 plants. Normally, an $F_2$ population should have a bare minimum of 100 plants and may have 300 or more plants. For an $F_3$ population, tomato breeders may grow 30-100 plants. Populations in the $F_4$ generation, or higher, usually consist of 20-50 plants.

Backcrossing is particularly useful when one parent (the recurrent parent) is well adapted, except for one or a very few deficiencies, and when another parent (donor, or nonrecurrent parent) is available that possesses the genes lacking in the recurrent parent. The backcross is a conservative method because a minimum performance is established for the derived lines by the recurrent parent. It is unlikely that the performance of the derived lines will be worse than the performance of the line to be improved. On the other hand, no line markedly better than the recurrent parent is likely to result.

The backcross method does not require the use of populations as large as those required by other techniques. Therefore, if the necessary screening can be accomplished in a greenhouse, as is often the case when disease resistance is involved, at least 2 or 3 generations can be grown a year. Also, use of the backcross technique requires only a limited amount of testing because the recurrent parent is usually a well-adapted line or variety.

Another breeding method sometimes used for tomatoes, especially when dealing with a multigenically controlled character, is recurrent selection. This is a milder form of inbreeding than the pedigree or backcross method.

Hybridization for the purpose of exploiting hybrid vigor has not been used extensively with tomatoes, although a few $F_1$ hybrids are available for home-garden and fresh-market use. Several male sterility genes have been reported in tomatoes, but these have not been successfully used to produce hybrids commercially.

Extensive effort has been devoted to improving tomatoes by breeding because they are grown throughout the world under a wide variety of environments. The hundreds of varieties that have been named and the breeding lines not named vary greatly genetically and provide excellent sources of germ plasm. The U.S. Seed Storage Laboratory in Fort Collins, Colorado, has the responsibility of maintaining seed of old varieties. Tomato introductions from foreign countries are maintained by the North Central Regional Plant Introduction Station, Ames, Iowa (Clark et al., 1968; Skrdla et al., 1968).

Wild relatives of the tomato are also extremely valuable sources of genetic variation. The Plant Introduction Station collections contain many accessions of wild species in addition to the *L. esculentum* material mentioned above.

Seven wild species will hybridize with *L. esculentum*. These are *L. pimpinellifolium* (Jusl.) Mill., *L. hirsutum* Humb. & Bonpl., *L. peruvianum* (L.) Mill., *L. glandulosum* C. H. Mull., *L. cheesmanii* Riley, *L. chilense* Dun., and *Solanum pennellii* Correll. Hybrids between *L. esculentum* Mill. and *L. peruvianum* (L.) Mill. may be only partially fertile, but with the aid of embryo culture, viable seed can be obtained and genetic material can be exchanged between the species (Smith, 1944). Two examples of the use of wild species as sources of disease resistance are the transfer of resistance to Fusarium wilt from *L. pimpinellifolium* (Bohn and Tucker, 1940) and the transfer of resistance to root knot nematodes from *L. peruvianum* (Harrison, 1960).

Radiation has not been a useful means for inducing desirable mutations to use in breeding new tomato varieties. The large amount of diverse germ plasm from all over the world will probably continue to be the best source of genes for improvement of tomatoes in the foreseeable future.

# Major Diseases and Disorders of Tomatoes

Tomato diseases are caused by fungi, bacteria, viruses, nematodes, or physiological agents. A list of the most important tomato diseases in the United States is found in Table 21-1, and sources of resistance to them are indicated.

In the United States, many sources of resistance were identified by the cooperative efforts of participants in a National Screening Program (Alexander and Hoover, 1955). The resistant cultivars or wild species of tomato have been donated to, or were originally collected by, the Office of Plant Exploration and Introduction (now, the New Crops Research Branch) of the U.S. Department of Agriculture. Walter (1967) summarized the progress made through 1966 in breeding disease-resistant tomatoes.

# Disease Evaluation Techniques

Historically, observations of resistance often are first made in the field. Later, as techniques for seedling evaluation in the greenhouse become available, plant introductions or cultivars can be rapidly evaluated in order to locate new sources of resistance, or segregating populations can be rapidly screened.

Techniques for evaluating tomato populations for disease resistance are outlined in Table 21-2. Survivors or resistant individuals from

TABLE 21-1  MAJOR TOMATO DISEASES, THEIR CAUSES, AND SOURCES OF RESISTANCE.

| DISEASE | CAUSE | SOURCE OF RESISTANCE[a] | GENE ACTION | REFERENCES[b] |
|---|---|---|---|---|
| Fusarium wilt | *Fusarium oxysporum* f. sp. *lycopersici* (Sacc.) Snyder & Hansen | Pan American | Dominant | Bohn and Tucker, 1940; Porte and Walker, 1941; Stall and Walter, 1965. |
| Verticillium wilt | *Verticillium albo-atrum* Reinke & Berth. | VR Moscow | Dominant | Clark et al., 1968; Walter, 1967. |
| Bacterial wilt | *Pseudomonas solanacearum* (E. F. Sm.) E. F. Sm. | P.I. 127805A | Recessive or Multigenic | Acosta et al., 1964; Kelman, 1953; Winstead and Kelman, 1952. |
| Bacterial canker | *Corynebacterium michiganense* (E. F. Sm.) H. L. Jens. | Bulgarian 12 | Multigenic | Thyr, 1968. |
| Late blight | *Phytophthora infestans* (Mont.) DBy. | West Virginia 63 | Dominant or Multigenic | Gallegly, 1960. |
| Early blight | *Alternaria solani* (Ell. & G. Martin) Sor. | (on foliage) 68B134 (collar rot) Southland | Dominant or partially dominant | Barksdale, 1969; Reynard and Andrus, 1945. |
| Septoria leaf spot | *Septoria lycopersici* Speg. | Targinnie Red | Dominant | Andrus and Reynard, 1945. |
| Leaf mold | *Cladosporium fulvum* Cke. | Waltham Mold Proof Forcing | Dominant | Guba, 1956; Kerr, 1955. |

[a] This list is not comprehensive; it simply gives examples.

[b] Several references list sources of resistance to more than one disease (Alexander and Hoover, 1955; Clark et al., 1968; Frazier et al., 1950; Harrison, 1960; Skrdla et al., 1968; and Walter, 1967).

TABLE 21-1 *(continued)*

| DISEASE | CAUSE | SOURCE OF RESISTANCE[a] | GENE ACTION | REFERENCES[b] |
|---|---|---|---|---|
| Gray leaf spot | *Stemphylium solani* Weber | Manalucie | Dominant | Hendrix and Frazier, 1949. |
| Anthracnose | [*Colletotrichum phomoides* (Sacc.) Chester] = *C. coccodes* (Wallr.) Hughes | P. I. 272636 | Multigenic | Barksdale, 1970; Hoadley, 1960. |
| Tomato mosaic | Tobacco mosaic virus | Ohio M-R9 | Dominant or Recessive | Pelham, 1966. |
| Spotted wilt | Spotted wilt virus | Pearl Harbor | Dominant and Multigenic | Frazier et al., 1950. |
| Curly top | Curly top virus | CVF4 | Dominant and Multigenic | Cannon and Waddoups, 1951; Martin, 1969. |
| Root knot | *Meloidogyne* spp. | Anahu | Dominant | Dropkin and Webb, 1967; Harrison, 1960. |
| Graywall (Internal browning or blotchy ripening) | Physiological? | | Obscure | Boyle and Bergman, 1967; Stall and Hall, 1969. |
| Blossom-end rot | Physiological | | Obscure | Geraldson, 1957; Young, 1942. |

seedling tests either can be transplanted to the field for necessary horticultural evaluation, or they can be used in making crosses.

Precise laboratory-greenhouse techniques can shorten the time required to screen large segregating populations; can permit more severe selection pressure than is possible in the field; and often can aid in determining segregation ratios more accurately than in the field. Planting survivors from greenhouse tests in the field can reduce the amount of field work, or, more importantly, when field acreage is limited, resistant individuals from a greater number of crosses can be examined.

Laboratory-prepared inoculum can be used to inoculate large populations of plants in the field. Such a technique has been suggested for bacterial wilt and tomato mosaic (McCarter and Jaworski, 1969). Epidemics of fungal leaf spots could be initiated early in the season with laboratory-grown inoculum applied by spraying it uniformly on a field planting; the timing of field inoculations should coincide with the start of a dew period or period of wet weather. Field inoculations have proved feasible for evaluating lines for anthracnose (Hoadley, 1960) and internal browning resistance (Boyle and Bergman, 1967).

Evaluation for resistance to fruit rots or fruit disorders requires that populations be grown to maturity. A fruit disorder (or, perhaps, three distinct disorders) variously called blotchy ripening, internal browning, or graywall has a confusing etiology. Symptom appearance in the field often grades from one disorder to the other. This presents a problem in applying selection pressure to populations: what pathogen or set of conditions should be used, and what symptoms are to be emphasized in rating? The disorder has been variously ascribed to bacteria (Stall and Hall, 1969) or tobacco mosaic virus (Boyle and Bergman, 1967). Nevertheless, some practical progress has been made by selecting on a line basis (Walter, 1967). Hopefully, precise techniques will be available in a few years.

Blossom-end rot severity differs among lines (Young, 1942), but breeding for resistance to it can be complicated by rather large differences within lines. Because blossom-end rot severity is increased by a calcium deficiency (Geraldson, 1957), populations or lines may be evaluated for resistance by growing them on soils low in calcium-supplying power and by encouraging droughty conditions as the first fruit mature.

Similar screening techniques can be employed for diseases not specifically mentioned. These techniques would depend on the disease and the available facilities. A good evaluation technique provides conditions that are both biologically necessary and convenient for the breeder and gives consistent results.

# Examples of Disease Resistance

The examples in this section represent situations where resistance is dominant (Fusarium wilt) or recessive (bacterial wilt), and where it is possible to screen populations for resistance to more than one disease at a time. See Table 21-2 for a summary of techniques used to evaluate resistance.

## Fusarium Wilt

The fungus causing Fusarium wilt lives in the soil and can survive almost indefinitely even when tomatoes are not grown. When fields become heavily infested with the fungus, they are no longer suitable for growing susceptible varieties of tomatoes. More effort has been made to develop varieties resistant to this disease than to any other tomato disease.

*Source of Resistance.*   Resistance was first found in a wild species of tomato, *L. pimpinellifolium,* which is variously designated in the literature as Porte's No. 2116, Missouri Accession 160, and P.I. 79532 (Bohn and Tucker, 1940; Harrison, 1960). Pan American was the first variety released with this resistance (Porte and Walker, 1941). Most varieties released in the past 20 years have carried this resistance, which is inherited as a single dominant gene.

Unfortunately, a new race of the fungus, called race 2, became an economic problem in parts of Florida during the mid-1960s (Stall and Walter, 1965). Efforts by breeders in that state have recently culminated with the release of the first variety, Walter, to carry genes for resistance to both races.

*Methods of Inoculation.*   The fungus is easily grown on potato dextrose agar (PDA). Stock cultures can be maintained on PDA under refrigeration.

Inoculum is grown on PDA in 10-cm petri dishes at 24 C. The cultures are comminuted in distilled water at the rate of one culture per 50 ml of water. Roots of seedlings whose cotyledons are fully expanded are dipped in this agar-mycelium-spore suspension and immediately transplanted into flats of sterilized sandy soil (Harrison, 1960). The flats should be held at about 24 C in a greenhouse. Wilting of the cotyledons of susceptible individuals will begin in about one week. Most susceptible individuals can be eliminated by the end of three weeks.

Older seedlings can be inoculated by the same root-dip method, or by

TABLE 21-2 TECHNIQUES FOR THE EVALUATION OF TOMATO POPULATIONS FOR DISEASE RESISTANCE.

| DISEASE | TEST LOCATION | PLANT GROWTH STAGE | INOCULATION TECHNIQUES | CONDITIONS DURING INCUBATION | SYMPTOM DEVELOPMENT ON SUSCEPTIBLES | |
|---|---|---|---|---|---|---|
| | | | | | Time Needed | Special Considerations |
| Fusarium wilt | GH[a] | Seedlings, cotyledons expanded. | Dip roots in inoculum and transplant to flats (Harrison, 1960). | Hold at 24 C. | 2-3 weeks | Sunny days; keep soil moist at first, then let dry in midday to encourage wilting. |
| Verticillium wilt | GH | Seedlings, cotyledons expanded. | Dip roots in inoculum and transplant to flats. | Hold at 21 C. | 3-4 weeks | Sunny days; keep soil moist at first, then let dry in midday to encourage wilting. |
| Bacterial wilt | GH or F | Seedlings, 5-6 leaves (in flats or seedbed). | Cut roots on one side. Pour bacterial suspension into soil trench (Winstead and Kelman, 1952). | Keep plants between 20 C (night) and 30 C (day); keep soil moist. | 1-3 weeks | Symptoms appear more rapidly at warm temperatures. Transplant survivors to field. |
| Bacterial canker | GH | Seedlings, 3-leaf stage. | Inoculate cut made by excising first true leaf at point of attachment with bacterial suspension (Thyr, 1968). | Provide conditions for good seedling growth. | 8 weeks | Plants with one or more unwilted shoots considered resistant. |
| Late blight Monogenic resistance | GH | Seedlings, 3-4 leaves. | Inoculate leaves with swarmspore suspension by atomization. | Place inoculated plants in a moist chamber at 21 C for 16 hr. | 4-7 days | Provide conditions for good growth. |

[a] The following abbreviations are used in this table: GH = greenhouse, L = laboratory, F = field.

TABLE 21-2 *(continued)*

| DISEASE | TEST LOCATION | PLANT GROWTH STAGE | INOCULATION TECHNIQUES | CONDITIONS DURING INCUBATION | SYMPTOM DEVELOPMENT ON SUSCEPTIBLES | |
|---|---|---|---|---|---|---|
| | | | | | Time Needed | Special Considerations |
| Multigenic resistance | GH | 6 weeks old, grown in shallow beds (Gallegly, 1960). | Inoculate leaves with swarmspore suspension by atomization. | Cover bench with plastic and introduce mist for 16 hr. | 1-2 weeks | Keep soil fertility low and give supplemental light. |
| Early blight Foliage | GH | Seedlings, 3-4 leaves. | Inoculate with atomized spore suspension (Barksdale, 1970). | Initial incubation for 24 hr at 21 C. | 7-10 days | Place plants in mist chamber nightly (Barksdale, 1970). |
| Collar rot | GH (Soil beds) | Seedlings, 4 weeks old. | Dunk tops and stems of seedlings into inoculum and transplant to a depth of 3 inches (Reynard and Andrus, 1945). | Keep soil damp at about 20 C. | 7-14 days | Water between rows to prevent washing inoculum from stems (Reynard and Andrus, 1945). |
| Septoria leaf spot | GH | Seedlings, 3-4 leaves. | Atomize conidial suspension onto plants (Andrus and Reynard, 1945). | 100% RH for 48 hr at 21 C. | 7-10 days | |
| Leaf mold | GH | Seedlings, 3-6 leaves. | Inoculate by spraying lower-leaf surfaces with spore suspension. | 100% RH for 48 hr at 21 C. | 2-3 weeks | Keep plants at 20-24 C and at 80% RH or above (Kerr, 1955). |
| Gray leaf spot | GH | Seedlings, cotyledons expanded, or 3-4 leaves. | Atomize spore suspension onto plants. | 100% RH for 24-28 hr at 21-24 C (Hendrix and Frazier, 1949). | 3-5 days on cotyledons; 7 days on true leaves. | Intergrade-resistant reactions sometimes occur on true leaves (Walter, 1967). |

TABLE 21-2 *(continued)*

| DISEASE | TEST LOCA- TION | PLANT GROWTH STAGE | INOCULATION TECHNIQUES | CONDITIONS DURING INCUBATION | SYMPTOM DEVELOPMENT ON SUSCEPTIBLES Time Needed | SYMPTOM DEVELOPMENT ON SUSCEPTIBLES Special Considerations |
|---|---|---|---|---|---|---|
| Anthracnose | L | Mature. | Inoculate fruit by atomization or puncture (Barksdale, 1970). | If inoculation is by atomization, 24 hr at 100% RH and 24 C. | 7-10 days | Hold fruit at about 24 C. |
| Tomato Mosaic | GH | Seedlings, start series of inoculations on first true leaf. | Apply expressed inoculum with an air brush; inoculate again 10 days later. | | Start roguing 10-14 days after inoculation. | Transplant survivors in field and continue roguing. |
| Spotted wilt | F | | Place seedlings in a disease nursery and encourage thrips (Frazier et al., 1950). | Transplant to field after 4 weeks. | 7-20 days after vector feeding. | Grow plants to maturity, and rogue on basis of plant and fruit symptoms. |
| Curly Top | GH | Seedlings 10 days old at start. | Release viruliferous leafhoppers into screened greenhouse 2 or 3 times at 1 week intervals (Cannon and Waddoups, 1951). | | 1 week after first feeding. | Classify plants after 1 month. |

TABLE 21-2  *(continued)*

| DISEASE | TEST LOCA-TION | PLANT GROWTH STAGE | INOCULATION TECHNIQUES | CONDITIONS DURING INCUBATION | SYMPTOM DEVELOPMENT ON SUSCEPTIBLES | |
|---|---|---|---|---|---|---|
| | | | | | Time Needed | Special Considerations |
| Root knot | GH | Seedlings, cotyledons expanded. | Transplant seedlings into flats containing soil infested with chopped, galled roots from a susceptible crop (Harrison, 1960). | Keep soil moist, and GH temperature about 24-27 C. | 4-6 weeks | Dig plants, wash roots, count plants with galled roots as susceptible. |
| Root knot | L | Germinated seedlings | Inoculate seedlings grown in test tubes with egg masses of nematodes (Dropkin and Webb, 1967). | Keep tubes at 28 C. | 7 days | Examine roots microscopically at 10 X for amount of necrosis and galling. |
| Blotchy ripening and Blossom-end rot | See text. | | | | | |

injecting spore suspensions into stems. However, when these methods are used, the time required to identify susceptibles will be longer.


## Bacterial Wilt

The bacteria causing bacterial wilt are soil residents and attack a wide range of crops (Kelman, 1953). The disease is most serious in warm climates and under moist soil conditions. Although no commercial varieties now have resistance to bacterial wilt, active breeding programs are being conducted in North Carolina and Hawaii. When varieties resistant to bacterial wilt are released from these efforts, the tomato industry in the southern part of the United States and in many tropical countries will benefit greatly.

*Sources of Resistance.* Several tolerant lines have been identified over the years, but P.I. 127805A seems to have a high level of resistance. Counts of mature survivors indicate that resistance is inherited as recessive genes, but data taken up to seven weeks after inoculation may support the idea of partial dominance or of complex inheritance (Acosta et al., 1964).

*Methods of Inoculation.* The bacteria may be maintained in sterile water blanks at 20 C and increased in a liquid nutrient medium. Seedling plants growing in flats in a greenhouse can be inoculated at about the five- to six-leaf stage, either by a needle puncture in the stem, or by cutting roots on one side of each plant with a knife and pouring a dilute bacterial suspension into the resulting soil trench. In either case, the soil should be kept moist after inoculation with day temperatures of 30 C and night temperatures of 20 C (Winstead and Kelman, 1952).

In the tropics, where conditions are warm and moist, the bacteria can be maintained in an outdoor experimental plot by routinely planting and plowing under a susceptible host plant. Healthy seedlings of a segregating population can be transplanted directly into the experimental field, and survivors can be counted at various time intervals (Acosta et al., 1964).


## Multiple Screening

Screening simultaneously for resistance to more than one disease is a time-saving device. It should not be used, of course, when data are desired on the mode of inheritance; but when the modes of inheritance are known, and one or both parents possess resistance to more than one disease, multiple screening can be very useful.

Seeds of individual $F_2$ populations are germinated and the roots of each seedling, with cotyledons expanded, are dipped into a mixture of both *Fusarium* and *Verticillium* inoculum prior to transplanting in flats in the greenhouse. After the first true leaf expands, the plants are inoculated with *Stemphylium* spores and the greenhouse benches containing the flats covered with plastic sheeting overnight to keep relative humidity near 100% during incubation.

About two weeks after transplanting, many of the wilt-susceptibles and all the gray leaf spot-susceptibles can be rogued. At this point, screening for resistance to late blight (*Phytophthora infestans* [ Mont.] DBy.) can also be done, again covering the benches with plastic and perhaps introducing mist from a fogging apparatus during an overnight incubation (Gallegly, 1960). It is dangerous to leave the plastic covering on during the day (unless it is cloudy) because of the danger of overheating the plants. One week later, the remaining *Fusarium*-susceptibles, most of the *Verticillium*-susceptibles, and the late blight-susceptibles can be rogued.

The remaining plants should be inoculated with bacterial wilt by pouring inoculum into trenches cut on one side of each plant. Survivors are planted in the field and selected for horticultural characters. Whether a program of inbreeding, backcrossing, or outbreeding is subsequently followed will depend partly on the desirability of the horticultural characters found in the plants with multiple resistance.

## Concurrent Considerations

During the 1960s, the emphasis on varieties suitable for machine harvesting gave new importance to searches for resistance to certain diseases. With once-over machine harvesting, some tomato fruit must remain on the vines in a ripe condition for long periods of time until the maximum percentage of fruit are ripe. Therefore, diseases of ripe fruit, such as anthracnose, become more serious than they were with the old system of multiple hand-harvests.

The advent of machine harvesting has also forced the tomato breeder to select for small determinate vines that set all their fruit in a short period of time, fruit that are firm and can withstand mechanical harvesting without breaking, fruit that separate readily from the vine, and fruit that will remain in a sound condition when left on the vine for a long period of time. Other factors that were important for hand-picked varieties and for which tomatoes must be selected continually include high yielding ability, adaptability to a wide range of environmental conditions, quality for specific uses in processing and fresh market, and insect resistance.

Today maximum progress in a tomato-breeding program requires complete cooperation among scientists of many disciplines: plant pathologists, geneticists, horticulturists, physiologists, entomologists, and biochemists.

## Unsolved Problems

With the advent of the wide commercial use of mechanical harvesting for tomatoes, the seriousness of several fruit-rot diseases has increased. In the East and Midwest, anthracnose, soil rot caused by *Rhizoctonia solani* Kuehn, buckeye rot caused by *Phytophthora parasitica* Dast., and freckle caused by *Alternaria tenuis* Auct. are important. In other parts of the country, *Pythium* sp., *Sclerotium rolfsii* Sacc., and *Botrytis cinerea* Pers. ex Fr. take an unknown toll of fruits in the field, as do bacteria. Techniques to evaluate lines for fruit-rot resistance are needed, as are sources of resistance.

To make mechanical harvesting successful, new cultural practices are being followed. Chief among these is direct seeding in the field. At this time no one knows which, if any, of the pathogens causing damping-off or seedling stem-rots may become serious. Resistance to these pathogens may be needed in the future.

Defoliation caused by early blight [ *Alternaria solani* (Ell. & G. Martin) Sor. ] in commercial varieties is still a problem, even though the resistance derived from lines with resistance to collar rot helps to some extent. Recently, lines derived from complex pedigrees have demonstrated exceptionally high levels of resistance to the early blight organism under highly favorable conditions in the field at Beltsville, Maryland, during a 3-year period. Evaluation techniques for resistance to the defoliation phase of this disease in the greenhouse are needed.

Resistance to some virus diseases of tomatoes is known (Clark et al., 1968; Walter, 1967), and resistance to tobacco mosaic virus is reasonably well advanced. However, resistance to other specific virus problems, such as Stolbur, is needed because they limit production in some parts of the world.

Today the tomato breeder has the challenge and the opportunity to solve these problems. He will work not only to develop new resistances, but also to keep old ones, while incorporating resistance or tolerance to insect damage and to such physiological disorders as atmospheric pollution.

# References

Acosta, J. C., J. C. Gilbert, and V. L. Quinon. 1964. Heritability of bacterial wilt resistance in tomato. Proc. Amer. Soc. Hort. Sci. 84:455-461.

Alexander, L. J., and M. M. Hoover. 1955. Disease resistance in wild species of tomato. Ohio Agr. Exp. Sta. Res. Bull. 752:1-76.

Allard, R. W. 1964. Principles of plant breeding. John Wiley & Sons, Inc., New York. 485 p.

Andrus, C. F., and G. B. Reynard. 1945. Resistance to *Septoria* leaf spot and its inheritance in tomatoes. Phytopathology 35:16-24.

Barksdale, T. H. 1969. Resistance of tomato seedlings to early blight. Phytopathology 59:443-446.

———. 1970. Resistance to anthracnose in tomato introductions. Plant Dis. Reporter 54:32-34.

Bohn, G. W., and C. M. Tucker. 1940. Studies on Fusarium wilt of the tomato. I. Immunity in *Lycopersicon pimpinellifolium* Mill. and its inheritance in hybrids. Missouri Agr. Exp. Sta. Res. Bull. 311. 82 p.

Boyle, J. S., and E. L. Bergman. 1967. Factors affecting incidence and severity of internal browning of tomato induced by tobacco mosaic virus. Phytopathology 57:354-362.

Cannon, O. S., and V. Waddoups. 1951. A greenhouse method for testing tomatoes for resistance to curly top. Phytopathology 41:936 (Abstr.)

Clark, R. L., J. L. Jarvis, S. W. Braverman, S. M. Dietz, G. Sowell, and H. F. Winters. 1968. A summary of reports on the resistance of plant introductions to diseases, nematodes, insects, and mites: *Lycopersicon* spp. New Crops Research Branch, Crops Research Division, Agricultural Research Service, U. S. Department of Agriculture. p. 1-62.

Dropkin, V. H., and R. E. Webb. 1967. Resistance of axenic tomato seedlings to *Meloidogyne incognita acrita* and to *M. hapla*. Phytopathology 57:584-587.

Frazier, W. A., R. K. Dennett, J. W. Hendrix, C. F. Poole, and J. C. Gilbert. 1950. Seven new tomatoes; varieties resistant to spotted wilt, Fusarium wilt, and gray leaf spot. Hawaii Agr. Exp. Sta. Bull. 103. 22 p.

Gallegly, M. E. 1960. Resistance to the late blight fungus in tomato. Proc. Plant Sci. Seminar, Campbell Soup Co. p. 113-135.

Geraldson, C. M. 1957. Control of blossom-end rot of tomatoes. Proc. Amer. Soc. Hort. Sci. 69:309-317.

Guba, E. F. 1956. Coordination in breeding tomatoes for resistance to *Cladosporium* leaf mold. Plant Dis. Reporter 40:647-653.

Harrison, A. L. 1960. Breeding of disease resistant tomatoes with special emphasis on resistance to nematodes. Proc. Plant Sci. Seminar, Campbell Soup Co. p. 57-78.

Hendrix, J. W., and W. A. Frazier. 1949. Studies on the inheritance of *Stemphylium* resistance in tomatoes. Hawaii Agr. Exp. Sta. Tech. Bull. 8. 24 p.

Hoadley, A. D. 1960. The development of anthracnose resistant tomatoes. Proc. Plant Sci. Seminar, Campbell Soup Co. p. 19-36.

Kelman, A. 1953. The bacterial wilt caused by *Pseudomonas solanacearum*. N. C. Agr. Exp. Sta. Tech. Bull. 99. p. 1-194.

Kerr, E. A. 1955. Breeding for resistance to leaf mold in greenhouse tomatoes. Rep. Hort. Exp. Sta., Ontario, 1953-1954. p. 76-80.

McCarter, S. M., and C. A. Jaworski. 1969. Field studies on spread of *Pseudomonas solanacearum* and tobacco mosaic virus in tomato plants by clipping. Plant Dis. Reporter 53:942-946.

Martin, M. W. 1969. Inheritance of resistance to curly top virus in the tomato breeding line CVF4. Phytopathology 59:1040. (Abstr.)

Pelham, J. 1966. Resistance in tomato to tobacco mosaic virus. Euphytica 15:258-267.

Porte, W. S., and H. B. Walker. 1941. The Pan American tomato, a new red variety highly resistant to Fusarium wilt. U.S. Dept. Agr. Circ. 611. 6 p.

Reynard, G. B., and C. F. Andrus. 1945. Inheritance of resistance to the collar-rot phase of *Alternaria solani* on tomato. Phytopathology 35:25-36.

Skrdla, W. H., L. J. Alexander, G. Oakes, and A. F. Dodge. 1968. Horticultural characters and reaction to two diseases of the World Collection of the genus *Lycopersicon*. Ohio Agr. Res. and Devel. Center Res. Bull. 1009. 110 p.

Smith, P. G. 1944. Embryo culture of a tomato species hybrid. Proc. Amer. Soc. Hort. Sci. 44:413-416.

Stall, R. E., and C. B. Hall. 1969. Association of bacteria with graywall of tomato. Phytopathology 59:1650-1653.

———, and J. M. Walter. 1965. Selection and inheritance of resistance in tomato to isolates of races 1 and 2 of the Fusarium wilt organism. Phytopathology 55:1213-1215.

Thyr, B. D. 1968. Resistance to bacterial canker in tomato, and its evaluation. Phytopathology 58:279-281.

Walter, J. M. 1967. Hereditary resistance to disease in tomato. Ann. Rev. Phytopathol. 5:131-162.

Winstead, N. N., and A. Kelman. 1952. Inoculation techniques for evaluating resistance to *Pseudomonas solanacearum*. Phytopathology 42:628-634.

Young, P. A. 1942. Varietal resistance to blossom-end rot in tomatoes. Phytopathology 32:214-220.

# 22 Apples

J. R. Shay,[1] L. F. Hough,[2]
E. B. Williams,[3] D. F. Dayton,[4]
Catherine H. Bailey,[2]
J. B. Mowry,[4] J. Janick,[5] and
F. H. Emerson[5]

In the communities where there is a relatively high standard of living, fruit is often used more for its aesthetic value than for its healthful and nutritional values. No other agricultural industry can afford so great an expense as that of the increasingly more sophisticated disease and pest control programs that are essential to commercial tree fruit production.

Commercial fruit growers would appreciate having disease-resistant cultivars which would lessen the cost of the production program. However, unlike most agronomic crops, the appearance and quality of the fruit are by far the most important criteria in determining selling price. Consequently, the fruit of any new disease-resistant cultivar must

1. Department of Botany and Plant Pathology, Oregon State University, Corvallis.

2. Department of Horticulture and Forestry, Rutgers—The State University, New Brunswick, New Jersey.

3. Department of Botany and Plant Pathology, Purdue University, Lafayette, Indiana.

4. Department of Horticulture, University of Illinois, Urbana.

5. Department of Horticulture, Purdue University, Lafayette, Indiana.

be fully competitive on the market or growers will continue to use costly sprays on the established cultivars.

The basic concepts of breeding for disease resistance with apples are the same as those for breeding for disease resistance with other crops, but some special characteristics of fruit trees influence the approach to the problem. Apple trees are long-generation plants, usually requiring 5-8 years from seed to first fruit. The individual plant units are large. These two biological facts mean that improvement through breeding requires many years per seedling and a high cost per seedling per year.

The ability to use clonal propagation makes it possible immediately to increase a superior phenotype, but we must recognize the special problems associated with clonally propagated crops. Each one of the few desirable cultivars of the modern apple industry is highly heterozygous, and therefore not particularly prepotent for the sum of characters that make it desirable!

Perhaps the most serious problem in breeding for disease-resistant apples is that, in order to be useful, a new disease-resistant orchard must maintain its resistance for many years after it has come into production. Yet it must be recognized that of all kinds of crops, disease-resistant fruit cultivars (which are propagated as clones) are the most likely to have their resistance destroyed by the occurrence and build-up of a new pathogenic race of the disease.

Finally, there is the problem common to all disease resistance breeding; that the original source of disease resistance is almost always in a relatively unadapted or commercially unacceptable form, so that several generations of hybridization and selection are necessary.

This combination of problems, which is essentially the same for all of the tree fruit species, makes breeding for disease resistance with tree fruits so formidable that few have undertaken it. Most of those who have done so have not been able to tackle the problem with enough resources to achieve appreciable results in a single researcher's lifetime. The apple scab resistance breeding reported in this chapter has been a cooperative program among three experiment stations. At each station it has been a major research effort and the total support from the three stations up to the time of the first resistant cultivar, 'Prima,' has been in excess of two million dollars. Yet even with this level of support there are not enough resources adequately to develop the additional objective of breeding for mildew resistance and to accept the challenge of using the resistance to mites that has recently been observed.

Not all disease-resistance breeding programs will be as costly as the apple scab resistance breeding program. And, with the increasing concern of society for maintaining a healthful environment, surely disease- and pest-resistance breeding programs will become more important, especially with those crops that are sustained with intensive chemical protection programs.

Disease resistance as a major breeding objective has led to modifications in methods and procedures formerly used in breeding tree fruits. Some of the more important of these include:

1. Identify and discard susceptible seedling segregates at the earliest possible growth stage to reduce costs.
2. Reduce to a minimum number of years the generation time from seed to fruiting of hybrid seedlings to speed progress, especially for the first few backcross generations.
3. Test seedlings selected from each generation that are used as parents for the next generation in as many regions as possible to sample pathogenic variants of disease organisms.
4. Maintain original sources of resistance and selected resistant parents of successive generations in holding blocks for future reference if new pathogenic races appear.
5. Plan each year to make crosses in different climatic zones to spread risks of loss of a year's progress by spring freeze damage and to extend the crossing season.

This chapter describes the problems and the progress of a cooperative disease-resistance apple-breeding program that has been successful (Hough et al., 1970). The goal is the ultimate control of apple scab caused by *Venturia inaequalis* (Cke.) Wint., a historic disease of continuing importance in all humid apple growing regions of the world. This account should help others contemplating a similar effort with a long-generation, clonally propagated crop to develop appropriate procedures and to anticipate problems — especially the need for *adequate sustained effort* by both pomologists and pathologists.

## Organization

A major cooperative program to breed commercially superior new apple cultivars resistant to apple scab began in 1945 following the observation of Hough that field immunity to this serious disease existed in some *Malus* species (Hough, 1944). His observations closely paralleled those made in Germany during the 1930s. He pointed out that the ratio of resistance to susceptibility that Schmidt (1938) reported for a progeny derived from *M. micromalus* was practically the same as that observed in Illinois for a progeny involving *M. floribunda* 821 as a parent. Since the ratio of resistance to susceptibility in both of these progenies approximated 1:1, Hough suggested that resistance could be simply inherited.

This program requires that scab resistance, coming in part from small-fruited *Malus* species, be added as an additional major objective to an apple-breeding program which must produce cultivars that surpass in fruit quality most of the existing commercial cultivars. At the outset it was realized that the objectives could be attained only if a high level of research activity by breeders and pathologists were maintained for many years. To provide this stability and team effort, contract agreements on research and release of ultimate products of the research were drawn in 1945 between the Department of Horticulture of the Illinois Agricultural Experiment Station and the Department of Botany and Plant Pathology of the Purdue University Agricultural Experiment Station. In 1948 this was extended to include the Department of Horticulture of the New Jersey Agricultural Experiment Station. Since then many other research workers in experiment stations in North America and other continents have participated in this program (Shay et al., 1955; Williams and Brown, 1968). In 1958 an organization named "Apple Breeders Cooperative" was formed which facilitates exchange and testing of selections and planning of crossing programs (Janick et al., 1966).

The hazard of the development of races of the fungus that would negate individual sources of resistance was recognized from the beginning. To assure early detection of such races, hybrids representing all important sources of resistance have been distributed globally to major apple production areas where the apple scab disease is a problem. The resistance trees have been grown at each location without fungicide protection to permit exposure to the full range of natural pathogenic variants. As a result of this effort unusual races have been detected in South Dakota (McCrory and Shay, 1951) and Indiana (Shay, Williams, and Janick, 1962) in the U.S.; in Nova Scotia, Canada (Shay and Williams, 1956); and in Norwich, England (Williams and Brown, 1968). An unanticipated benefit from this distribution of scab-resistant hybrids was the dramatic demonstration of the presence of latent viruses that resulted from the sensitivity of many of the *Malus* species breeding lines to these viruses, which were widely prevalent in old-line rootstocks and commercial apple cultivars (Mink and Shay, 1962). Fortunately, the modified backcross method used in the breeding program has permitted selection of virus-tolerant segregates.

Initially the program had two phases: (1) to survey *Malus* species and hybrids for resistance to apple scab and other diseases, and then to study the inheritance and the mode of action of such sources of resistance; and (2) to breed for commercially desirable cultivars (Shay and Hough, 1952).

## Sources and Types of Resistance

Early in the program 25 apparently different sources of resistance were identified (Dayton, Shay, and Hough, 1953; Hough, Shay, and Dayton, 1953; McCrory and Shay, 1951; Shay, Dayton, and Hough, 1953; Shay and Hough, 1952). The breeding behavior with respect to disease resistance of many of these sources of resistance was summarized by Williams and Kuc (1969) in their Table 1. Several of the backcross progenies from different sources of resistance showed phenotypically similar resistant reactions following greenhouse inoculations with the fungus. This suggested that several of the genes might be identical. To simplify and facilitate future hybridization work, it was necessary to determine the relationships among the resistant genes from different lines. Six loci have been identified (Dayton and Williams, 1968, 1970; Dayton, Williams, and Shay, 1966; Williams and Dayton, 1968). These loci and the resistant source or sources carrying these qualitative genes are:

$V_f$   *M. floribunda* 821          Morton Arboretum #4

    *M. prunifolia* 19651          Morton Arboretum #8

    *M. prunifolia xanthocarpa* 591-25          Morton Arboretum #16

    *M. prunifolia microcarpa* 782-26          Morton Arboretum #1255

    *M. atrosanguinea* 804 (3 type)          Hansen's baccata #1

    *M. micromalus* 245-38 (3 type)

$V_m$ *M. micromalus* 245-38 (pit type)

    *M. atrosanguinia* 804 (pit type)

$V_r$ *M. pumila* Russian seedling R12740-7A (non-differential)

$V_b$ Hansen's baccata #2

$V_{bj}$ *M. baccata jackii*

$V_a$ Antonovka PI 172623 (pit type)

There may be still other qualitative sources of resistance that can be identified, especially from crabapples derived from *Malus* species or species hybrids.

Resistance of still others of the original sources of resistance seems to be controlled by multiple genes. These include Antonovka and certain of its seedlings, and selected seedlings of the following species: *M. baccata, M. sargenti* 843, *M. sieboldii* 2972-22, *M. toringo* 852, and *M. zumi calocarpa*. A number of old apple cultivars of Irish, English, and European origin also carry multigenic scab resistance, but our knowledge of their breeding behavior is limited (Shay, Williams, and Janick, 1962; Williams and Kuc, 1969). In the modified backcross breeding program the original level of resistance is diluted each generation, and sibbing or intercrossing is required to restore full resistance. This delays progress toward improvement, but several selections of improved quality with a high level of resistance have been obtained.

## Host-Pathogen Interaction

A part of the total effort in breeding for resistance has been to understand the mechanical, physiological, and biochemical aspects of the host-pathogen interaction. Williams and Kuc (1969) reviewed the research through 1968.

Much careful work has been done during the past 15 years to describe the biochemical differences of the leaf cell and/or cell surfaces of susceptible (that is, tolerant) versus resistant (hypersensitive) hosts. It is difficult to generalize beyond Kuc's statement (1968) that abnormal metabolites, probably capable of inducing metabolic stress, induce resistance. Hopefully the range of variation in the expression of resistance in the host and the wealth of different genotypes in the host, together with the different genotypes for pathogenicity in the fungus, will encourage continued research in this area. Certainly a more complete understanding of the biochemical processes that result in resistance of apples to attack by *V. inaequalis* should help us to develop and to recognize apple cultivars that will remain resistant.

In young leaves of both susceptible and resistant hosts, spore germination and the penetration of the cuticle are similar. All the work indicates that resistance results from a hypersensitive response of the tissue of young leaves to the fungal hypha after penetration of the cuticle is accomplished (Biehn, Williams, and Kuc, 1966; Enochs, 1964; Nusbaum and Keitt, 1938; Shay and Hough, 1952).

Recently, using electron microscopy, Maeda (1970) studied the mechanism of penetration and early development of the fungus. She observed that appressorial formation and penetration appear to be influenced by the proximity of pectin-rich areas in anticlinal wall junctions. Furthermore, changes and dissolution in leaf cuticular materials indicate that penetration is in part accomplished in response to a diffusible substance produced by the germinating spore.

Biehn et al. (1966) noted that, although spores germinated as readily on mature leaves as on young leaves of a resistant host, the hypersensitive response was not observed on mature leaves. The inference is that this increased resistance of mature leaves may be largely mechanical; for example, increased resistance of the cuticle to spore penetration. Maeda's observations (1970) give further support to this idea that the increased thickness of the cuticle of mature leaves and the consequent greater separation of the spore from the epidermal cell walls may indeed be a factor in preventing establishment of infection.

In the hypersensitive resistant reaction of the epidermal leaf cells to the fungus, four phenotypically different classes of resistance have been recognized (Hough, Shay, and Dayton, 1953; Shay and Hough, 1952). Our experience during more than 20 years indicates that these classes are still valid.

*Class 1:* pin-point pits with no sporulation. This may be considered the extreme hypersensitive reaction. The epidermal cells of the host immediately adjacent to the infection peg collapse within 40-72 hours after penetration. The fungus may be killed shortly thereafter (Enochs, 1964).

*Class 2:* irregular chlorotic or necrotic lesions with no sporulation. These result from a somewhat greater extension of the fungus affecting a larger volume of host cells that are ultimately destroyed. It takes 6-9 days before the host reaction may be observed macroscopically. The fungus may remain viable for as long as 21 days (Enochs, 1964).

*Class 3:* few restricted sporulating lesions. This is a lesser level of hypersensitivity. Even under optimum conditions for infection and fungal development the fungus is still very restricted in development but may persist, and limited sporulation may be observed after 2-3 weeks.

*Class M:* may be considered as intermediate between Class 2 and Class 3. It is recognized as a mixture of necrotic, nonsporulating lesions and of sparsely sporulating lesions. Apparently it represents a slightly more hypersensitive condition than Class 3, so that fewer of the fungal infections develop sufficiently to permit limited sporulation.

*Class 4* (susceptible): extensive, abundantly sporulating lesions. These symptoms are evident in 9-14 days under the conditions of artificial inoculation used in our tests.

It is significant that the restricted sporulation in both Class M and Class 3 occurs later than abundant sporulation in the case of the susceptible Class 4 reaction.

To date, four new unusual physiologic races of the fungus have been recognized (McCrory and Shay, 1951; Shay and Williams, 1956; Shay, Williams, and Janick, 1962; Williams and Brown, 1968) that can infect certain of the breeding lines that have been developed from the *Malus* sources resistant to the common race (Race 1). We are optimistic that we can recombine scab resistance that will persist. At the outset it seemed that *Malus floribunda* 821, *M. atrosanguinea* 804, the *M. pumila* selection R12740-7A, and Antonovka had had the greatest continuing exposure to infection in forests, arboreta, or unsprayed orchards, and yet had never been attacked. At the present time, our knowledge of the genes for scab resistance supports this early perspective. For example, Antonovka (PI 172623) carried a single locus for pitting together with quantitative genes for resistance. *M. atrosanguinea* 804, and *M. micromalus* 245-38 each had a pit gene as well as a gene for Class 3 resistance (Williams and Kuc, 1969).

After the discovery of race 5 of the fungus, which could attack the *micromalus* pit resistance, we found selections with the pit resistance which were not attacked by race 5 because of another closely linked gene which alone gives a Class M or 3 reaction. It is also interesting that one backcross selection carrying Class 3 resistance from *M. micromalus* 245-38 exhibited a Class 1 resistance when infected with race 5. This indicates that the susceptible backcross parent contributed something that bolstered the Class 3 type resistance which came from *M. micromalus* (Williams and Brown, 1968).

With *M. floribunda* 821 there was a change in reaction type with succeeding backcross generations. The original *M. floribunda* 821 gave a Class 1 resistant reaction, while $F_226829$-2-2 and $F_226830$-2 (selections obtained from sibbing $F_1$'s of 'Rome Beauty' x *M. floribunda* 821) gave Class 2 reactions (Shay and Hough, 1952). In subsequent generations of modified backcross progenies the resistant classes ranged from 2 to M to 3 (Williams and Kuc, 1969). When all four resistant classes (1, 2, M, and 3) are considered resistant, we have obtained a 1:1 ratio through 5 backcross generations involving a total of many tens of thousands of seedlings with this $V_f$ gene from *M. floribunda* 821. In other laboratories, where the sparsely sporulating classes M and 3 are considered susceptible, the resistant fraction of the backcross progenies is much less, rarely as much as a third (Lamb and Hamilton, 1969; Spangelo et al., 1956).

This change in reaction type of the resistance derived from *M. floribunda* 821 indicates that the original level of resistance is not due solely to the $V_f$ gene. The genetic basis for the difference between our parent selections carrying the $V_f$ resistance and the original expression of the $V_f$ gene in *M. floribunda* 821 is not obvious. Probably we are dealing with some general resistance (Caldwell, 1968) which fortifies the specific resistance of $V_f$ Since we have many of the intermediate parents, as well as the original clone, it may be possible to get some additional information about the genetic controls involved in this phenomenon. This would be especially desirable, since the $V_f$ source of resistance has been used so extensively in our breeding program to date.

## Breeding

A satisfactory screening procedure was developed at the outset for identifying scab susceptibility or resistance in very young seedlings in the greenhouse (Shay and Hough, 1952).

As soon as an apparently new source of scab resistance has been identified, an effort has been made to use this scab-resistant selection in hybridization with large-fruited, superior-quality apple cultivars or selections. In order to incorporate a greater variety of characters for commercial fruit quality, and at the same time to avoid the undesirable results of inbreeding, we have usually modified the strict concept of backcrossing by using a different susceptible parent in each successive generation (Shay et al., 1955). This use of different susceptible backcross parents has also allowed us to incorporate additional characteristics which we anticipate will be of value for the orchards of the future, such as spur type trees, adaptability to mechanical harvesting, and so forth.

With each source of resistance the emphasis at first was placed on moving through generations as quickly as possible. With *M. zumi calocarpa,* in seven years we were able to see two generations of fruit and have three generations of seed from controlled hybridization (Shay, Dayton, and Hough, 1954). This was accomplished by using the first seedlings to blossom for each of the backcross generations. These seedlings were used both as seed parents and as pollen parents the first spring. Then, at the time of the first fruit harvest, hybrid seed progenies were kept only from the selections that had the largest and best quality fruit.

The greatest progress has been made with the $V_f$ gene from *M. floribunda* 821. One selection with Class 3 resistance with quality similar to 'Jonathan' has been named 'Prima' (Dayton et al., 1970). It ripens earlier than 'Jonathan,' is larger, and is more resistant to mildew. 'Prima' promises to have international acceptance very quickly.

In addition to 'Prima' we have many other selections with the $V_f$ gene from *M. floribunda* 821 that have acceptable commercial fruit size, appearance, and quality with different seasons of ripening ranging from early mid-summer to late keeping winter types.

Very good progress has been made in obtaining satisfactory commercial size, appearance, and quality with many of the other sources of the $V_f$ gene, as well as with each of the qualitative genes. There are also improved selections in some of the lines carrying quantitative genes for scab resistance.

Attention has also been given to resistance of the scab-resistant selections to other diseases, such as powdery mildew (*Podosphaera leucotricha* [ Ell. & Ev.] Salm.), fireblight (*Erwinia amylovora* [ Burr.] Winslow), and the prevalent leaf spot and limb canker pathogens (Shay, Williams, and Janick, 1962). For these diseases we have relied on natural infections in the nursery or seedling orchards to help us identify unusual susceptibility or resistance. Since the scab-resistant seedlings do not require fungicide protection in the field, the conditions during most seasons are excellent for selecting for general good health.

# Perspective

At the very beginning of this breeding program we realized that it would require a long and expensive program to recover commercial size and superior quality in scab-resistant cultivars. Plant breeding experience assured that recovery of size and quality could be accomplished, whereas disease-resistance breeding experience was not so reassuring with respect to obtaining permanent disease resistance. We have made very satisfactory progress, but the durability of scab resistance is just as fragile or questionable as it was 25 years ago.

We have kept each breeding line distinct while improving fruit quality, since in most cases it would not be possible phenotypically to identify combined genotypes for resistance in the absence of differential races of the fungus. We hoped to achieve an advanced stage in the breeding program before many differential races arose. We are now at a stage where we have excellence in several lines, and we have begun to combine genetically different sources of resistance.

We anticipate that the most rapid progress will be made in combining the several single loci for resistance that have already been identified. There is also the possibility that we can fortify the $V_f$ gene and recover the original level of resistance of *M. floribunda* 821. This may be done through recombining $V_f$ selections with the Class 2 resistance, for example, or by combining $V_f$ lines with improved multigenic resistance lines.

We have been aware of the importance of working with quantitative resistance, and this aspect of the program should be expanded. In view of the considerable work that has been done in Europe with the quantitative resistance from Antonovka and other old European varieties (Blaja, 1968; Cociu, 1970; Visser and de Vries, 1970; Zwintzscher, 1954, 1970), there would be great mutual advantage in close coordination and active collaboration among all scab-resistance breeding programs.

# References

Biehn, W. L., E. B. Williams, and J. Kuc. 1966. Resistance of mature leaves of *Malus atrosanguinea* 804 to *Venturia inaequalis* and *Helminthosporium carbonum*. Phytopathology 56:588-589.

Blaja, D. 1968. Contributii la ameliorarea soiurilor de mar rezistente la fainare, patarea cafenie si ripan. Ph.D. Thesis, Facultatea de Horticultura, Institut Agronomie "N. Balescu," Bucuresti.

Caldwell, R. M. 1968. Breeding for general and/or specific plant disease resistance. Proc. 3rd Int. Wheat Genet. Symp., Canberra 1968, Aust. Acad. Sci., Canberra. p. 263-272.

Cociu, V. 1970. L'amélioration des arbres fruitiers à pépins en Roumanie. *In* Proc. Angers Fruit Breeding Symp., Eucarpia, Sept. 14-18, 1970. p. 59-68.

Dayton, D. F., J. B. Mowry, L. F. Hough, Catherine H. Bailey, E. B. Williams, J. Janick, and F. H. Emerson. 1970. Prima — an early fall red apple with resistance to apple scab. Fruit Vars. Hort. Dig. 24:20-22.

———, J. R. Shay, and L. F. Hough. 1953. Apple scab resistance from R12740-7A, a Russian apple. Proc. Amer. Soc. Hort. Sci. 62:334-340.

———, and E. B. Williams. 1968. Independent genes in *Malus* for resistance to *Venturia inaequalis*. Proc. Amer. Soc. Hort. Sci. 92:89-94.

——— ———. 1970. Additional allelic genes in *Malus* for scab resistance of two reaction types. J. Amer. Soc. Hort. Sci. 95:735-736.

——— ———, and J. R. Shay. 1966. Gene pools for apple scab resistance (Abstr.). Proc. XVII Int. Hort. Cong. 1:13.

Enochs, N. J. 1964. Histological studies on the development of *Venturia inaequalis* (Cke.) Wint. in susceptible and resistance selections of *Malus*. Master's Thesis. Purdue Univ., Lafayette, Ind.

Hough, L. F. 1944. A survey of the scab resistance of the foliage on seedlings in selected apple progenies. Proc. Amer. Soc. Hort. Sci. 44:260-272.

————, J. R. Shay, and D. F. Dayton. 1953. Apple scab resistance from *Malus floribunda* Sieb. Proc. Amer. Soc. Hort. Sci. 62:341-347.

————, E. B. Williams, D. F. Dayton, J. R. Shay, Catherine H. Bailey, J. B. Mowry, J. Janick, and F. H. Emerson. 1970. Progress and problems in breeding apples for scab resistance. *In* Proc. Angers Fruit Breeding Symp., Eucarpia, Sept. 14-18, 1970. p. 217-230.

Janick, J., E. B. Williams, J. R. Shay, D. F. Dayton, Catherine Bailey, and L. F. Hough. 1966. Breeding apples for resistance to scab. Proc. XVII Int. Hort. Congr. 1:14 (Abstr.)

Kuc, J. 1968. Biochemical control of disease resistance in plants. World Rev. Pest Control 7:42-55.

Lamb, R. C., and J. M. Hamilton. 1969. Environmental and genetic factors influencing the expression of resistance to scab (*Venturia inaequalis* Cke. Wint.) in apple progenies. J. Amer. Soc. Hort. Sci. 94:554-557.

Maeda, Karen M. 1970. An ultrastructural study of *Venturia inaequalis* (Cke.) Wint. infection of *Malus* hosts. Masters Thesis. Purdue Univ., Lafayette, Ind.

McCrory, S. A., and J. R. Shay. 1951. Apple scab resistance survey of South Dakota apple varieties and breeding stocks. Plant Dis. Reporter 35:433-434.

Mink, G. I., and J. R. Shay. 1962. Latent viruses in apple. Purdue Univ. Agr. Exp. Sta. Res. Bull. 756. 24 p.

Nusbaum, C. J., and G. W. Keitt. 1938. A cytological study of host-parasite relations of *Venturia inaequalis* on apple leaves. J. Agr. Res. 56:595-618.

Schmidt, M. 1938. Venturia inaequalis (Cooke) Aberhold. VIII. Weitere Untersuchungen zur Züchtung schorf widerstandsfähiger Apfelsorten. (Erste Mitteilung.) Züchter 10:280-291.

Shay, J. R., D. F. Dayton, and L. F. Hough. 1953. Apple scab resistance from a number of *Malus* species. Proc. Amer. Soc. Hort. Sci. 62:348-356.

———— ———— ————. 1954. Foredling av skurvresistente eplesorter. Frukt og Baer 7:33-36.

———— ———— ————, E. B. Williams, and J. Janick. 1955. Apple scab resistance. Rept. XIV Inter. Hort. Cong., Wageningen, Netherlands 1:735-739.

————, and L. F. Hough. 1952. Evaluation of apple scab resistance in selections of *Malus*. Amer. J. Bot. 39:288-297.

————, and E. B. Williams. 1956. Identification of three physiologic races of *Venturia inaequalis*. Phytopathology 46:190-193.

———— ————, and J. Janick. 1962. Disease resistance in apple and pear. Proc. Amer. Soc. Hort. Sci. 80:97-104.

Spangelo, L. P. S., J. B. Julien, H. N. Racicot, and D. S. Blair. 1956. Breeding apples for resistance to scab. Canad. J. Agric. Sci. 36:329-338.

Visser, Tyjis, and D. P. deVries. 1970. Personal communication. Institute for Horticultural Plant Breeding, Wageningen, Netherlands.

Williams, E. B., and A. G. Brown. 1968. A new physiologic race of *Venturia inaequalis,* incitant of apple scab. Plant Dis. Reporter 52:799-801.

——, and D. F. Dayton. 1968. Four additional sources of the $V_f$ locus for *Malus* scab resistance. Proc. Amer. Soc. Hort. Sci. 92:95-98.

—— ——, and J. R. Shay. 1966. Allelic genes in *Malus* for resistance to *Venturia inaequalis.* Proc. Amer. Soc. Hort. Sci. 88:52-56.

——, and J. Kuc. 1969. Resistance in *Malus* to *Venturia inaequalis.* Ann. Rev. Phytopathology 7:223-246.

Zwintzscher, M. 1954. Über die Widerstandfähigkeit von $F_2$ Bastarden des Apfels gegenüber dem Schorferreger, *Venturia inaequalis* (Cooke) Aderhold. Gartenbauwissenshaft' 19:22-35.

——. 1970. Die Schorfresistenz des Apfels: Ausgangsmaterial, Infektions and Selektionsmethoden. *In* Proc. Angers Fruit Breeding Symp., Eucarpia, Sept. 14-18, 1970. p. 199-216.

# 23 Forest Trees
## Henry D. Gerhold[1]

The idea of breeding trees is staggering to many biologists, and some plant pathologists consider it preposterous even to propose this method for controlling forest diseases. Several features of trees are responsible for such skepticism: their long life, the advanced age at which they start flowering, and the difficulty of reaching their flowers and seeds. Pathogens that kill or damage trees obviously can reproduce much faster than the trees, so in theory they could overcome genetic resistance mechanisms more rapidly than they were developed.

Are trees at a hopeless disadvantage, then? Doesn't a tree breeder have any chance of creating varieties with durable, useful types of resistance? Simple logic indicates that such a gloomy prognosis cannot be correct, or trees would have disappeared from the earth long ago. The hundreds of tree species that have coexisted with their pathogens for many generations stand in mute witness to this conclusion. In fact, such balanced host-pathogen situations probably represent conditions of genetic equilibria that can guide breeders in setting goals. The immunity

1. Professor of Forest Genetics, The Pennsylvania State University.

or high resistance of some species to pathogens that severely damage close relatives is another type of evidence that supports the resistance breeder's confidence.

Substantial progress already has been made in creating disease-resistant tree varieties and in making them available on a practical basis. In 1968 a survey (Gerhold, 1970) revealed that over 50 forest tree-breeding projects in 19 countries were concentrating on disease resistance. There are many additional projects in which disease resistance receives some attention but is not a primary objective of improvement. Numerous resistant varieties and clones of several genera have already been released, including *Castanea, Larix, Pinus, Populus, Pseudotsuga, Salix, Thuja,* and *Ulmus.* The parasitic pathogens to which they are resistant include *Aplanobacterium, Ceratocystis, Cronartium, Didymascella, Diplodia, Dothichiza, Marssonina, Melampsora,* and *Phytophthora.* Most of the tangible accomplishments have been realized since 1960.

Two of these projects are described in the following sections. They illustrate some of the variations in techniques, improvement objectives, and breeding strategies that are associated with different species. The underlying principles upon which genetic improvement of forest trees is based are, of course, the same as those employed for agricultural crops.

# Breeding Blister-Rust Resistant Western White Pine[2]

White pine blister rust, caused by *Cronartium ribicola* J. C. Fisch. ex Rabenh., has devastated many forest stands of western white pine, *Pinus monticola* Dougl., since serious damage was discovered in western North America in 1928. The commercial importance of this valuable timber species is greatest in northern Idaho and adjacent parts of Washington, Montana, and British Columbia, where it is native on some 3 million acres. Attempts to control the extensive damage by eradicating the obligate alternate host, *Ribes* species, or by chemotherapy, have proven unsuccessful (Ketcham et al., 1968), even though the former method had been relied upon for many years. In contrast, resistance breeding is regarded as the only promising means of

2. This section has been reviewed by Richard T. Bingham, Raymond J. Hoff, and Geral I. McDonald, whose helpful comments are gratefully acknowledged.

safeguarding this tree in the future. Several breeding programs have been started in North America with the object of producing blister-rust resistant varieties of *P. monticola, P. strobus, P. lambertiana,* or interspecies hybrids that utilize the greater resistance of certain Eurasian white pines.

The U.S. Forest Service project at Moscow, Idaho, started in 1950, at a time when very little was known about the prospect for developing disease-resistant trees. Initial emphasis was on finding resistant phenotypes in severely infected forest stands for intraspecific breeding and on defining the heritability of resistance. Canker-free individuals were rare, and it was a time-consuming task to find them in the rough, mountainous terrain. Mass selection under these conditions proved to be very effective, though, and over 400 candidates at 25 locations had been selected by 1965. Most of the resistant phenotypes were found to be genetically resistant as grafts or in progeny tests, and about one-fourth exhibited fairly high breeding values, estimated as general combining ability. After artificial inoculation of seedlings, an average of 35% of the $F_1$ progenies of selected parents survived, compared to 5% survival in reference populations. This degree of improvement in resistance is considered sufficient for practical use in developing seed orchards. It is likely that damage would be lower in forest plantations, as there would be less inoculum and because older trees are less susceptible. Three distinct varieties are being developed for low, middle, and high elevation zones because of the diverse environments they will encounter from 2,000 to 5,000 feet (Bingham, 1968; Bingham et al., 1969).

Fairly standard techniques have been devised for mating, inoculation, and evaluation. Controlled pollinations of phenotypically resistant parents are made in the forest. Female flowering of this monoecious species occurs in the upper branches, while pollen is produced in the lower part of the crown. Bagging, pollen collection, and pollination must be completed within a few weeks during the spring (see Bingham and Squillace, 1957; and Wright, 1962, for detailed descriptions of flowering phenology and pollination techniques). Periodicity of years with adequate flowering, time requirements for travel and climbing, and differences in flowering date severely limit the number of trees that can be mated by a small staff. Lately, four resistant male testers have been used to estimate general combining ability of selected trees that usually are represented in progeny tests as maternal parents. A pollen mixture from 10 male parents is considered nearly as efficient for selection and could be used to reduce costs, which have averaged about $600 per selected candidate. Sausage-casing pollination bags are removed by midsummer and replaced by protective cloth bags in April and May in the following spring. Cones are collected about September in the year following pollination. Each tree must be climbed at least four times, and commonly five or six.

Rust resistance is evaluated in a special nursery after artificial inoculation. Seeds presown on strips of paper toweling are planted in a randomized complete block design. Each plot contains 16 seedlings, and 10 replications are used. Large tents of polyethylene shaded by canvas are erected over the nursery beds to provide cool, moist conditions during the three- to five-day inoculation period. *Ribes* leaves are laid on screens above the seedlings, so that sporidia shed from telia on the undersides of leaves will fall on them. The wild currant leaves or branches are collected from naturally infected plants at several locations in northern Idaho.

Seedlings ordinarily are inoculated at the end of their second growing season. Foliar symptoms are rated during the following late spring, and bark symptoms after two or more years following inoculation. One-year-old seedlings have been inoculated in some experiments, but it is more difficult to interpret their reaction because they are more susceptible and they seldom have secondary, five-needle foliage. Seedlings that lack active bark cankers at the second annual inspection are classified as "survivors." The proportion of surviving trees per 16-tree plot is used for calculating genetic parameters.

It is assumed that the relative resistance ranking of various families measured in the nursery will be similar in the forest, and early results of field tests tend to confirm this. In two forest plantations exposed 11-15 years to high rust levels, the number of infected trees among the progenies of parents selected for general combining ability was 38-52% below controls and wild seedlings (Steinhoff, 1971).

Seed orchards are being developed to mass-produce improved $F_2$ varieties. Orchards for three elevational zones will each contain 24-30 full-sib $F_1$ seedling lines. Parents of the $F_1$s will be selected from the original 400 candidates on the basis of high general combining ability, as estimated in nursery tests. Selected parents from the same elevational zone are then mated, $F_1$ seedlings inoculated, and the survivors planted into seed orchards. By utilizing additive genetic variance in this manner, a gain of 10-30% in survival above the base population is expected by the $F_2$ generation (Bingham et al., 1969; Becker and Marsden, 1972). The selection and testing require about seven years.

Little had been learned about the genetical or physiological basis of resistance until 1969, despite a number of careful studies. More recently, evidence of several major genes was reported (McDonald, 1969; McDonald and Hoff, 1970; R. J. Hoff and G. I. McDonald, *unpublished data;* G. I. McDonald and R. J. Hoff, 1971). The abscission of infected needles and a fungicidal reaction in the short-shoot at the base of the needle fascicle were proposed as two resistance mechanisms governed by independent, recessive genes. Genes at three loci have been proposed that govern resistance to yellow or red lesions, normal or large size, that occur on needles; resistance is conferred by a homozygous recessive condition in one case, and by dominant genes in the others. Another gene is believed to control reduced needle-lesion frequency. Several bark-lesion types have been recognized, and these may lead to the identification of additional genes.

"Total resistance" has been described recently as a composite of independent mechanisms (Hoff and McDonald, 1972). The controlling genes that have been identified are all inherited in a simple Mendelian pattern, though polygenic inheritance may be involved, too. Two genes for virulence in *C. ribicola* have been hypothesized, plus a third that apparently acts as a modifer (G. I. McDonald and R. J. Hoff, *unpublished data*). These new findings may lead to revisions in the goals and methods of this breeding project. A central question is how best to utilize the genes that confer resistance on the host in a manner that minimizes the risk of consequent increases in virulence of the pathogen.

Various problems need to be investigated as progress continues, several of which are related to the stability of resistance. The capacity for genetic change in the pathogen has received scant attention, though the need for investigating genetic variation in *C. ribicola* clearly has been recognized. To guard against increased virulence, resistance genes will be sought in other white pine species (R. J. Hoff and G. I. McDonald, *unpublished data*). Those native to Asia are of particular interest because *C. ribicola* is assumed to have an Asian origin. The techniques that led to the discovery of major genes governing resistance may be applied in studies of possible gene-for-gene relationships in the host-parasite systems of American or Asian white pines. It would be of practical and scientific interest to compare balanced equilibria where the pathogen is endemic with evolution under artificial or natural conditions (Hoff and McDonald, 1972; Leonard, 1969; Hattemer, 1967; Watson, 1970). The insight that may be obtained about the most useful types and levels of resistance would be especially useful to tree breeders, who must be concerned with maintaining resistance of populations for periods of 30-80 years or longer.

## Breeding Elms Resistant to Dutch Elm Disease[3]

Dutch elm disease, caused by the fungus *Ceratocystis ulmi* (Buism.) C. Moreau, has killed uncounted millions of elms in extensive portions of North America, Europe, and Asia. Circumstantial evidence indicates that the disease originated in Asia. After being discovered in Europe in 1919 and in the United States in 1930, it spread rapidly through most regions where various elm species grow. Many towns lost most of their shade trees within a few years, and stands of timber have been devastated. Long distance transport from Europe to America has been attributed to immature bark beetles carried under the bark of elm burl logs or rough lumber. Locally, spores are spread by these insects as they breed in dead or weakened trees and feed on the twigs of healthy trees. Transmission to neighboring trees can occur through root grafts.

Some protection can be provided to valuable shade trees by annual spraying to control the insect vector. Various internal chemotherapy treatments have been tried, but so far they have not proven effective. Either chemical control program is very costly and would not be practical where trees are inaccessible to equipment. Furthermore, insecticides such as DDT carry with them considerable risks to animals. Thus, the protection of existing elms remains a partly unsolved problem for phytopathologists and entomologists.

In the meantime, there is a large demand for ornamental trees in new plantings, not only in gaps left by dead elms, but also in new housing developments, parks, and highway borders. Disease-resistant elms would be preferred in many localities, not only because of their shape, but also because they would be adapted to endure various environmental stresses better than most other species. Breeding projects have been started in recent years in several countries (Lester and Smalley, 1968; Ouellette, 1964; Ouellette and Pomerleau, 1965; Wright, 1968), partly because of the encouraging progress in Holland (Heybroek, 1966), in an improvement program which started much earlier. The various projects differ somewhat in their breeding goals, plant materials, and procedures.

Selection of resistant trees in the Netherlands was started in 1928 by Christine Buisman for the purpose of replacing susceptible native elms (Heybroek 1957, 1962, 1966, 1969a, 1969b). Over 99% of the planted trees at that time belonged to a single susceptible clone, *U. hollan-*

---

3. This section has been reviewed by Hans M. Heybroek, whose helpful comments are gratefully acknowledged.

*dica* Belgica. Already in 1936 a resistant clone named Christine Buisman, derived from a *U. carpinifolia* seedling from Spain, was released to nurserymen, but later it was considered a failure in the Netherlands mainly because of its poor growth habit and susceptibility to coral spot disease caused by *Nectria cinnabarina* (Tode) Fr. Hybridization of selected trees was started in 1936 (Went, 1954). Other clones that have been released are Bea Schwartz, Commelin, and Groeneveld. None of these are immune, but Groeneveld in particular is highly resistant to both *Ceratocystis* and *Nectria.* One quarter million grafted Commelin elms have been planted in Holland in the first ten years after the release of this clone.

The Dutch breeding project has multiple objectives. Its aim is to produce several varieties with different combinations of properties, each designed for a particular use. Some trees are planted only for ornamental qualities, others solely for timber, while roadside trees may be utilized for both purposes. A coastal zone is recognized, where the risk of contracting Dutch elm disease is lower than farther inland, but where high wind resistance is required. Accordingly, selection criteria include resistance to both *Ceratocystis* and *Nectria,* desirable crown form and growth rate, resistance to wind and cold winters, and good timber quality. These criteria are applied sequentially and somewhat informally as independent culling levels as soon as they can be judged. Disease resistance is the first to receive attention, and the overall value of each clone to be released must be superior to clones already in use. North American elm enthusiasts should be aware that most Dutch selections have a compact crown quite different from the vase-shaped *U. americana,* and that many may not be hardy enough for regions with severe winters.

The breeding system employed by Hans Heybroek (1962) combines several traditional methods. There are four steps in each cycle, involving sexual and vegetative reproduction. (1) Diverse inter- and intra-specific hybrids are made, to search for parents with high breeding values. Most of these come from $F_1$ and $F_2$ generations, and they are mated to original selections and to each other. Additional species, natural hybrids, and provenances are chosen through judgment and intuition gained by accumulated experience and data. There are relatively few barriers to crossing, especially among the species that are generally used and which belong to section *Ulmus* (syn. *Madocarpus* Dum.). *U. americana* is not used in the Dutch program, and it seems very difficult to cross with any other species. It belongs to the small section *Blepharocarpus* and apparently is the only natural tetraploid elm species, with $2n = 56$ chromosomes. (2) Concurrently, crosses that produced promising seedlings earlier are repeated on an expanded scale to produce a larger number of offspring for more intensive selection. These are supplemented with related crosses using full- or half-siblings of the parents. (3) The few individuals having the best combination of qualities are cloned by grafting for further testing. (4) Clones that meet the rigorous standards for commercial release are generally reproduced by grafting using rootstocks that will not produce rootsuckers. Layers of *U. hollandica* Belgica are preferred for this reason, and seedlings of *U. glabra* Huds. are also utilized. Rooting of cuttings seems to be feasible and may replace grafting after techniques have been perfected, though it is unknown yet whether the recent selections will grow suckers from rooted cuttings.

Artificial crossing is usually done in the field, commonly in a breeding arboretum or in test plantings. Young grafts with flowerbuds are sometimes potted and pollinated in a greenhouse if no more than 4 m high. Seeds can be produced on cut branches after pollination in a greenhouse, but this technique has not been reliable enough for a large scale breeding project. Pollen is obtained in this manner from cut branches shortly before flowers become receptive, generally in March. Nonwoven terylene bags with transparent plastic windows are tied over branches with flower buds, and these are opened briefly when pollen is applied with a small paintbrush. Emasculation is unnecessary except for crossability experiments. Protogyny is common, but does not invariably prevent selfing. A small proportion of selfed progeny can be tolerated, however, as selfing results in greatly reduced seed set and marked reduction in growth of seedlings. Seeds are collected in May or June and planted soon afterwards.

Each cycle of selection lasts for 12 or more years, and subsequent cycles may start annually or whenever sufficient seed is on hand. Seedlings grown without replication in a special nursery are inoculated typically at ages 3, 4, 7, and 9. Every tree that develops symptoms following inoculation is cut out. After the second inoculation and an initial selection for other traits, only 5-10% are left from an initial group of 6,000-10,000 seedlings. The few hundred trees that survive four inoculations are further evaluated according to the various selection criteria. Only those with the very best combination of traits are utilized for further testing and breeding. These are cloned by grafting, reinoculated several times, and set out in several experimental plantations for performance testing. Less than one clone from the 10,000 original seedlings will be released eventually to the nursery trade, after a lapse of up to 20 years after mating.

The success of the inoculation procedure with a spore suspension of

*C. ulmi* depends to a considerable extent on proper timing. Susceptibility is highest during the formation of spring wood; in Holland this is between mid-May and early July, but in North America the situation is more variable (Smalley and Kais, 1966). Warm, sunny weather shortly after inoculation results in optimum symptom development, while much lower damage results if inoculation is preceded by a drought or followed by a cold, wet period. Transplanting or other suboptimal growing conditions also lower the susceptibility of trees. Inoculum is prepared by growing some selected isolations of the fungus in a shaking culture for a few days (Holmes, 1965; Tchernoff, 1965). Several drops of spore suspension held on a knife blade are sucked into xylem vessels through horizontal cuts in the stem about 90 cm above the ground. The first symptoms may appear two weeks after infection, so there is ample time to evaluate trees in late summer.

The artificial selection process bypasses the insect vector responsible for a majority of natural inoculations. Thus the possibility of utilizing one type of resistance has been ignored purposely, for several reasons (Heybroek, 1969a). The disease is caused by just one fungus, but it is spread by several insect species. A reliable method of screening for resistance to these insects has not been developed. Genetic variation of insect resistance in elms remains largely unknown. Therefore, the simpler and more reliable approach using artificial inoculation has been used.

Much remains to be learned about the mechanisms of resistance (Elgersma, 1969). There are indications (Heybroek, 1966, 1969b) of different kinds of resistance, possibly involving numerous minor genes or cytoplasmic inheritance. Different patterns of wood discoloration associated with differences in resistance have been recognized (Tchernoff, 1965). Two types of resistance have been proposed (McNabb et al., 1970). One is a form of tolerance in which the pathogen spreads systemically inside the wood but with few external symptoms. The second is a localizing of pathogen establishment with very limited internal and external symptoms. The degree of resistance of resistant clones and various experimental clones is highly correlated with the size and distribution of xylem vessels. This relationship might explain the high susceptibility of *U. americana* because of its large vessels. Furthermore, it may imply that polyploids in general may be more susceptible if an increase in ploidy typically causes larger vessels to be formed.

Variation in pathogenicity among *C. ulmi* strains has been studied by Holmes (1965) and others. Differences in virulence have been found, but the consequent implications concerning durability of resistance remain unclear. Evidence that clone-specific strains of the fungus may occur either has been very limited, or has been contradicted by other experiments (Heybroek, 1969*b*). Observations that Christine Buisman elms have in some cases contracted Dutch elm disease could be explained in various ways. Those who developed this clone never claimed that its resistance approached the immune level, and they actually reported a few slight disease symptoms. Occasional infections are to be expected if trees happen to be exposed to massive infections during weather conditions optimal for disease development.

The large scale planting of elm clones, each represented by thousands of trees in the Netherlands, can provide a very sensitive test of the permanence of their resistance. Permanence can never be proved in advance, of course, but precautions have been taken to ensure that resistance will last as long as possible. An essential part of the program is to collect resistance genes from many different sources and species of elm, and to combine these through hybridization. Repeated testing in different locations and in different years is also important in this respect, so that there will be adequate exposure to a sampling of environments, fungus populations, and their interactions with the elm clones.

A message from H. M. Heybroek and H. S. McNabb, Jr., to elm breeders in September, 1971, serves as a postscript to underscore the need for continuing vigilance. Having noted that Dutch elm disease had become much more serious in Great Britain recently, Heybroek used two British isolates and four Dutch isolates for field inoculations of clones Belgica, Commelin, and highly resistant #496. The British isolates caused much greater defoliation, indicating that a more pathogenic race of the fungus probably has developed. Clone #496 had the highest resistance to both the Dutch and British isolates. This finding confirms the wisdom of having taken precautions against increased virulence. It also represents an opportunity to learn more about the genetic systems that control resistance and virulence.

# References

Becker, W. A., and M. A. Marsden. 1972. Estimation of heritability and selection gain for blister rust resistance in western white pine. *In* R. T. Bingham et al. [ ed.] Biology of rust resistance in forest trees. USDA Forest Service Misc. Publ. 1221:397-409.

Bingham, R. T. 1968. Breeding blister rust resistant western white pine. IV. Mixed-pollen crosses for appraisal of general combining ability. Silvae Genetica 17(4):133-138.

———. 1969. Rust resistance in conifers: present status, future needs. Proc. 2nd World Consulta. on Forest Tree Breeding. FO-FTB-69-5/2:1-11.

————, R. J. Hoff, and G. I. McDonald [ eds.] 1972. Biology of rust resistance in forest trees. USDA Forest Service Misc. Publ. 1221. 681 p.

————, R. J. Olson, W. A. Becker, and M. A. Marsden. 1969. Breeding blister rust resistant western white pine. V. Estimates of heritability, combining ability, and genetic advance based on tester matings. Silvae Genetica 18(1-2):28-38.

————, and A. E. Squillace. 1957. Phenology and other features of the flowering of pines, with special reference to *Pinus monticola* Dougl. U.S.D.A. Forest Service, Intermountain For. & Range Exp. Sta. Res. Pap. 53. 26 p.

Elgersma, D. M. 1969. Resistance mechanisms of elms to *Ceratocystis ulmi*. Phytopath. Lab. "Willie Commelin Scholten," Baarn, Neth., Med. No. 77. 84 p.

Gerhold, H. D. 1970. A decade of progress in breeding disease-resistant forest trees. Unasylva 24(2-3) No. 97-98:37-44.

Hattemer, H. H. 1967. Genetic mechanisms allowing an equilibrium between a parasitic fungus and a forest species. Proc. IUFRO Congress 14(3):413-425.

Heybroek, H. M. 1957. Elm-breeding in the Netherlands. Silvae Genetica 6(3-4):112-117.

————. 1962. Ulmen, *Ulmus*. Handbuch per Pflanzenzüchtung, end Ed. 6:819-824.

————. 1966. Aims and criteria in elm breeding in the Netherlands. *In* H. D. Gerhold et al [ ed.] Breeding pest-resistant trees. Pergamon Press, New York. p. 387-389.

————. 1969*a*. Host plant resistances in the elms. Proc. North Central Branch, Ent. Soc. Am. 24(2):69-74.

————. 1969*b*. Three aspects of breeding trees for disease resistance. Proc. 2nd World Consulta. on Forest Tree Breeding. FO-FTB-69-5/4:1-12.

Hoff, R. J., and G. I. McDonald. 1971. Resistance to *Cronartium ribicola* in *Pinus monticola*: short shoot fungicidal reaction. Canad. J. Bot. 49(7):1235-39.

———— ————. 1972. Stem rusts of conifers and the balance of nature. *In* R. T. Bingham et al. [ ed.] Biology of rust resistance in forest trees. USDA Forest Service Misc. Publ. 1221:525-535.

————, ————. 1971. Resistance to *Cronartium ribicola* in *Pinus monticola*: genetic control of needle-spots-only resistance factors. Canad. J. For. Res. 1(4): 197-202.

Holmes, F. W. 1965. Virulence in *Ceratocystis ulmi*. Neth. J. Plant Pathol. 71:97-112.

Ketcham, D. E., C. A. Wellner, and S. S. Evans, Jr. 1968. Western white pine management programs realigned on northern Rocky Mountain National Forests. J. For. 66(4):329-332.

Leonard, K. J. 1969. Genetic equilibria in host-pathogen systems. Phytopathol. 59:1858-1863.

Lester, D. T, and E. B. Smalley. 1968. Prospects for elm breeding in Wisconsin. Proc. 6th Central States Tree Improv. Conf. p. 37-41.

McDonald, G. I. 1969. Resistance to *Cronartium ribicola* J. C. Fisch. ex Rabenh. in *Pinus monticola* Dougl. seedlings. Ph.D. Thesis. Wash. State U. 74 p.

————, and R. J. Hoff. 1970. Resistance to *Cronartium ribicola* in *Pinus monticola*: early shedding of infected needles. USDA Forest Service Intermountain For. & Range Exp. Sta. Res. Note INT-24. 8 p.

McNabb, H. S., Jr., H. M. Heybroek, and W. L. MacDonald. 1970. Anatomical factors in resistance to Dutch elm disease. Neth. J. Plant Pathol. 76:196-204.

Ouellette, C. E. 1964. A review of the Dutch elm disease: resistant varieties. Canada Dept. Forestry, Forest Ent. & Path. Branch, Prog. Rep. 20(4):6-8.

―――, and R. Pomerleau. 1965. Recherches sur la resistance de l'orme d'Amerique au Ceratocystis ulmi. Can. J. Bot. 43:85-96.

Smalley, E., and A. G. Kais. 1966. Seasonal variations in the resistance of various elm species to Dutch elm disease. In H. D. Gerhold et al. [ ed.] Breeding pest-resistant trees. Pergamon Press, New York. p. 279-287.

Steinhoff, R. J. 1971. Field levels of infection of progenies of western white pines selected for blister rust resistance. USDA Forest Service Res. Not INT-146. 4 p.

Tchernoff, V. 1965. Methods for screening and for the rapid selection of elms for resistance to Dutch elm disease. Acta Bot. Neerl. 14:409-452.

Watson, I. A. 1970. Changes in virulence and population shifts in plant pathogens. Ann. Rev. Phytopathol. 8:209-230.

Went, J. C. 1954. The Dutch elm disease. Summary of fifteen years hybridization and selection work (1937-1952). Tijds. O. Plantenziekten 60:109-127.

Wright, J. W. 1962. Genetics of forest tree improvement. Food & Agr. Org. U.N. Rome. p. 330-352.

―――. 1968. A first step in breeding resistant elms. Proc. 6th Central States Tree Improv. Conf. p. 25-28.

# Index of Names

# Index of Diseases
# and Causal Agents

# Subject Index